Associative Memory Cells: Basic Units of Memory Trace

Jin-Hui Wang

Associative Memory Cells: Basic Units of Memory Trace

Springer

Jin-Hui Wang
University of Chinese Academy of Sciences
Beijing, China

ISBN 978-981-13-9503-1 ISBN 978-981-13-9501-7 (eBook)
https://doi.org/10.1007/978-981-13-9501-7

This Springer imprint is published by the registered company Springer Nature Singapore Pte Ltd.
The registered company address is: 152 Beach Road, #21-01/04 Gateway East, Singapore 189721, Singapore

Preface

Associative learning is a common approach to acquire multiple associated signals including knowledge, experiences, and skills from the natural environment or social interaction. Associative memory stands for the integrative storage and reciprocal retrieval of these associated signals, which are essential for cognitive processes, emotional reactions, and behaviors. The revelation of cellular and molecular mechanism underlying associative memory is important for us to understand the principles of memory formation and memory-relevant behaviors as well as to develop the therapeutic strategies that enhance memory capacity in health individuals and improve memory deficit in patients suffering from neurological disease and psychiatric disorders. Neural substrates for associative memory have been studied for centuries. A series of hypotheses is given, such as memory traces, engrams, cell assemblies, and neural plasticity for short-term memory and long-term memory consolidation. However, numerous questions need to be addressed, especially basic units and their working principles in engram circuit specifically for various memory patterns. In addition to memories to the associated signals learned from natural and social environments, the contents about associative thinking, logical reasoning, imagination, computation, and emotional reactions can be memorized for subsequent recall and recollection. Two sets of basic units in memory traces have been proposed for memories to exogenous signals from environments and endogenous signals generated from cognitions and emotions. By a series of experiments, basic units in memory traces have been discovered to be recruited during associative learning. These basic units in engram circuits include primary associative memory cells in sensory cortices, secondary associative memory cells in brain areas in relevance to cognition, emotion, and behaviors, as well as memory output cells in the motor cortex. Their characters include the following. The coactivation of cortical neurons recruits them as associative memory cells through the formation of their mutual synapse innervation. These associative memory cells receive new synapse innervation from coactivated brain areas alongside innate synapse input. Associative memory cells can encode new and innate associated signals for their integrative storage and reciprocal retrieval. Their axons innervate brain areas relevant to cognition, emotion, and behaviors to recruit secondary associative memory cells. The number and functional upregulation of associative memory cells influence memory strength and maintenance. The activation of associative memory cells grants logical reasoning and associative thinking. Their recruitment is influenced by epigenetics-regulated genes and proteins that manage axon prolongation and synapse formation. The working principle of these associative memory cells is based on their mutual synapse innervation, their reception strength of synapse innervations, their ability to convert synaptic analogue signals into digital spikes for encoding associated signals, as well as their ability to output sequential spikes for driving behavior and cognition. The synapse innervations to associative memory cells determine the specificity of memory contents. The number and activity level of associative memory cells and the number and activity strength of synapse innervations set up the power and persistence of memory and memory-relevant behaviors.

In this book, the author expects to provide comprehensive diagrams about associative memory cells and their working principles that may be applied to associative memories and memory-relevant cognitions for neuroscientists, psychologists, and students, based on historical and current literatures in memoriology. In the meantime, the author will present personal perspectives about pathology and therapeutic strategies for memory deficits in patients suffering from neurological diseases and psychiatric disorders.

Beijing, China
05-31-2019

Jin-Hui Wang

Acknowledgement

This study is funded by the National Key R&D Program of China (2016YFC1307100) and the National Natural Science Foundation of China (81671071) to JHW. The author thanks Shan Cui for drawing some illustrations.

Contents

About the Author

Wang Jin-Hui received his PhD from Shanghai Institute of Physiology, Chinese Academy of Sciences. After his postdoctoral training in the State University of New York and the University of Texas, he joined the University of Kansas as tenure-track assistant professor. Subsequently, he is appointed by the Chinese Academy of Sciences as a distinguished professor. After being an independent researcher, he has found working principles of neurons and synapses about the conversion of inactive into active synapses, the conversion of irregular to regular transmission pattern at unitary synapse, homeostatic plasticity among subcellular compartments, spike amplification at axons, as well as the functional compatibility between axonal branches and their postsynaptic partners. In terms of cellular architecture for learning and memory, his group has discovered and identified associative memory cells as basic units in memory trace.

Chapter 1
History in the Study of Learning and Memory

Abstract Learning is the acquisition of information, knowledge, and experiences from natural environments as well as social interactions. Memory is termed to be the storages and the subsequent retrievals of these acquired messages through recall, recollection and representation. The acquisition of exogenous signals is mainly based on associative learning despite an assumption of non-associative learning. The storage of associated signals is integrative in nature, which is essential for cognitive events and emotional reactions. In the meantime, the processes of cognition, emotion and behavior can be memorized. Associative memory includes the integrative storage of exogenous and endogenous associated signals. Learning and memory have been studied many centuries. Neuronal substrates potentially relevant to learning and memory have been presented in the following names, such as memory trace, engram, cell assemblies, associative memory cells, and so on. In terms of mechanisms underlying these processes, most of the studies indicate that these neuronal substrates are widely distributed in the cerebral brains. Although the natures of memory traces or engrams have not well been documented, cell assemblies are hypothetically to be interconnected neurons. Recently, associative memory cell has been detected. Associative memory cells are neurons with mutual synapse innervations in coactivated brain regions and are able to encode multiple signals carried by these innervations. The recruitment of associative memory cells as well as the refinement of these interconnected neurons and their synapses are thought to be basically for memory formation. In this chapter, author intends to review the histories in the study of learning and memory as well as the trend of revealing neural substrates for information storage and memory retrieval.

Keywords Learning · Memory · Engram · Memory trace and associative memory cells

J.-H. Wang, *Associative Memory Cells: Basic Units of Memory Trace*,
https://doi.org/10.1007/978-981-13-9501-7_1

1.1 Introduction

The acquisitions and storages of exogenous signals from natural environments and social interactions as well as of endogenous signals during cognitive processes and emotional reactions are essential for humans to treat the forthcoming events and solve various issues. In other words, memory constitutes the intelligent basis of human life. With this recognition, learning and memory have been the hot topic to be investigated centuries [1–4]. However, human intelligence to reveal intelligence-related activities appears difficult since the balance of their equal strength makes contradictory complications. In this period, numerous terms, concepts, and neural processes in the fields of physiology, psychology, and neuroscience are given to interpret mechanisms underlying memory. For instance, memory can be classified into declarative (explicit) memory and nondeclarative (implicit) memory. The explicit form includes episodic memory and semantic memory [5]. Based on temporal feature, memory can be short-term and long-term. Neuronal substrates for memory formation are called as memory traces, engrams, cell assemblies, memory encoding neurons, and associative memory cells [6–8]. Plasticity as synapses and neurons presumably plays a major role in memory formation [9, 10]. More than hundreds of molecules are found to be related to memory formation [11]. Although these enriched data are collected, the cellular and molecular processes have not been linked to the features of various memory events well. Most of these molecules works for signaling transductions in cells and communications among cells or the building elements for neurons and synapses, such that there appear no specific molecules for specific memory events except these molecules work for the normal structure and function of neurons and synapses. Instead, specific neurons and their synapse-interconnected routes constitute the storage of specific information [12].

With the advance of technologies in biology and biophysics, the studies of memory traces or engram have been paid attention recently. Memory traces or engrams are widely distributed in the central nervous system because fear memory can be evoked by stimulating many places and synaptic plasticity related to fear memory is seen in many areas in the brain. These engrams are regulated by neurotransmitters from several sources. The plasticity in synaptic transmission is detected in the hippocampus and cerebral cortices. It is likely that the memory traces or engrams are neurons and their interconnected synapses whose activities can carry the flow of specific signals memorized in the brain [12–16]. Currently, associative memory cells are presumably the basic units in memory traces. Specific associative memory cells and their interconnected pathways may work for the integrative storages and reciprocal retrievals of associated signals. Primary associative memory cells in sensory cortices and secondary associative memory cells in brain areas relevant to cognition and emotion are featured by mutual synapse innervations among them, convergent innervations onto their downstream neurons as well as functionally encoding to the associated signals [17, 18]. In this chapter, the historical progress and the milestones in the study of learning and memory are summarized.

1.2 Memory Traces

The term "memory trace" was issued in ancient Greek, which has been defined with various means, such as a postulated biochemical change presumably in nerve tissues that represents a memory, a hypothetically permanent change in the nervous system brought about by memorizing something, or biophysical and biochemical alternations induced by a given form of learning and memory. In spite of these definitions, the core meaning of memory traces is the changes of the nerve system in response to information acquisition. Based on current views in biological science, these changes can be morphological, biochemical, and functional, and basic units in memory traces may be neurons, glia cells, and/or their synapses. There are more than ten billion of neurons in the brain and equal number of glia cells. Synapses for the communication of these nerve cells may be 10^3 higher than neurons. Neural networks composed by interconnected neurons may constitute the foundation of memory traces for information storage. Thus, the number of these basic units is sufficient to be used for working correspondently to specific events experienced in lifetime.

In order to have the comprehensive view of memory traces, the following different but interrelated features should be elucidated. The precise sites with memory-relevant alternations should be localized in the brain. Neuronal circuits that specifically encode the storage and retrieval of the given signals are expected to be detected at the levels of neurons and their circuits. Once the location related to memory storage and retrieval is identified, the basic units of memory traces need to be figured out at the levels of individual cells that are able to encode the learned signals and to be actively functional during memory retrieval. In addition, biophysical changes in neuronal circuits and memory cells correlated with memory formation should be profiled, e.g., alternations in the strength of synapse connections among network neurons. These processes are regulated by the expressions of memory-relevant genes and proteins. Furthermore, the memory-specific activity patterns of neurons and synapses changed from the conventional activity patterns in memory traces should be identified during memory storage and retrieval. The correlation of the neuronal circuits to memory formation and retrieval can be delineated by the following experiment strategies, such as parallel changes between memory formation and neural circuit that consists of neurons and synapses, the downregulation of neurons and synapses leading to lack of memory formation, as well as the upregulation of such neurons and synapses strengthening memory formation. As long as all of these criteria are met in experiment results, neurons and synapses would be considered as the relevance to specific memory, i.e., they are basic units of memory traces.

In addition to the strategy of experiment study, the theoretical modeling of memory traces has been widely used by the computational simulation of neural substrates for learning and memory. Data for theoretical modeling is based on experimental results. In the study of neuronal and synaptic architectures for memory traces and memory-related events, there are clear indications that neuronal ensemble and synaptic plasticity are involved in processes of learning and memory.

However, it remains lack of evidences for synapses, neurons and their plasticity specifically correlated to memory formation [19–24].

In the identification of memory traces, numerous experiment studies have presented that the recruitment of memory cells as well as the refinement of neurons and synapses in wide areas of the brain are essential for memory formation [12, 25, 26]. For instance, the location of memory traces related to specific memory behaviors has been thought to be widely distributed in the brain, since the surgical removal of brain tissues and the pharmacological blockade of neuronal activity cannot erase all of memory-relevant events despite their partial impairment [27, 28]. The downregulation of axon prolongation and synapse formation to attenuate the recruitment of associative memory cells will weaken memory formation [29, 30]. Blockades of intracellular or intercellular signaling molecules inhibit synaptic plasticity presumably related to memory formation [31–33].

1.3 Engrams

In addition to memory traces as a term of neural substrates underlying the information storage, "engrams" in mneme theory have been suggested by Richard Wolfgang Semon [7, 34], a zoologist and evolutional biologist, especially as renowned theologist in the field of learning and memory. Engrams are proposed to be biophysical and biochemical changes in response to external stimulations. Engram and ecphory are relevant to memory traces and memory retrieval, respectively [13, 35]. Based on Semon's view about engrams and ecphory for memory retrievals, further hypothesis is that the interactions between the stored engram and retrieval cues can generate new engrams. As long as a stimulus that awakes engrams is closely similar to an original stimulus, this incomplete retrieval cue is sufficient to make the memorized engrams to be retrieved. In the meantime, awakening the originally stored engram may generate new engrams related to this given event. During strengthening original memory by the contiguity, these old retrieved engrams and new engrams may be associated. The simultaneous retrieval of multiple engrams that store similar contents and their subsequent associations, likely a resonance among engrams, may lead to even more complicated cognitive processes, such as the generalization, abstraction, and organization of memory contents [23]. These processes can be considered as a model of the conversion of episodic memory to semantic memory. This proposal suggests the coordination between awakening engrams and recruiting engrams in dynamic and use-dependent manner. Although engrams coined by Semon lack experimental evidences in that period, his frameworks about engrams are compatible to the characters of memory formation and memory-relevant behaviors. For instance, more representations induce impressive memory. Repeated simultaneous recalls of similar memory contents will make them to be summarized, generalized, reorganized, and conceptualized. Thus, Semon's theoretical model about the engram and its extensions has constituted the bases of memory formation and memory-relevant cognition. It should be clearly emphasized that engrams stand for memory traces equally in many aspects of current view about learning and memory.

Donald Hebb, another well-known theologist, describes engrams to be cell assemblies relevant to memory. In his and Penfield's observations, the destructing to large amount of cerebral cortices in human beings leads to little effect on memory-relevant behavior [27, 36]. Lashley's experiments show that the ablation of wide-spread cortices in the animals does not induce the parallel change in memory-relevant behaviors [37, 38]. Based on these results, Hebb has proposed "cell assemblies" in the engram that are the widely distributed neural substrates for memory. Each cell ensemble is a group of interconnected cells, and their connections are strengthened in their simultaneous activities [6, 39]. As these cells are interconnected, the activity of cell assemblies in this circuit is maintained briefly after the events, i.e., short-term memory. Recurrent activities of this cell ensemble with sufficient time will induce the growth and the metabolic alternation that will strengthen those interconnections among assembly cells, so that short-term memory is converted to longer-term memory [39]. The strengthening of interconnections between presynaptic and postsynaptic neurons in their simultaneous activities confers these neuronal units to be a rule, firing together and strengthening together, which has been hypothetically extended to neuron connection together. The strengthening of neuronal connection has been indicated in long-term potentiation of synaptic transmission [9, 40]. The high interconnections of neurons may allow entire ensemble to be active if a subset of neurons is activated by the process of pattern completion that evokes memory retrieval. As Hebb's cell assemblies are widely distributed across brain regions, the destruction in a small proportion of neural tissues may not lead to catastrophic memory traces or graceful network degradation. This view may account for Lashley's experiment results below. In brief, Hebb's theory has overlapped multiple spatial scales from the integrated synaptic strengthening (a microscale level) to cell ensemble formation (a mesoscale level).

The study of memory formation by theoretical models has generated great frameworks that provide useful guideline for addressing cellular mechanisms underlying learning and memory. Yet, these hypotheses about engrams and cell assemblies have not indicated the insights about the integrative storage and reciprocal retrieval of associated signals as well as need be proved by experimentations, since associative learning and memory are the most common approaches to acquire and store the information. Similarly to the goal of defining memory traces, experimental studies of engrams should include the address of their distributions, basic units, biophysical/biochemical changes, and specific patterns related to particular memory events by approaches in morphology, neurophysiology, and molecular biology.

Karl Lashley started identifying engrams experimentally, in which he used the surgical ablation to search the distribution of engrams. Although he was unable to localize engrams, his studies indicate the wide distribution of engrams in the cerebral brain [37, 38, 41]. Furthermore, the removal of the temporary lobe in human beings leads to the loss of recent memory due to the impairment of the hippocampus [27, 42–44]. In the study of engram distribution by pharmacological reagents, Josefa Flexner has observed that recent memory in learned maze behaviors can be blocked by the intracerebral injections of puromycin into the neocortex and the hippocampus [28, 45, 46]. In addition to an indication of the widespread distribution of memory traces in cerebral brain, protein synthesis is required for memory formation since puromycin blocks protein synthesis. With his studies, Flexner is thought of as a pioneer to investigate the molecular mechanism underlying learning and memory. These studies reveal the causal relationship between memory traces in wide brain areas and memory formation and retrieval, though engrams specific for memory content-related events remain to be identified. With the advance of molecular biology, the downregulation of gene expressions by gene knockout [47] and optogenetics [48, 49] successfully shows the inverse correlation among molecules, memory cells, and memory events. These studies provide strong evidences for causal relationships among molecular substrates, cellular architectures, and memory formation.

There are three strategies in experimental studies of learning and memory which can be used to identify the causal relationships between engrams and memory-relevant events. (1) Engrams with specific activity patterns are detected in memory formation and cue-induced memory retrievals through the recording system. Cell electrophysiology and two-photon cell imaging can be used to record engram's activities and to profile activity patterns in relevance to the information storage. Fluorescent protein-based neural tracing carried by adeno-associated virus (AAV) can be used to track the location and distribution of engrams. (2) The downregulation of engrams or memory cell assemblies can inhibit memory-related events through their surgical removal, pharmacological blockade, and gene knockout. (3) The upregulation of engrams can facilitate memory-relevant behaviors by pharmacological, electrical, optogenetic, or chemogenetic stimulation to the potential engrams as well as by gene overexpression in engrams [17, 18, 29, 30]. Two commonly used approaches have been associated with three strategies above to track engrams or cell assemblies, such as the detection of memory cells in vivo during learning and memory as well as the activation of memory cells to retrieve memory-relevant behaviors. The detection of memory cell assemblies is to observe their responses to memory cues by electrophysiological recording and two-photon cell imaging. The activation of memory cells can be performed by electrical, pharmacological, optogenetic, or chemogenetic stimulations to induce the emergence of memory-relevant events [18]. However, if engrams are widely distributed in the brain and brain areas are mutually connected, these stimulations may cause the antegrade and retrograde activation of neural pathways to indirectly activate engrams, such that primary versus secondary allocations for memory formation cannot be localized [12].

1.4 The Location and Distribution of Memory Traces or Engrams

To fully understand mechanisms underlying memory formation, the location of memory traces or engrams should be addressed at the first step. With a possibility of many locations involving memory-relevant behaviors, the distribution of engrams in the brain should be figured out in order to further manipulate memory processes in either the enhancement of memory or the erasing of aversive memory. The strategies to track engrams described above can be used.

Initially, physiologists and psychologists have paid attention to localize the distribution of memory trace or engrams by the combination of surgical lesions or pharmacological block with behavioral tasks of memory-relevant events quantitatively. Their studies are based on a hypothesis that each specific memory is represented as the change of one particular site in the brain. Karl Lashley has started searching for the memory trace by the combination of the surgical ablation of brain tissues with rat's searching food in the maze. In this experiment, it is necessary to find out memory traces with an intention to subsequently analyze natures of memory traces [50]. After many studies have been done to search engrams, he concludes that the removal of brain tissues in the piece-by-piece manner cannot completely erase memory-relevant behaviors in mazes, though partial impairment in such behaviors [37, 38, 41, 51–53]. Lashley further suggests that memories in response to the complex mazes, at least for rats, do not reside in any particular location in cerebral cortices, though the more cortices remove and the more memory impairs. Based on his studies, a wide location and distribution of engrams in the central nervous system are generally cited and trusted. In the meantime, Walter Hunter points out that the removal of more cerebral cortices may impair more sensory information including visual, auditory, pain, and kinesthetic signals, such that the number of available cues is reduced in their effects, such as animals blinded do not know mazes well [54]. Whether memory traces or sensory systems are impaired in these classic experiments has never really been resolved, at least for complex maze learning in the rats, although we now know that the hippocampus and most of the cerebral cortex are important for these memories.

After Lashley is unable to localize engrams, some scientists have adopted the view that engrams are widely distributed throughout the brain or broadly across certain brain areas, such as the cerebral cortices and hippocampus. With more and more accumulated knowledge about anatomical and functional organizations of the brain, it becomes clear that the brain isn't the diffusely organized tissues; instead it is characterized as regularly structured and highly interconnected tissues. Donald Hebb as a theologist in physiology has proposed the resolution for this issue. He has assumed that the organization of any specific memory trace is a complex and includes a number of brain areas in that memory trace can involve specific connection in particular areas [6, 55, 56]. Hebb further proposes a possible mechanism underlying the formation of engrams, which has been known as Hebb's synapses. In his theoretical hypothesis, the neurons and synapses are locations where memory traces are formed, i.e., cell assemblies. There must be active inputs from a source ready to be learned (conditioned stimulus in Pavlovian terms) to a neuron while this neuron fires action potentials. The coordinated activities of the presynaptic and postsynaptic neurons strengthen their interconnections. Hebb's synapse has become a more general view that one synapse input onto one neuron will be strengthened or weakened if this input is active simultaneously with the activation of this neuron from another input [6], i.e., active together and strengthening together or weakening together. Although heterosynaptic depression has not been observed, heterosynaptic potentiation is really present at the synapses in the central nervous system [57–60]. In typical form of associative learning and memory, coactivations among cortical neurons induce their mutual synapse innervations and in turn recruit associative memory cells for the integrative storage and reciprocal retrieval of associated signals [8, 61, 62].

To the end of the twentieth century, the research focus has been shifted away from the localization of engrams in complex tasks toward more specific and discrete learning and memory tasks. Scientists in the field of learning and memory emphasize to identify the distribution of entire neural circuits essential for each of particular memory forms. The term "essential" stands for the requirement to meet the necessary and sufficient conditions in the logical analysis. Once the identification and distribution of engrams for each particular form of memory have been accomplished, the memory traces for entire memory tasks can be mapped and analyzed expectedly. The following approaches are well-established experimentally and are often used to identify engrams in particular memory. These approaches include the surgical lesion of nerve tissues, the functional detection of neuronal activity by electrical recording and cellular imaging, the structural identification of engram cells by neural tracing, as well as the stimulation of brain tissues by electrical apparatus and optogenetic method. With these approaches, the identification and distribution of engrams or essential memory circuits in the brain have led to much progress for simpler forms of learning, though experiment challenges are formidable. Other new methods, such as probes to genetic expression and the local infusion of highly specific receptor antagonists or agonists, have also become widely applicable tools to search engrams [12, 18]. Once the complete circuitries for a particular form of learning and memory have been identified, the next step of revealing basic units in the engram circuit is needed in spite of more difficult. It should be noteworthy that there are no universal methods generally accepted for all scientists to do in searching engram cells. Animal models currently used to study the specific form of learning and memory, such as fear memory by electrical shocks or aversive taste, are not simple, since animal behaviors in response to these cues include cognitive activities and emotional reaction. Nonspecific behaviors cause these studies to be complicated [12], though the aspect of searching memory traces has become the conceptual center of the field [63].

To localize the distribution of engrams relevant to memories, the types of learning and memory should be clearly classified in mind. There are different types of information acquisition, such as associative learning and non-associative learning. The classification of memory is even more complicated, such as declarative (explicit) memory versus nondeclarative (implicit) memory, episodic (event-relevant) memory versus semantic (knowledge-relevant) memory, working memory (operant conditioning) versus perceptual (visual, auditory, olfactory, gustatory, and tactile) memory, eidetic memory versus false memory, spatial memory versus temporal memory, and short-term memory versus long-term memory (please refer to Chap. 2 for details). For instance, learning to ride a bicycle, in which individuals learn how procedures are handled, differs from learning a list of contents about concepts and/

or events, in which people learn what contents are taken into account. Memories formed in learning procedures and operations refer to nondeclarative memory. Memories formed in learning concepts and knowledge fall into declarative memory. In animal models, animal's memory to its recent experiences is declarative, and Pavlovian classical conditioning, in which specific conditioned responses like salivation, eyeblink, or fear behavior are learned to a given conditioned stimulus, is nondeclarative. Although both types of memory formation involve many brain areas with partial overlap, brain structures and functional systems essential for these two types of memories are quite different. By experiment studies, a few memory circuits and systems are found in the mammalian brain, such as the hippocampus for spatial memory, associative cortices for perceptual memory, the prefrontal cortex for memory-relevant cognition, and the temporal lobe for emotion memory [12, 18].

The hippocampus presumably plays a key role in recent experiential memory or declarative memory since extensive hippocampal lesions can substantially impair the newly formed memory in humans and monkeys [27, 64–68]. The lesion of the hippocampus appears to influence more in the establishment of memories than in memory retrievals. The influence of hippocampal lesion on memory retrieval impairment takes weeks in monkeys and years in humans, whereas the memory formation impairs shortly after a hippocampal lesion. Recent working memory and spatial memory in rodents impair in the hippocampal lesion. Short-term memory in monkeys involves the prefrontal cortex. The thalamus, the largest subdivision of the human diencephalon, plays an important role in recent memory [4, 68, 69]. That is, the formation of short-term declarative memory is based on neural circuits formed among the thalamus, prefrontal cortex, and hippocampus. However, long-term memories relevant to knowledge (semantic memory) and life experiences (episodic memory) are not impaired by lesioning the hippocampus, prefrontal cortex, or thalamus. Broad areas in cerebral cortices are often suggested as the allocations for long-term memories though definitive evidence is lacking [70–72].

To a conditioned eyeblink response in humans and mammalians thought to be nondeclarative memory, the association of the neutral stimulus of a brief tone with another stimulus of air-puffing to the eye cornea will lead to that subjects express a conditioned tone inducing a specific eyeblink. By using stimulations, lesions, and recordings, engrams that mediate this motor learning are located in the cerebellum and its associated brain stem. Engrams are further localized in the interpositus nucleus of the cerebellum [72]. In addition, there appear to be the additional storage areas in the cerebellar cortex [71]. As there are interconnections between the cerebral cortices and the cerebellum, it is not surprising that engrams relevant to this motor learning are certainly distributed into population of neurons in these areas [18]. However, that the cerebral cortices or the cerebellum is primarily for the integrative storage and retrieval of signals tone and air-puff remains not known.

In the model of fear memory conditionally formed by pairing a tone or light with a painful electric shock to feet, the conditioned responses including tone- or light-induced upregulation of heartbeat and blood pressure as well as the freezing or shaking of body and arms involve the hypothalamus and amygdala. In particular, critical areas for memory formation in conditioned fear appear to be localized to the basolateral area in the amygdala [73–81]. Stimulations to the amygdala or hippocampus by optogenetic approach can activate engrams relevant to fear memory, which supports the role of amygdala in fear memory [82–85]. To the results why the lesion to the amygdala or the activation to the amygdala and hippocampus leads to similar results in fear memory, the interpretation may be the following that fear memory is influenced by many areas in the central nervous system, the brain in matured human beings and animals is highly wired, and its different regions are interconnected [86]. Stimulations to potential engrams by electrical and optogenetic approaches to a site of the brain may indirectly activate other regions connected with this location to induce memory-relevant behaviors indirectly or those behaviors across or similarly to the memory retrieval. The replay of "memory-relevant behaviors" may not be directly or realistically controlled by memory engrams [18]. It is unclear whether the amygdala is the place for long-term storage of fear information. It is noteworthy that the body freeze and involuntary or voluntary shake used to signify fear memory can be induced by extreme fear, anxiety, emotional reactions (angry and fighting), and physiological processes (hypothermia and hypoglycemia). Electrical shock may activate the whole brain by spreading electrical current in the body, such that the association is not region-specific in the brain [17, 18]. Compared to electrical stimulations used in the study of fear memory, physical and psychological stress in social interaction is closer to the life [87–89].

In the current model of associative learning, the coactivations of sensory cortices including barrel, piriform, and S1-tail cortices lead to their mutual innervations through axon projection and synapse formation as well as the integrative storage of these associated signals in all of these sensory cortices [8, 29, 61] and in the prefrontal cortex and the hippocampus [90, 91]. Primary associative memory cells in sensory cortices and secondary associative memory cells in brain areas related to cognitive processes, spatial localization, and emotional reactions, basic units of engrams, are broadly distributed in the brain through their mutual synapse innervations in convergent and divergent manners [18].

Taken these together, the localization of a specific engram and the distribution of basic units in this engram circuit are not simply restricted to a single site. The conclusion from experimental studies above well explains the early observation by Karl Lashley, who has claimed the wide distribution of engrams in the brain. Physiological impacts for the broad distribution of a specific engram in the brain are given below. The storage of a specific knowledge or experience in many areas of the brain prevents the complete loss of this memory trace from various influences, such as neurological diseases and psychological disorders. Convergent and divergent synapse interconnections among the areas confer the reciprocal retrieval of associated signals, automatic conversions in the retrieval of memories among different modalities, the integrative cognitions and emotions (logical reasoning, associative thinking, and associative emotional reactions), as well as the applications of knowledge and experiences for multiple neural behaviors [12].

1.5 The Identification of Basic Units in Engrams: Memory-Relevant Cells

After memory trace or engram relevant to a specific memory content is localized, basic units in this engram circuit are expected to be identified at the level of individual cells that are able to encode the learned information and to actively execute memory retrieval through behaviors. As the central nervous system is composed of neurons, glia cells, and their interconnected synapses, theoretical basic units in engrams are these nerve cells and synapses. Similar to those strategies used to localize engrams, the correlation of neurons and synapses in a specific engram to memory formation and retrievals can be delineated by the following experiment strategies. For instance, parallel changes are seen between memory formation and neurons or synapses, the downregulation of neurons or synapses leads to lack of memory formation in a prevention manner or to interrupt memory previously formed, and the upregulation of neurons or synapses facilitates memory formation or strengthens memory previously formed. With the meet of all of these criteria in experiment studies, those neurons and synapses identified would be considered as basic units in engram circuits relevant to specific memory [18].

In terms of the identification of neurons in a particular engram, their activities in response to specific cues relevant to this memory are expected to be detected. Neuronal activities are featured by electrical signals generated on cell membrane as well as calcium signals elevated in cells, such that the recording of electrical signals and the imaging of intracellular calcium dynamics can be applied to track cell assemblies relevant to memory formation and retrieval. This approach is called as the detection of the replays of memory-relevant cells [12, 92]. The electrophysiological recording by electrodes or electrode array in vivo has been used to observe the replay of neurons in relevance to particular memory, in which O'keefe and Delong as pioneers use this technique to record memory-relevant neuronal replays in the hippocampus and basal ganglia [93, 94]. Subsequently, this kind of experiments has been done in the hippocampus, sensory cortices, ventral tegmental areas, and amygdala under the distinct conditions including retrieval cue, wakefulness, and sleeping state [95–109]. For example, the coordinate interactions from neurons in the hippocampus to neurons in the prefrontal cortex and association cortices are involved in spatial memory [110, 111]. The cortical-hippocampal-cortical neuronal circuits are important for memory consolidation [112]. Hippocampal cell assemblies can trigger neuronal activities in the ventral striatum during the replay of place-reward messages [113]. The acquisition of associative memory in the hippocampus initiates a gradual-to-stable encoding process at the neurons in the medial prefrontal cortex after the training stops [114]. The representation of emotional memory is associated with the reactivation of the hippocampus-amygdala system during sleeping [115]. These results from functional studies are supported by anatomical evidences, in which the connections are established among the prefrontal cortex, hippocampus, and thalamic nucleus [116].

Memory-relevant neurons in a particular engram have been recently identified in cerebral cortices by using two-photon cell calcium imaging in vivo [117–119]. For instance, the gradual emergence of neuronal activities relevant to spatial memory in the retrosplenial cortex, which mainly receives synaptic inputs from the hippocampus, depends on the intact hippocampus. The indirect connections between the retrosplenial cortex and the hippocampus indicate the mutual influence between the hippocampus and neocortex via polysynaptic neural pathways, i.e., the hippocampus and cerebral cortices include memory-relevant neurons in the widely distributed engrams [120]. Repetitive motor learning induces the formation of new dendritic spines in vivo [121]. Associative memory cells in response to retrieval cues have been identified in the primary sensory cortices, the prefrontal cortex, the hippocampus, and the motor cortex [8, 29, 90, 91]. Therefore, memory-relevant neurons or cell assemblies can be tracked by electrophysiological and imaging recordings based on their activities in response to the retrieval cues and during memory-relevant events.

The functional identification of memory-relevant neurons or cell assemblies should be proved and validated by the morphological approach, i.e., their morphology is quantified and their distribution is localized. Two ways can be used for this purpose, the trace of their synapse innervations from axon inputs that carry the learned signals as well as the labeling of memory cell assemblies by molecules specifically relevant to memory. In the identification of associative memory cells about their mutual synapse innervations, the expression of fluorescent proteins mediated by adeno-associated viruses (AAV) in memory-relevant neurons and their axons has been used. One of these AAVs which are tagged with genes for encoding these fluorescent proteins is injected into the source site of the predicted memory traces. With its transferring and expression at neurons in the source site, the tracing of their axon terminals on neurons in other areas, or they target onto associative memory cells genetically labeled by another fluorescent protein, is performed. The injection and detection in neural tracing can be conducted the other way around to examine their mutual synapse innervations [29, 61, 122]. Associative memory cells are characterized by receiving both innate and new synapse innervations. It is noteworthy that the combination of tracing new synapse contacts and labeling memory cell assemblies with memory-relevant molecules would be an ideal way to identify memory cell assemblies.

The activity of neurons and synapses can induce the alternation of their included molecules [123–125]. These molecules may be relevant to recruit neurons as memory cells during learning. Thus, these molecules may be used to label memory cells to indicate the allocation of memory cell assemblies. The stimulations to neurons couple with the expression of immediate early genes [126]. Their expression in dendrites is regulated by synapse activities [127]. The immediate early gene Arc is specifically linked to neuronal encoding [128]. Some immediate early genes are widely expressed in the brain after fear memory, in which the number of labeled cells is positively correlated to fear memory [129, 130]. There may be the association between the expression of immediate early genes and the active strength of memory cells. Their parallel change leads to a thought that cells labeled by immediate early gene expression are likely engram cells, in which their structures and functions can be studied [16, 119, 131–133]. However, an upregulated expression of immediate early genes is also observed in those neurons with hyperactivity, such as seizure discharge in epilepsy [134–137] and neuron toxicity in brain ischemia [138–140]. Therefore, immediate early genes may be suitable to identify all of the neurons that are highly active. Genes and proteins specifically linked to memory cell assemblies and their memory contents remain to be explored [141, 142]. It is noteworthy that there are around 10,000 types of proteins in living cells [143], which is much less than unit signals remembered in individual's life, such as words, unitary images, odorants, and so on. On the other hand, more than ten billion of neurons reside in the central nervous system; those neurons with synapse interconnection, i.e., associative memory cells, should be basic units in memory traces, instead of the possibility in a specific protein for a given content of memory.

Because particular memory retrieval may be based on engram cells recruited during learning, the activation of memory cells to induce the emergence of memory-relevant behaviors should be used to study memory formation. This strategy is based on a positive correlation between memory cells and memory behaviors. If some neurons store particular memory contents, the activation of these cells by electrical, pharmacological, and optogenetic approaches should induce the expression of memory-relevant behaviors. The electrical stimulations to engram cells in the brain have been used by Penfield who originally expect to localize the source of epilepsy [144]. The stimulations to engram cells in the temporal lobe induce memory recalls in wakeful epileptic patients [145, 146]. Pharmacological stimulations to activate serotonin or norepinephrine systems by using these transmitters facilitate memory formation successfully [147, 148]. Optogenetic stimulations to activate memory engrams induce the fear memory and false memory [82–85, 149]. There is a positive correlation between engram cells and memory-relevant behaviors. It is noteworthy that direct optogenetic activation to neurons without increases of synaptic strength and dendritic spine density leads to memory retrievals [150], implying nonspecific neuron activation. As engram cells are widely distributed in the brain and the brain areas are interconnected, stimulations may lead to the antegrade and retrograde activation of neuronal pathways as well as the indirect activation of memory cells. The approach in stimulations to those engram cells is unable to localize primary versus secondary allocations for memory formation.

If particular memory contents depend upon memory cells recruited during learning, the downregulation of certain molecules critical for memory cells by pharmacological blockers, gene knockout, or optogenetic approaches should prevent or attenuate the emergence of memory-relevant behaviors. This is a common strategy to address the causal relationship among molecular substrates, cellular architectures, and memory events. The first use of surgical ablation to search the distribution of engrams was done by Lashley. Although he failed to localize memory traces, his studies imply the wide distribution of engram cells in the cerebral brain [37, 38, 41]. In subsequent studies, the removal of the temporary lobe in human beings leads to the loss of recent memory due to the impairment of the hippocampus [27, 42–44]. In the study of memory cells by pharmacological reagents, recent memory can be blocked by the intracerebral injection of puromycin [28, 46]. These studies reveal a causal relationship between memory cells in wide brain areas and memory formation though these memory cells specific for content-relevant events remain to be tracked. With the advanced molecular biology, the downregulation of gene expressions by gene knockout [47] and optogenetics [48, 49] has been successfully used to find out the negative correlations among molecules, memory cells, and behaviors. These studies provide strong evidences for the causal relationships among molecular substrates, cellular architectures, and memory formation.

These strategies and approaches to identify memory cells need be evaluated and validated in terms of their advantage and disadvantage. In logical analyses, parallel change, negative correlation, as well as positive correlation between functions and changeable factors should be met to ensure their causal relationship. What the manipulations of molecules and/or cells lead to the changes of memory-relevant events in these three criteria should be combinedly used to figure out memory cells recruited during learning, through which consistent results are expected to be obtained for the conclusion. However, inconsistent results may occur in these studies. For instance, silencing and stimulating patriate cortical neurons lead to inconsistent results during memory retrieval. Parietal lesions do not normally yield severe episodic memory deficits, whereas parietal activations are seen frequently in functional neuroimaging studies of episodic memory [151]. These two categories of evidence suggest that an answer to this puzzle requires to distinguish the contributions of dorsal versus ventral parietal areas and the influence of top-down versus bottom-up attention on memory. The natures of memory trace cells identified in these studies include the following Engram cells can encode the trained signals, receive synapse inputs, and undergo synaptic plasticity [2, 13, 15, 152]. The activation of memory cells induces strong memory retrieval. Memory events are upregulated by norepinephrine and serotonin. How the memory traces encode multiple signals associatively learned remains to be reveal (please refer to associative memory cells).

1.6 Biophysical Changes in Memory-Relevant Cells: Neuronal and Synaptic Plasticity

In line with identifying basic units including neurons and synapses in engrams, another line in the study of learning and memory is to profile biophysical alternations in nervous tissues relevant to memory events, such as the plasticity or refinement of synapses and neurons. As hypothesized by Donald Hebb, neuronal interconnections may be strengthened if both presynaptic and postsynaptic neurons are coincidently activated [6, 39]. This hypothesis is supported by long-term potentiation of synaptic transmission discovered between hippocampal neurons [40] and long-term depression of synaptic transmission in the hippocampus [153]. In addition, the plasticity has been observed in the intrinsic excitability of neurons [10] and the patterns of neuronal spiking [154].

Long-term potentiation (LTP) at synaptic transmission can be induced by high-frequency stimulations to presynaptic axons [9]. Based on Hebb's hypothesis that the connection strengthening induced by presynaptic and postsynaptic coincidence is relevant to memory formation, synapses showing LTP in the cortex and hippocampus have been called as Hebb's synapses. LTP is presumably one of the cellular substrates underlying learning and memory [15, 155, 156]. In terms of the natures LTP induction, firstly, long-term potentiation of synaptic transmission is induced at the specific axon-innervated synapses that have received high-frequency stimulation, which is called as specificity of LTP expression [9]. Secondly, the cooperativity among synapses in a given pathway is thought to be critical for LTP induction. The high-frequency stimulation with weak intensity to a bundle of presynaptic axons, which presumably activates only an axon-made or a few axon-made synapses, is usually unable to induce long-term potentiation at these synapses. However, the high-frequency stimulation with strong intensity to this bundle of presynaptic axons induces LTP expression well [9]. It is possible that tetanus-induced LTP may not be expressed at single synapse. This possibility is also granted by experiment results that the high-frequency stimulation in maximal intensity cannot induce LTP [157], in which all axons in this bundle are presumably activated likely as a unitary synapse. In this regard, tetanus-induced LTP at the bundle of axons may not be due to the potentiation expressed at synapses but due to the decrease of axonal threshold to fire action potentials [10, 157]. Thirdly, LTP shows the associativity of its induction, in which LTP can be induced at synapses in a pathway receiving weak stimulation when a strong stimulation is given to its neighboring pathway that may activate postsynaptic neurons [9]. This associativity of LTP induction can support the cooperativity among synapses, as well as granted by an observation that the association of postsynaptic neuron depolarization for NMDA receptor activation with stimuli to presynaptic axons [158, 159]. As specificity, cooperativity, and associativity are seen in memory formation, the induction and maintenance of long-term potentiation of synaptic transmission have been linked with the neuronal substrates relevant to memory formation and consolidation. It is noteworthy that the associativity of LTP induction differs from the learning of associated signals since associative memory is featured by two independent pathways with the simultaneous activation in the intensity that is sufficient to activate reflexes, respectively [8].

Moreover, the correlation of LTP expression at synapses to memory formation is granted by the following studies [9]. The blockade of LTP induction by pharmacological and genetic approaches can inhibit behavioral tasks to examine memory retrieval. The facilitation of LTP induction by pharmacological and genetic upregulations of gene expression and molecular activation can strengthen behavioral tasks to show memory retrieval. This causal relationship between memory and LTP induction support a possibility that LTP is involved in learning and memory [9]. The following points should be emphasized. As discussed above, the expression of tetanus-induced LTP occurs in a bundle of axons and their innervated synapses, due to the decrease of their spiking threshold, instead of the expression at individual synapses. If the strengthening of synaptic transmission expresses in a given neuronal pathway, this potentiation cannot explain the integrative storage of associated signals carried from their correspondent inputs. Moreover, tetanus-induced LTP usually cannot express at synapses in unmatured animals, such that it is not relevant to learning and memory in early life when learning is efficient and memory is impressive. In addition, long-term depression at synapses is induced by low frequency stimulation, in which 1 Hz stimuli do not mimic any physiological spiking in the neurons, as well as is generated from the extensive depletion of presynaptic vesicles, especially at synapses in young animals. The association of LTP in some pathways and depression in other pathways may be relevant to the contrast in memories.

Biophysical changes in the potentiation of synaptic transmission may result from these factors, such as the conversion of developmentally unmatured synapses into natured synapses, the conversion of silent synapses into function synapses, the conversion of inactive synapses into active synapses, as well as the upregulated compatibility between presynaptic and postsynaptic partners [12]. During postnatal development, the formation of glutamatergic synapses is featured by an emergence of NMDA receptors in the center of postsynaptic density and subsequent insertion of AMPA receptors around [160–162]. The complete formation of glutamatergic synapse includes the embedding of NMDA and AMPA receptors in the postsynaptic density, i.e., matured glutamatergic synapses that are crucial for a fulfillment of synapse plasticity. Even though both NMDA receptors and AMPA receptors are localized in the postsynaptic density, their functions appear sequentially to be different in synaptic transmission. AMPA receptors are activated by glutamates for fast synaptic transmission. The membrane depolarization facilitates the activation of NMDA receptors to mediate slow synaptic transmission [163]. With this complete form of glutamatergic synapses, they demonstrate the feature about the conversion of silent synapses into functional synapses during synaptic plasticity, in which the membrane depolarization activates NMDA receptors and subsequently AMPA receptors to have synapses be converted from silence to function. In fact, even though synapses include both kinds of receptors, they remain to be inactive. Once calmodulin-dependent protein kinase II is activated in postsynaptic neurons, these inactive synapses are converted into active synapses; especially this process is matured during postnatal development [159]. Because this protein kinase plays an important role in LTP induction and maintenance [164–166], the conversion of inactive synapses into active synapses constitutes the mechanism for synaptic potentiation. Finally, there is the functional compatibility between presynaptic and postsynaptic partners [167]. The upregulated compatibility between presynaptic and postsynaptic partners may be critical for the refinement of synapses.

In addition to synapse plasticity, the refinement of neurons in their excitability and spine features is likely involved in learning and memory. The upregulation of neuronal excitability includes an increased capability to fire sequential spikes and a decreased threshold to initiate action potentials [168–172]. In this regard, the upregulated excitability of memory-relevant cells may facilitate memory retrievals induced by cues and spontaneous memory recall. The upregulated excitability may also facilitate the coactivation of ready-recruited neurons for the conversion of neurons into memory-relevant cells [18]. There are two forms of factors that can induce the upregulation of neuronal excitability, i.e., neuronal activities and synaptic activities coordinated by presynaptic and postsynaptic neurons. For instance, the intensive activities in cortical neurons are able to induce themselves to possess an increased spiking capability and a decreased spiking threshold in a multigrade manner [10]. The synapse activations by coordinated activities between presynaptic and postsynaptic neurons enhance the excitability at presynaptic neurons [173, 174]. Moreover, the stronger activation of synapses in the cerebellar nuclear neurons can induce themselves to express the increases of intrinsic excitability and spiking patterns [154, 171, 175–177]. In addition, the neuronal excitability can be upregulated in a compensatory manner after the blockade of voltage-gated sodium channels for 48 h is removed [178]. These results indicate that the activity of neurons undergoes the upregulated plasticity induced by their intensive activity or driven by the strong activity of synapses, though these neurons have not been identified to memory-relevant cells. Currently, the cortical and hippocampal memory cells in vivo are functionally upregulated by input cues learned during the acquisition of associated signals [91]. Furthermore, the enhancement of synapse function and neuron excitability has been detected in associative memory cells in vitro slices [61, 62, 122, 179–181].

As synapses are presumably formed on dendritic spines, spine density and morphology may undergo plastic changes correspondent to synaptic plasticity. Numerous studies have indicated that the density and size of spines are upregulated by the processes relevant to learning and memory [182–197]. In addition, the function of spines is upregulated in an experience-dependent manner [198–200]. On the other hand, extinction procedure induces the pruning of dendritic spines in CA1 hippocampal field depending on strength of training in rats [201]. These studies lead to a conclusion that new spines are recruited and upregulated for new memory [202]. It is noteworthy that spine density in associative memory cells appears not increased though their size is enlarged [62, 180] and that the elimination of dendritic spines with long-term memory is specific to active circuits [203]. These data indicate the enlargement of spine size from filopodia to mushroom shapes which may play an important role in the memory formation. As the increased number and size of dendritic spines in ipsilateral barrel field cortex following unilateral whisker trimming in postnatal rat [204], dendritic spines are also regulated the strength of sensory inputs in an inverse manner. Therefore, the number of spines may be responsible for raising the sensitivity to sensory inputs, whereas the size of spines may be responsible for increasing the formation of functional synapses to ensure the consolidation storage of learned signals.

It is noteworthy that the neuronal and synaptic plasticity occurs under various conditions, such as intensive physiological activities and pathological conditions including anxiety, depression, and epilepsy. Therefore, searching the plasticity at neurons and synapses that is specifically characterized for memory formation is critically important for issuing the conclusion neural plasticity for learning and memory [12].

1.7 Memory-Specific Activity Patterns as Basic Units of Memory Traces

With the profiles about the distribution of memory traces, the identification of memory-relevant cells, and the biophysical change of memory-relevant cells, the activity patterns of these memory cells in relevance to specific memory should be documented. The characters of learned signals and memory contents include the strength and the specificity, such that the basic units of memory traces in relevance to memory content specificity and signal strength should be addressed. Theoretically, the memory-relevant cells are mainly neurons, which express general features of nerve cells and their interconnected synapses, such as electrical activity in synaptic transmission and neuronal action potentials. As the strength of synaptic transmission is influenced by the density and responsiveness of postsynaptic receptors as well as the quantity of presynaptic released transmitters in any of neural pathways [205], this functional feature of synapses may play a major role in the power and maintenance of memory formation and retrieval, but not a specific memory. In addition, neuronal excitability including the number of sequential spikes and the spike threshold in neurons is proportional to the strength of their activities [172], such that this function feature of neurons plays a major role in the power and maintenance of memory formation and retrieval, but not a specific memory. It seems to us that the activity patterns of memory-relevant cells for particular memory contents may not be present.

Many studies have been performed to record neuronal activities relevant to particular memory, based on the fact that neuronal activities are featured by electrical signals generated on cell membrane and calcium signals raised in the cell. The recording of electrical signals and the imaging of intracellular calcium dynamics in the neurons have been used to correlate neuronal activity patterns with particular memory contents [95–109]. For instance, the electrophysiological recording was used to monitor the replays of neurons in the hippocampus, sensory cortices, amygdala, and ventral tegmental areas under the conditions of retrieval cues, wakefulness, and sleep state. The coordinate interactions among hippocampal neurons, prefrontal cortical neurons, and associative cortical neurons were found in relevance to spatial memories by recording neural activity [110–112]. The activation of hippocampal cell assemblies to neuronal activities in the ventral striatum was observed during the replay of place-reward message [113]. In the acquisition of associative memory, the gradual-to-stable encoding process in prefrontal cortical neurons was instigated by the hippocampus without continued training [114]. The retrieval of emotion memory is associated to the reactivation of in the hippocampus-amygdala system in the sleep state [115]. Although these studies have indicated the correlation between these neurons and different types of memories, precise relationships between neuronal activity patterns and specific memory contents have not been quantified successfully.

What mechanisms are influencing the specificity in the storage of various learned signals? Memory cells and their network architectures in sensory cortices as well as associative cortices may play a critical role in specific memory contents [18], since these learned signals come originally from different sensory systems [92]. For instance, visual images including specific features in photon beams and colors as well as their spatial distribution are detected by distinct cone and/or rod cells in the retina and integrated in the visual cortex. Auditory signals in specific tones or sound frequencies are detected by distinct hair cells on the basement membrane of cochlea and integrated by the auditory cortex. Olfactory signals from specific odorants are detected by different olfactory epithelia cells and encoded in the olfactory cortex. Gustatory signals in specific taste sensation are detected by gustatory receptor cells in the tongue and encoded in the gustatory cortex. Somatosensory signals in temperature, tactile, and pain sensation are detected by various somatosensory receptors and encoded by the somatosensory cortex. Memory contents always include more or less the sensory signals, i.e., individuals can tell the sources of the learned and stored signals inputted from the visual system or others. Memory cells may be primarily recruited in the sensory cortices [18]. With these exogenous signals learned from external environments, the signals stored in sensory cortices can be used for cognitive processes, emotional reactions, and other behaviors. Endogenous signals generated in these cognitive and emotional processes specifically, such as logical reasoning and associative thinking, can be memorized. It is likely that memory cells may be secondarily recruited in the brain areas related to cognitions, emotion, and behaviors [17]. These primary and secondary memory cells recruited during learning the specific signals work for encoding specific memory contents.

1.8 The Recruitment of Memory-Relevant Neurons by Their Coactivation: Associative Memory Cells

Memory-relevant nerve cells can be thought to be the different stores for preserving specific exogenous and endogenous signals. In order to be the storeroom, there should be different pathways that connect to these rooms for inputting various merchandises. Similarly, in order to store some new signals into neurons in the brain, new synapse innervations onto these neurons are expected to be established to carry and input these signals. In other words, the neurons with receiving new synapse innervations that carry on exogenous and endogenous signals would be memory cells [17]. The neurons that receive new synapse innervations as well as encode their carried-in signals and relevant cues would be identified as memory cells recruited during learning specific signals and contents. As learning-associated signal is the most common approach of information acquisition, in which a signal previously learned or innately formed is associated with a new signal, memory-related neurons that receive both innate and new synapse innervations as well as encode their carried signals are termed to be associative memory cells [8, 17, 18].

In order to memorize newly learned signals, the formation of new synapse innervations onto neurons that are ready to be recruited as memory traces is essential for the recruitment of associative memory cells [18]. The synapses on ready-recruited neurons, which are formed in previous learning or innate development, should be activated by previously learned signals and innate signal during current learning to drive their membrane potentials toward a threshold of producing sequential spikes. These action potentials on ready-recruited neurons can activate a chain reaction from the epigenetic processes to the expression of their regulated genes and proteins. These molecules participate in building up the axon prolongation and synapse formation for mutual synapse innervations among ready-recruited neurons [29, 30]. In other words, the coactivations of ready-recruited neurons among cerebral cortices across distinct sensory modalities or in a sensory cortex within intramodal sensory modality will lead them to form mutual synapse innervations. With these new synapse innervations plus the synapses formed during previous learning and innate development on ready-recruited neurons, associative memory cells will be recruited [8, 17]. Based on these exogenous signals be memorized, cognitive processes, emotional reactions, or behaviors coactivate the neurons in brain areas that encode these events. Mutual synapse innervations are formed among these neurons, and secondary associative memory cells are recruited [90, 91]. The principle for the recruitment of associative memory cells is active together, connecting together and strengthening together [12, 17, 18].

After cortical neurons are recruited to be associative memory cells in the sensory cortex, they have specific characters for associative memory and general features for neurons. Their specific characters are used as criteria to identify whether the neurons seen in experiments are associative memory cells. The synchronous activity of cortical neurons, i.e., their coactivations, triggers their mutual synapse innervations and recruits them as primary associative memory cells [17, 18]. In terms of their specific characters, primary associative memory cells receive new synapse innervations from the coactivated neurons within or across sensory cortices, such that they are interconnected alongside the innervations from the innate sensation and the arousal system (Fig. 1.1). Associative memory cells are able to encode new and innate signals for their integrative storage and reciprocal retrieval. Associative memory cells include glutamatergic, GABAergic, and glia cells. Their axons convergently and divergently project to and make synapses onto the neurons in brain regions that execute for the cognition, emotion, and behaviors. The recruitment of neurons to be associative memory cells is influenced by microRNA-regulated genes and proteins for axon projections and synapse formations [29, 30, 122]. Interconnections among associative memory cells allow the reciprocal retrieval of associated signals by relevant cues or automatic conversion among various modalities. For instance, the recall of image signals can be stated by verbal language. The recall of verbal signals can be represented by visual diagrams. Their synapse convergence onto downstream neurons plus their coactivations by mutual synapse innervation permits associative thinking, logical reasoning, and computation in forward and backward manners. Their synapse divergence onto other brain regions to activate associative memory cells allows their memorized signals for multiple uses. Regarding general characters for the neurons, the number and encoding ability of associative memory cells influence memory strength and maintenance. The number of recruited associative memory cells is influenced by the weight of their mutual synapse innervations. The synapse weight depends on neuronal coactivation

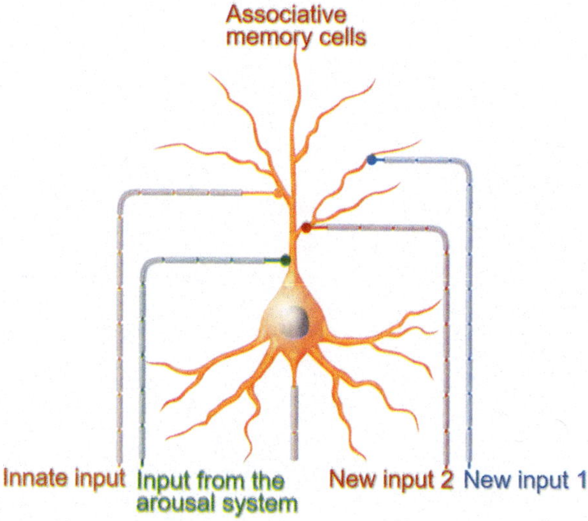

Fig. 1.1 Associative memory cell and its synapse innervation. This primary associative memory cell receives its innate synapse input (orange), synapse input from the arousal system (green), as well as new synapse innervations one (blue) and two (red). During associative learning, this cortical neuron is coactivated with other neurons that are either neighbors in intramodality or other brain regions (cross-modality) by their innate inputs under the condition of wakefulness based on the arousal system. Their coactivity and connection together make these receiving new synapse innervations from coactivated neurons

strength, repetitive coactivations, developmental stages, and axon growth locally to form en passant synapses [8, 12, 29]. The encoding ability of associative memory cells is regulated by the potency of innate/new synapse inputs, their capability to convert synaptic analogue signals into digital spikes and their axon capability to propagate sequential action potentials out [167, 172, 206, 207]. Glutamatergic associative memory cells restrain their neighboring associative memory cells via GABAergic associative memory cells and lateral inhibition, such that they are dominantly active for the memory to be contrast and retrieval [62, 208].

In terms of secondary associative memory cells recruited in the association cortices as well as brain regions related to cognition and emotion, the characters include the following. Secondary associative memory cells receive new synapse innervations convergently from primary associative memory cells across coactivated sensory cortices in cognitive events and emotion reactions. They are able to encode endogenous signals associated across the sensory cortices for the integrative storage (Fig. 1.2). The association of cognitive processes and emotion reactions evokes mutual synapse innervations among these secondary associative memory cells. Their axons will project to memory-output cells in behavior-relevant brain regions for memory retrieval via language, countenance, gesture, and writing. The number of recruited secondary associative memory cells is regulated by mutual synapse innervations induced by coactivation strength and repetitive coactivations in cognition and emotion. The encoding capability of secondary associative memory cells is

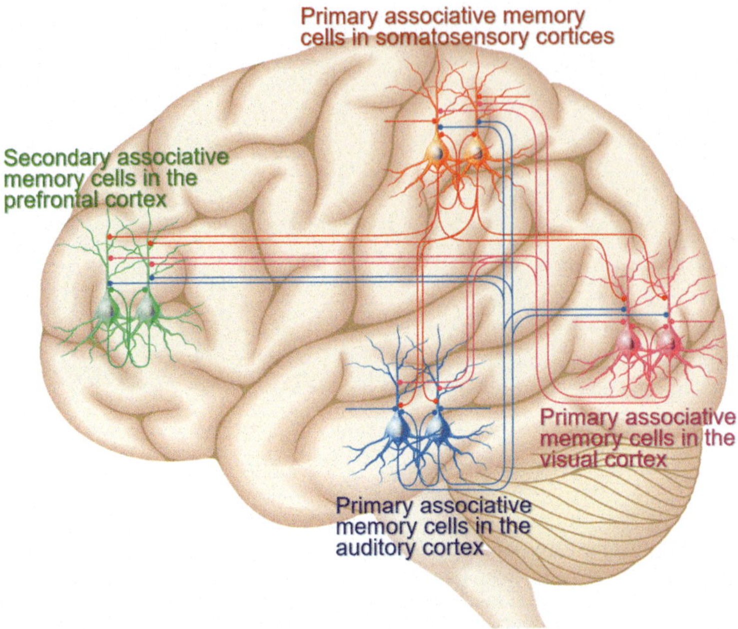

Fig. 1.2 Associative memory cells in the sensory cortices and the frontal lobe of neocortex as well as connections. During associative learning, cortical neurons are coactivated by their innate synapse inputs. Their coactivations will initiate a chain reaction from intensive neuronal spikes to axon prolongation and synapse formation via microRNA-regulated expression of genes and proteins. There cortical neurons are synaptically interconnected and recruited as associative memory cells in intramodal and cross-modal manners. Primary associative memory cells in the auditory cortex (blue), the sensorimotor cortex (orange), and the visual cortex (pink) are mutually innervated. The axons of these primary associative memory cells project convergently toward the prefrontal cortex and make synapses on its neurons to recruit them as secondary associative memory cells (green)

set by synapse inputs, their capability to convert synaptic analogue signals into digital spikes and their axon capability to propagate spikes and drive memory-output cells. The synapse innervations onto secondary associative memory cells determine the specificity of memory contents in the cognition and emotion. The population size, connection strength, and excitation of secondary associative memory cells set up the persistence and power of memory formation and retrieval. Activities of secondary associative memory cells permit the rehearsals of associative thinking, logical reasoning, and emotional reactions. Secondary associative memory cells innervate the hypothalamus and extrapyramidal system to regulate sympathetic and parasympathetic balance, temperature set point, food ingestion, and hormones associated with emotional reactions and behaviors [12, 17, 18].

In addition, primary and secondary associative memory cells are interconnected in memory traces. Basic architectures for their union include the following. Mutual synapse innervations form among primary and/or secondary associative memory cells. The primary associative memory cells in sensory cortices convergently and divergently innervate the secondary associative memory cells in brain regions relevant to cognitions, emotions, and behaviors. Each set of primary associative memory cells reciprocally connects one set of secondary associative memory cells, whose functions are closely relevant. In addition to feedforward innervations from primary to secondary associative memory cells, there may be feedback connection from secondary associative memory cells to primary associative memory cells (Fig. 5.2). The axons from all of these associative memory cells terminate to memory-output cells in the motor cortex. Mutual synapse innervations among primary associative memory cells build up the interactive circuits for the reciprocal retrieval of associated signals by each of sensory cues as well as the automatic conversion retrieval among distinct modalities [8, 29, 61, 62]. Convergent synapse innervations from primary to secondary associative memory cells confer logical reasoning, associative thinking, and other integrative cognitions induced by one of the cues [90]. For instance, one of the secondary associative memory cells is convergently innervated by multi-sets of primary associative memory cells that carry these types of signals, which maintain basic activities in this secondary associative memory cell. When an input cue activates these sets of primary associative memory cells by their mutual synapse innervation, such primary associative memory cells can convergently activate this secondary associative memory cell, in addition to its activation through the dominantly innate chain from one set of primary associative memory cells onto one set of secondary associative memory cells. In other words, these types of signals triggered by one of the cues drive this secondary associative memory cell to achieve the integration of these associated signals for associative thinking and logical reasoning. This integration is also facilitated by mutual synapse innervations among secondary associative cells that contribute to interactions of the higher-order cognitions and emotions. The divergent synapse innervations from primary associative memory cells in sensory cortices to secondary associative memory cells in the prefrontal cortex, the hippocampus, and other areas [90, 91, 209] make associated signals to be stored in several brain areas for long-term maintenance with less chance being loss, as well as to be used for different cognitive processes and emotional reactions [18, 90]. Functional compatibility between presynaptic and postsynaptic partners [167] plays an important role in their coordination for memory formation and decay in these multiple uses. By feedforward and feedback connection among primary and secondary associative memory cells, the learned exogenous signals will automatically initiate cognitions and emotions, and endogenous signals from cognitive and emotional events usually contain sensory signal sources [18].

1.9 Major Milestones in the Study of Mechanisms Underlying Learning and Memory

The milestones for the study of mechanisms underlying learning and memory include the following aspects such as theoretical models for the neural substrates of memory formation and retrieval, animal models for searching cellular and molecular mechanisms, the detection of locations and distributions of these neural substrates, cellular architectures of neural substrates, and molecular mechanisms. In this section of this chapter, the author tries to organize the first and major movement in these aspects [12, 13].

1.9.1 Theoretical Modeling

Before the introduction of experimental approaches into the study of learning and memory, major efforts to figure out mechanisms for memory formation and retrieval are theoretical model. The first term describes the neural substrates of memory formation are memory traces by ancient Greeks around the eighteenth century. During that period, memory traces are thought of as biochemical changes in nerve tissues to represent a memory, a hypothetically permanent change in the nervous system for memorizing something, or biophysical and biochemical alternations induced by a given form of learning and memory. To the early of the twentieth century, Richard Semon issues a term engram in mneme theory [7, 34]. Engrams are proposed to be biophysical and biochemical changes in response to external stimulation. In mneme theory, engram and ecphory are relevant to memory traces and memory retrieval, respectively. Subsequently, Donald Hebb proposes cell assemblies to be the core of engrams or memory traces. Based on experimental observations, he suggests that cell assemblies in the engram are the widely distributed neural substrates for memory. Each cell ensemble is a group of interconnected neurons, and their connections are strengthened in their simultaneous activities [6, 39].

In an ancient Chinese book, *Inner Canon of Yellow Emperor* memories are based on three factors, i.e., "髓 (shui)" that can be translated as core material (such as the brain, spinal, and medulla), "精 (jing)" that can be translated as the spirit, and "血 (xue)" that can be translated as the blood. This theory comes from the fact that these three factors influence memory capability and capacity. Although this theory in Chinese medicine does not tell neural substrates and structural identities for memory formation many centuries ago, we understand that the brain and wakefulness are essential for memories and blood supply influences the brain functions including memories.

1.9.2 Animal Model

To study the natures of memory traces or engrams, animal models are needed to fill the gap of impossible study in human being. The first animal model to study associative memory is issued by Ivan Pavlov in 1897. In this model, conditioned (irrelevant) and nonconditioned (innate) stimulations are associated in dogs. This pairing makes conditioned stimulation to induce nonconditioned responses [210]. Edward Thorndike introduces another model for studying associative memory, called as operant conditioning, in which cats tried to escape from the homemade puzzle boxes [211]. With initially constraining in a puzzle box, cats spend long time to get out by pulling a cord or pushing a pole. After the repeated trials with the frequent success, these cats escaped more quickly. Thorndike generalizes this result to be a rule that behaviors with satisfied outcome tend to be repeated and behaviors with unpleasant consequences are less likely repeated. Similarly to the classical and operant conditioning, other models for associative learning include fear conditioning [212–214] and eyelid-blinking conditioning [215, 216]. Animal models of these conditionings become behaviorism basis in school psychology and psychological therapy.

In the study of mechanisms underlying associative learning and memory in complicated situations, animal models about spatial memory are developed. David Olton designs a radial arm maze to test spatial memory abilities of rats or mice [217, 218]. The maze includes a center platform and the number of arms branching off with food placed at the ends. A rat or mouse is placed in the maze's center to explore each arm individually and retrieve food while simultaneously remembering which arms it has already pursued. Richard Morris designs a water maze to examine rat's or mouse's water navigation task for studying spatial learning and memory [219] (please see Chap. 3 in detail).

After associative memory is formed, the retrieval of associated signals is featured to be reciprocal, i.e., one signal induces the retrieval of its associated signals or the other way around [17]. The animal models of associative learning and memory above nerve indicate the reciprocal retrieval of conditioned and nonconditioned stimulation. A new mouse model from author's lab shows that the pairings of whisker and odor stimulations induce odorant-induced whisker motion and whisker-induced olfactory response [8, 61, 62, 220, 221]. In this mouse model, the recruitment of associative memory cells including glutamatergic neurons, GABAergic neurons, and astrocytes is detected in the coactivated cortices after the formation of associative memory [8, 29, 30, 62].

1.9.3 Location and Distributions

In terms of the distribution of the neural substrates for memory formation, or engram or memory trace, two approaches have been used in the early time, the removal of nerve tissue by surgical ablation and the activation of nerve tissue by stimulation. Karl Lashley searches engrams by the combination of the surgical ablation of brain tissues with rats seeking food in the maze [50]. He is not able to see the complete disappearance of memory-relevant behaviors in mazes through removing brain tissues in the piece-by-piece manner. His conclusion is that memories, at least memories for rats in response to complicated mazes, do not reside in any particular location in cerebral cortices. Furthermore, the removal of the temporary lobe in human being leads to a loss of recent memory [27]. On the other hand, the electrical stimulation to specific brain areas to find out the seizure source in epileptic patients is used to localize the location of engrams by Penfield [144]. He has observed that the stimulation to engram cells in the temporal lobe induces memory recalls in wakeful epileptic patients [145, 146]. The searching of location for engrams has been done by pharmacological reagents, in which recent memory can be blocked by intracerebral injections of puromycin [28].

1.9.4 Cellular Architectures in Engram Circuits

Activity patterns relevant to memory behaviors and memory retrieval can be detected by electrical recording and cellular imaging. As the hippocampus is thought of as a critical area for information storage, O'Keefe records the activities of unitary neurons in the hippocampus from free-moving rats and finds correlation between the hippocampus and space movement [93]. In the meantime, Mahlon Delong shows the activity of pallidal neurons in vivo from basal ganglia in relevance to encode behaviors of operant conditioning [94, 222, 223]. These two pioneers detect neuronal activities in relevance to memory-relevant movement. After two-photon cell imaging is developed, this technique has been used to study the activity of memory-relevant cells, such as associative memory cells [8, 29, 62, 220, 221]. The structure of memory-relevant cells and their networks has been investigated by neural tracing, such as AAV-tagged fluorescent proteins to trace associative memory cells that encode triple associated signals [29].

1.9.5 Molecular Mechanisms

Molecular mechanisms underlying the formation of memory traces or engrams are studied by pharmacological approach and genetic manipulation. Flexner has injected puromycin into cerebral cortices to examine the role of protein synthesis in memory formation, for instance, recent memory can be blocked by intracerebral injections of puromycin [28, 45, 46]. In addition, genetic manipulation has been done by Susumu Tonegawa's lab, in which the knockout of α-calmodulin-dependent protein kinase II leads to the impairment of spatial learning and memory [47]. These two lines of studies constitute the foundation for addressing molecular profiles underlying learning and memory.

References

1. Eichenbaum H. Still searching for the engram. Learn Behav. 2016;44(3):209–22.
2. Josselyn SA, Kohler S, Frankland PW. Finding the engram. Nat Rev Neurosci. 2015;16(9):521–34.

3. McGaugh JL. Memory – a century of consolidation. Science. 2000;287(5451):248–51.
4. Squire LR, Knowlton B, Musen G. The structure and organization of memory. Annu Rev Psychol. 1993;44:453–95.
5. Squire LR. Declarative and nondeclarative memory: multiple brain systems supporting learning and memory. J Cogn Neurosci. 1992;4(3):232–43.
6. Hebb DO. The organization of behavior, a neuropsychological theory. New York: Wiley; 1949.
7. Semon RW. Chapter II Engraphic action of stimuli on the individual. In: Semon RW, editor. The mneme. London: George Allen & Unwin; 1921.
8. Wang D, et al. Neurons in the barrel cortex turn into processing whisker and odor signals: a cellular mechanism for the storage and retrieval of associative signals. Front Cell Neurosci. 2015;9:320.
9. Bliss TVP, Lynch MA. Long-term potentiation of synaptic transmission in the hippocampus: properties and mechanisms. In: Landfield PW, Deadwyler SA, editors. Long-term potentiation: from biophysics to behavior. New York: Alan R. Liss; 1988. p. 3–72.
10. Zhang M, et al. Calcium signal-dependent plasticity of neuronal excitability developed postnatally. J Neurobiol. 2004;61:277–87.
11. Kandel ER, Dudai Y, Mayford MR. The molecular and systems biology of memory. Cell. 2014;157(1):163–86.
12. Wang J-H. Searching basic units of memory traces: associative memory cells. F1000Research. 2019;8(457):1–28.
13. Josselyn SA, Kohler S, Frankland PW. Heroes of the engram. J Neurosci. 2017;37(18):4647–57.
14. Kandel ER. Nerve cells and behavior. In: Kandel ER, Schwartz JH, Jessell TM, editors. Principles of neural science. 4th ed. New York: McGraw-Hill; 2000. p. 19–35.
15. Poo MM, et al. What is memory? The present state of the engram. BMC Biol. 2016;14:40.
16. Tonegawa S, et al. Memory engram cells have come of age. Neuron. 2015;87(5):918–31.
17. Wang JH, Cui S. Associative memory cells: formation, function and perspective. F1000Res. 2017;6:283.
18. Wang JH, Cui S. Associative memory cells and their working principle in the brain. F1000Res. 2018;7:108.
19. Buzsaki G. Network properties of memory trace formation in the hippocampus. Boll Soc Ital Biol Sper. 1991;67(9):817–35.
20. Katkov M, Romani S, Tsodyks M. Memory retrieval from first principles. Neuron. 2017;94(5):1027–32.
21. Jahans-Price T, et al. Computational modeling and analysis of hippocampal-prefrontal information coding during a spatial decision-making task. Front Behav Neurosci. 2014;8:62.
22. Lansner A. Associative memory models: from the cell-assembly theory to biophysically detailed cortex simulations. Trends Neurosci. 2009;32(3):178–86.
23. McClelland JL, McNaughton BL, O'Reilly RC. Why there are complementary learning systems in the hippocampus and neocortex: insights from the successes and failures of connectionist models of learning and memory. Psychol Rev. 1995;102(3):419–57.
24. Schwindel CD, McNaughton BL. Hippocampal-cortical interactions and the dynamics of memory trace reactivation. Prog Brain Res. 2011;193:163–77.
25. Fields RD. Imaging learning: the search for a memory trace. Neuroscientist. 2011;17(2):185–96.
26. Russell IS, Ochs S. Localization of a memory trace in one cortical hemisphere and transfer to the other hemisphere. Brain. 1963;86:37–54.
27. Penfield W. Bilateral frontal gyrectomy and postoperative intelligence. Res Publ Assoc Res Nerv Ment Dis. 1948;27(1):519–34.
28. Flexner JB, Flexner LB, Stellar E. Memory in mice as affected by intracerebral puromycin. Science. 1963;141(3575):57–9.

29. Feng J, et al. Barrel cortical neuron integrates triple associated signals for their memory through receiving epigenetic-mediated new synapse innervations. Cereb Cortex. 2017;27(12):5858–71.
30. Lei Z, et al. Synapse innervation and associative memory cell are recruited for integrative storage of whisker and odor signals in the barrel cortex through miRNA-mediated processes. Front Cell Neurosci. 2017;11(316):1–11.
31. Wang JH, Feng DP. Postsynaptic protein kinase C essential to induction and maintenance of long-term potentiation in the hippocampal CA1 region. Proc Natl Acad Sci U S A. 1992;89(7):2576–80.
32. Wang JH, Kelly PT. Postsynaptic injection of Ca2+/CaM induces synaptic potentiation requiring CaM-KII and PKC activity. Neuron. 1995;15(2):443–52.
33. Wang JH, Ko G, Kelly PT. Cellular and molecular bases of memory: synaptic and neuronal plasticity. J Clin Neurophysiol. 1997;14:264–93.
34. Semon RW. In: Semon RW, editor. Mnemic psychology. London: Allen, Unwin; 1923.
35. McGaugh JL. The search for the memory trace. Ann N Y Acad Sci. 1972;193:112–23.
36. Hebb DO. Spontaneous neurosis in chimpanzees; theoretical relations with clinical and experimental phenomena. Psychosom Med. 1947;9(1):3–19.
37. Lashley KS. Mass action in cerebral function. Science. 1931;73(1888):245–54.
38. Lashley KS. Structural variation in the nervous system in relation to behavior. Psychol Rev. 1947;54(6):325–34.
39. Hebb DO. Animal and physiological psychology. Annu Rev Psychol. 1950;1:173–88.
40. Bliss T, Lomo T. Long-lasting potentiation of synaptic transmission in the dentate area of the anaesthetized rabbit following stimulation of the perforant path. J Physiol Lond. 1973;232:331–56.
41. Lashley KS. Cerebral organization and behavior. Res Publ Assoc Res Nerv Ment Dis. 1958;36:1–4; discussion 14–18
42. Milner B, Penfield W. The effect of hippocampal lesions on recent memory. Trans Am Neurol Assoc. 1955;(80th Meeting): 42–8.
43. Penfield W, Milner B. Memory deficit produced by bilateral lesions in the hippocampal zone. AMA Arch Neurol Psychiatry. 1958;79(5):475–97.
44. Penfield W. Engrams in the human brain. Mechanisms of memory. Proc R Soc Med. 1968;61(8):831–40.
45. Flexner LB, Flexner JB, Roberts RB. Memory in mice analyzed with antibiotics. Antibiotics are useful to study stages of memory and to indicate molecular events which sustain memory. Science. 1967;155(3768):1377–83.
46. Flexner JB, Glexner LB. Studies on memory: evidence for a widespread memory trace in the neocortex after the suppression of recent memory by puromycin. Proc Natl Acad Sci U S A. 1969;62(3):729–32.
47. Silva AJ, et al. Impaired spatial learning in alpha-calcium-calmodulin kinase II mutant mice. Science. 1992;257:206–11.
48. Liu X, et al. Identification and manipulation of memory engram cells. Cold Spring Harb Symp Quant Biol. 2014;79:59–65.
49. Ramirez S, Tonegawa S, Liu X. Identification and optogenetic manipulation of memory engrams in the hippocampus. Front Behav Neurosci. 2013;7:226.
50. Lashley KS. In: Lashley KS, editor. Brain mechanisms and intelligence. Chicago: University of Chicago; 1929.
51. Lashley KS. The mechanism of vision; effects of destroying the visual associative areas of the monkey. Genet Psychol Monogr. 1948;37(2nd Half):107–66.
52. Lashley KS, Chow KL, Semmes J. An examination of the electrical field theory of cerebral integration. Psychol Rev. 1951;58(2):123–36.
53. Lashley KS, Clark G. The cytoarchitecture of the cerebral cortex of Ateles; a critical examination of architectonic studies. J Comp Neurol. 1946;85(2):223–305.

54. Hunter WE, Rolf BB. The psychosomatic aspect of dysmenorrhea; a sensory conditioning process. Am J Obstet Gynecol. 1947;53(1):123–31.
55. Hebb DO. Cerebral organization and consciousness. Res Publ Assoc Res Nerv Ment Dis. 1967;45:1–7.
56. Hebb DO. Physiological learning theory. J Abnorm Child Psychol. 1976;4(4):309–14.
57. Bailey CH, et al. Is heterosynaptic modulation essential for stabilizing Hebbian plasticity and memory? Nat Rev Neurosci. 2000;1(1):11–20.
58. Lynch G, Dunwiddie T, Gribkoff V. Heterosynaptic depression: a postsynaptic correlate of long-term potentiation. Nature. 1977;266:737–9.
59. Muller D, Hefft S, Figurov A. Heterosynaptic interactions between LTP and LTD in CA1 hippocampal slices. Neuron. 1995;14:599–605.
60. Wigström H, Gustafsson B. Heterosynaptic modulation of long-lasting potentiation in the hippocampal slice. Acta Physiol Scand. 1983;119:455–8.
61. Liu Y, et al. Piriform cortical glutamatergic and GABAergic neurons express coordinated plasticity for whisker-induced odor recall. Oncotarget. 2017;8(56):95719–40.
62. Yan F, et al. Coordinated plasticity between barrel cortical glutamatergic and GABAergic neurons during associative memory. Neural Plast. 2016;2016(ID5648390):1–20.
63. Thompson RF, Krupa DJ. Organization of memory traces in the mammalian brain. Annu Rev Neurosci. 1994;17:519–49.
64. Milner B. The memory defect in bilateral hippocampal lesions. Psychiatr Res Rep Am Psychiatr Assoc. 1959;11:43–58.
65. Penfield W. Neurosurgery; yesterday, today and tomorrow. J Neurosurg. 1949;6(1):6–12.
66. Penfield W. Observations on the anatomy of memory. Folia Psychiatr Neurol Neurochir Neerl. 1950;53(2):349–51.
67. Penfield W. Some observations on amnesia. Am J Psychiatry. 1954;110(11):834.
68. Squire LR. Mechanisms of memory. Science. 1986;232(4758):1612–9.
69. Squire LR, Wixted JT, Clark RE. Recognition memory and the medial temporal lobe: a new perspective. Nat Rev Neurosci. 2007;8(11):872–83.
70. Thompson RF. In search of memory traces. Annu Rev Psychol. 2005;56:1–23.
71. Thompson RF. An essential memory trace found. Behav Neurosci. 2013;127(5):669–75.
72. Thompson RF, Kim JJ. Memory systems in the brain and localization of a memory. Proc Natl Acad Sci U S A. 1996;93(24):13438–44.
73. Agren T, et al. Disruption of reconsolidation erases a fear memory trace in the human amygdala. Science. 2012;337(6101):1550–2.
74. LeDoux JE, et al. The lateral amygdaloid nucleus: sensory interface of the amygdala in fear conditioning. J Neurosci. 1990;10(4):1062–9.
75. LeDoux J. Emotion circuits in the brain. Annu Rev Neurosci. 2000;23:155–84.
76. LeDoux J. The emotional brain, fear and the amygdala. Cell Mol Neurosci. 2003;23(4–5):727–38.
77. Lee JH, Lee S, Kim JH. Amygdala circuits for fear memory: a key role for dopamine regulation. Neuroscientist. 2017;23:542–53.
78. Ehrlich I, et al. Amygdala inhibitory circuits and the control of fear memory. Neuron. 2009;62:757–71.
79. Johansen JP, et al. Molecular mechanisms of fear learning and memory. Cell. 2011;147(3):509–24.
80. Maren S. The amygdala, synaptic plasticity, and fear memory. Ann N Y Acad Sci. 2003;985:106–13.
81. Morena M, et al. Upregulation of anandamide hydrolysis in the basolateral complex of amygdala reduces fear memory expression and indices of stress and anxiety. J Neurosci. 2019;39:1275–92.
82. Liu X, et al. Optogenetic stimulation of a hippocampal engram activates fear memory recall. Nature. 2012;484(7394):381–5.

83. Liu X, Ramirez S, Tonegawa S. Inception of a false memory by optogenetic manipulation of a hippocampal memory engram. Philos Trans R Soc Lond Ser B Biol Sci. 2014;369(1633):20130142.
84. Ramirez S, et al. Creating a false memory in the hippocampus. Science. 2013;341(6144):387–91.
85. Ramirez S, et al. Activating positive memory engrams suppresses depression-like behaviour. Nature. 2015;522(7556):335–9.
86. Cohen Kadosh R, Walsh V. Cognitive neuroscience: rewired or crosswired brains? Curr Biol. 2006;16(22):R962–3.
87. Martinez M, Calvo-Torrent A, Pico-Alfonso MA. Social defeat and subordination as models of social stress in laboratory rodents: a review. Aggress Behav. 1998;24:241–56.
88. Tsankova NM, et al. Sustained hippocampal chromatin regulation in a mouse model of depression and antidepressant action. Nat Neurosci. 2006;9(4):519–25.
89. Vasconcelos M, Stein DJ, de Almeida RM. Social defeat protocol and relevant biomarkers, implications for stress response physiology, drug abuse, mood disorders and individual stress vulnerability: a systematic review of the last decade. Trends Psychiatry Psychother. 2015;37(2):51–66.
90. Wang J-H, et al. Prefrontal cortical neurons are recruited as secondary associative memory cells for associative memory and cognition. Biophys J. 2018;114(3):155a.
91. Wang JH, et al. Secondary associative memory cells and their plasticity in the prefrontal cortex. Biophys J. 2019;116(3):427a.
92. Zhao J, Wang D, Wang JH. Barrel cortical neurons and astrocytes coordinately respond to an increased whisker stimulus frequency. Mol Brain. 2012;5:12.
93. O'Keefe J, Dostrovsky J. The hippocampus as a spatial map. Preliminary evidence from unit activity in the freely-moving rat. Brain Res. 1971;34(1):171–5.
94. DeLong MR. Activity of pallidal neurons during movement. J Neurophysiol. 1971;34(3):414–27.
95. Carr MF, Jadhav SP, Frank LM. Hippocampal replay in the awake state: a potential substrate for memory consolidation and retrieval. Nat Neurosci. 2011;14(2):147–53.
96. Harris KD, et al. Organization of cell assemblies in the hippocampus. Nature. 2003;424(6948):552–6.
97. Ji D, Wilson MA. Coordinated memory replay in the visual cortex and hippocampus during sleep. Nat Neurosci. 2007;10(1):100–7.
98. Jadhav SP, et al. Awake hippocampal sharp-wave ripples support spatial memory. Science. 2012;336(6087):1454–8.
99. Kay K, et al. A hippocampal network for spatial coding during immobility and sleep. Nature. 2016;531(7593):185–90.
100. Kudrimoti HS, Barnes CA, McNaughton BL. Reactivation of hippocampal cell assemblies: effects of behavioral state, experience, and EEG dynamics. J Neurosci. 1999;19(10):4090–101.
101. McNaughton BL, Barnes CA, O'Keefe J. The contributions of position, direction, and velocity to single unit activity in the hippocampus of freely-moving rats. Exp Brain Res. 1983;52(1):41–9.
102. Yu JY, et al. Specific hippocampal representations are linked to generalized cortical representations in memory. Nat Commun. 2018;9(1):2209.
103. Penfield W, Perot P. The Brain's record of auditory and visual experience. A final summary and discussion. Brain. 1963;86:595–696.
104. Sirota A, et al. Communication between neocortex and hippocampus during sleep in rodents. Proc Natl Acad Sci U S A. 2003;100(4):2065–9.
105. Skaggs WE, McNaughton BL. Replay of neuronal firing sequences in rat hippocampus during sleep following spatial experience. Science. 1996;271(5257):1870–3.
106. Wilson MA, McNaughton BL. Dynamics of the hippocampal ensemble code for space. Science. 1993;261(5124):1055–8.

107. Wilson MA, McNaughton BL. Reactivation of hippocampal ensemble memories during sleep. Science. 1994;265(5172):676–9.
108. Wirth S, et al. Single neurons in the monkey hippocampus and learning of new associations. Science. 2003;300(5625):1578–81.
109. Yokose J, et al. Overlapping memory trace indispensable for linking, but not recalling, individual memories. Science. 2017;355(6323):398–403.
110. Jones MW, Wilson MA. Theta rhythms coordinate hippocampal-prefrontal interactions in a spatial memory task. PLoS Biol. 2005;3(12):e402.
111. Khodagholy D, Gelinas JN, Buzsaki G. Learning-enhanced coupling between ripple oscillations in association cortices and hippocampus. Science. 2017;358(6361):369–72.
112. Rothschild G, Eban E, Frank LM. A cortical-hippocampal-cortical loop of information processing during memory consolidation. Nat Neurosci. 2017;20(2):251–9.
113. Lansink CS, et al. Hippocampus leads ventral striatum in replay of place-reward information. PLoS Biol. 2009;7(8):e1000173.
114. Takehara-Nishiuchi K, McNaughton BL. Spontaneous changes of neocortical code for associative memory during consolidation. Science. 2008;322(5903):960–3.
115. Girardeau G, Inema I, Buzsaki G. Reactivations of emotional memory in the hippocampus-amygdala system during sleep. Nat Neurosci. 2017;20(11):1634–42.
116. Varela C, et al. Anatomical substrates for direct interactions between hippocampus, medial prefrontal cortex, and the thalamic nucleus reuniens. Brain Struct Funct. 2014;219(3):911–29.
117. Nikolenko V, Poskanzer KE, Yuste R. Two-photon photostimulation and imaging of neural circuits. Nat Methods. 2007;4(11):943–50.
118. Stosiek C, et al. In vivo two-photon calcium imaging of neuronal networks. Proc Natl Acad Sci USA. 2003;100(12):7319–24.
119. Wang KH, et al. In vivo two-photon imaging reveals a role of arc in enhancing orientation specificity in visual cortex. Cell. 2006;126(2):389–402.
120. Mao D, et al. Hippocampus-dependent emergence of spatial sequence coding in retrosplenial cortex. Proc Natl Acad Sci U S A. 2018;115(31):8015–8.
121. Fu M, et al. Repetitive motor learning induces coordinated formation of clustered dendritic spines in vivo. Nature. 2012;483(7387):92–5.
122. Gao Z, et al. Associations of unilateral whisker and olfactory signals induce synapse formation and memory cell recruitment in bilateral barrel cortices: cellular mechanism for unilateral training toward bilateral memory. Front Cell Neurosci. 2016;10(285):1–16.
123. Sugiura H, et al. Transducing neuronal activity into dendritic spine morphology: new roles for p38 MAP kinase and N-cadherin. Neuroscientist. 2009;15(1):90–104.
124. Thiagarajan TC, Piedras-Renteria ES, Tsien RW. Alpha- and beta-CaMKII. Inverse regulation by neuronal activity and opposing effects on synaptic strength. Neuron. 2002;36:1103–14.
125. Wong-Riley MT. Cytochrome oxidase: an endogenous metabolic maker for neuronal activity. Trends Neurosci. 1989;12(3):94–101.
126. Morgan JI, Curran T. Stimulus-transcription coupling in neurons: role of cellular immediate-early genes. Trends Neurosci. 1989;12(11):459–62.
127. Link W, et al. Somatodendritic expression of an immediate early gene is regulated by synaptic activity. Proc Natl Acad Sci U S A. 1995;92(12):5734–8.
128. Guzowski JF, et al. Environment-specific expression of the immediate-early gene Arc in hippocampal neuronal ensembles. Nat Neurosci. 1999;2(12):1120–4.
129. Reijmers LG, et al. Localization of a stable neural correlate of associative memory. Science. 2007;317(5842):1230–3.
130. Tayler KK, et al. Reactivation of neural ensembles during the retrieval of recent and remote memory. Curr Biol. 2013;23(2):99–106.
131. Bissiere S, et al. Electrical synapses control hippocampal contributions to fear learning and memory. Science. 2011;331(6013):87–91.
132. Grundemann J, Luthi A. Ensemble coding in amygdala circuits for associative learning. Curr Opin Neurobiol. 2015;35:200–6.

133. Tanaka KZ, McHugh TJ. The hippocampal engram as a memory index. J Exp Neurosci. 2018;12:1179069518815942.
134. Kiessling M, Gass P. Immediate early gene expression in experimental epilepsy. Brain Pathol. 1993;3(4):381–93.
135. Meldrum BS. Concept of activity-induced cell death in epilepsy: historical and contemporary perspectives. Prog Brain Res. 2002;135:3–11.
136. Wang X, et al. Persistent hyperactivity of hippocampal dentate interneurons after a silent period in the rat pilocarpine model of epilepsy. Front Cell Neurosci. 2016;10:94.
137. Simonato M, et al. Differential expression of immediate early genes in the hippocampus in the kindling model of epilepsy. Brain Res Mol Brain Res. 1991;11(2):115–24.
138. Abe H, Nowak TS Jr. Induced hippocampal neuron protection in an optimized gerbil ischemia model: insult thresholds for tolerance induction and altered gene expression defined by ischemic depolarization. J Cereb Blood Flow Metab. 2004;24(1):84–97.
139. Bokesch PM, et al. Dextromethorphan inhibits ischemia-induced c-fos expression and delayed neuronal death in hippocampal neurons. Anesthesiology. 1994;81(2):470–7.
140. Kiessling M, et al. Differential transcription and translation of immediate early genes in the gerbil hippocampus after transient global ischemia. J Cereb Blood Flow Metab. 1993;13(6):914–24.
141. Asok A, et al. Molecular mechanisms of the memory trace. Trends Neurosci. 2019;42(1):14–22.
142. Sossin WS. Molecular memory traces. Prog Brain Res. 2008;169:3–25.
143. Milo R. What is the total number of protein molecules per cell volume? A call to rethink some published values. BioEssays. 2013;35(12):1050–5.
144. Penfield W. Some mechanisms of consciousness discovered during electrical stimulation of the brain. Proc Natl Acad Sci U S A. 1958;44(2):51–66.
145. Rasmussen T, Penfield W. The human sensorimotor cortex as studied by electrical stimulation. Fed Proc. 1947;6(1 Pt 2):184.
146. Penfield W, Welch K. Instability of response to stimulation of the sensorimotor cortex of man. J Physiol. 1949;109(3–4):358–65, illust
147. Ferry B, Roozendaal B, McGaugh JL. Role of norepinephrine in mediating stress hormone regulation of long-term memory storage: a critical involvement of the amygdala. Biol Psychiatry. 1999;46(9):1140–52.
148. Meneses A, Liy-Salmeron G. Serotonin and emotion, learning and memory. Rev Neurosci. 2012;23(5–6):543–53.
149. Roy DS, et al. Memory retrieval by activating engram cells in mouse models of early Alzheimer's disease. Nature. 2016;531(7595):508–12.
150. Ryan TJ, et al. Memory. Engram cells retain memory under retrograde amnesia. Science. 2015;348(6238):1007–13.
151. Cabeza R, et al. The parietal cortex and episodic memory: an attentional account. Nat Rev Neurosci. 2008;9(8):613–25.
152. Dubnau J, Chiang AS, Tully T. Neural substrates of memory: from synapse to system. J Neurobiol. 2003;54(1):238–53.
153. Stanton PK, Sejnowski TJ. Associative long-term depression in the hippocampus induced by hebbian covariance. Nature. 1989;339(6221):215–8.
154. Daoudal D, Debanne D. Long-term plasticity of intrinsic excitability: learning rules and mechanisms. Learn Mem. 2003;10:456–65.
155. Bliss TVP, Collingridge GL. A synaptic model of memory: LTP in the hippocampus. Nature. 1993;361:31–9.
156. Malenka RC, Nicoll RA. Long-term potentiation – a decade of progress? Nat Neurosci. 1999;285:1870–4.
157. Chavez NLE, Halliwell JV, Bliss TV. A decrease in firing threshold observed after induction of the EPSP-spike (E-S) component of long-term potentiation in rat hippocampal slices. Exp Brain Res. 1990;79(3):633–41.

158. Collingridge GL, et al. Involvement of excitatory amino acid receptors in long-term potentiation in the Schaffer collateral-commissural pathway of rat hippocampal slices. Can J Physiol Pharmacol. 1991;69(7):1084–90.
159. Wang J-H, Kelly PT. Ca2+/CaM signalling pathway up-regulates glutamatergic synaptic function in non-pyramidal fast-spiking neurons of hippocampal CA1. J Physiol Lond. 2001;533(2):407–22.
160. Liao D, et al. Regulation of morphological postsynaptic silent synapses in developing hippocampal neurons. Nat Neurosci. 1999;2:37–43.
161. Petralia RS, et al. Selective acquisition of AMPA receptors over postnatal development suggests a molecular basis for silent synapses. Nat Neurosci. 1999;2:31–6.
162. Takumi Y, et al. Different modes of expression of AMPA and NMDA receptors in hippocampal synapses. Nat Neurosci. 1999;2:618–24.
163. Liao D-Z, Hessler NA, Malinow R. Activation of postsynaptic silent synapses during pairing-induced LTP in CA1 region of hippocampal slice. Nature. 1995;375:400–4.
164. Malinow R, Schulman H, Tsien RW. Inhibition of postsynaptic PKC or CaMKII blocks induction but not expression of LTP. Science. 1989;245:862–6.
165. Nichols RA, et al. Calcium/calmodulin-dependent protein kinase II increases glutamate and noradrenaline release from synaptosomes. Nature. 1990;343(6259):647–51.
166. Wang J-H, Kelly PT. Balance between postsynaptic Ca2+–dependent protein kinase and phosphatase activities controlling synaptic strength. Learn Mem. 1996;3:170–81.
167. Yang Z, et al. Functional compatibility between Purkinje cell axon branches and their target neurons in the cerebellum. Oncotarget. 2017;8(42):72424–37.
168. Chen N, et al. The refractory periods and threshold potentials of sequential spikes measured by whole-cell recordings. Biochem Biophys Res Commun. 2006;340:151–7.
169. Chen N, et al. Sodium channel-mediated intrinsic mechanisms underlying the differences of spike programming among GABAergic neurons. Biochem Biophys Res Commun. 2006;346:281–7.
170. Chen N, et al. After-hyperpolarization improves spike programming through lowering threshold potentials and refractory periods mediated by voltage-gated sodium channels. Biochem Biophys Res Commun. 2006;346:938–45.
171. Chen N, Chen X, Wang J-H. Homeostasis established by coordination of subcellular compartment plasticity improves spike encoding. J Cell Sci. 2008;121(17):2961–71.
172. Wang JH, et al. The gain and fidelity of transmission patterns at cortical excitatory unitary synapses improve spike encoding. J Cell Sci. 2008;121(17):2951–60.
173. Ganguly K, Kiss L, Poo M-M. Enhancement of presynaptic neuronal excitability by correlated presynaptic and postsynaptic spiking. Nat Neurosci. 2000;3(10):1018–26.
174. Nick TA, Ribera AB. Synaptic activity modulates presynaptic excitability. Nat Neurosci. 2000;3(2):142–9.
175. Aizenmann C, Linden DJ. Rapid, synaptically driven increases in the intrinsic excitability of cerebellar nuclear neurons. Nat Neurosci. 2000;3:109–11.
176. Campanac E, Debanne D. Plasticity of neuronal excitability: Hebbian rules beyond the synapse. Arch Ital Biol. 2007;145(3–4):277–87.
177. Sourdet V, et al. Long-term enhancement of neuronal excitability and temporal fidelity mediated by metabotropic glutamate receptor subtype 5. J Neurosci. 2003;23(32):10238–48.
178. Desai NS, Rutherford L, Turrigiano GG. Plasticity in the intrinsic excitability of cortical pyramidal neurons. Nat Neurosci. 1999;2(6):515–20.
179. Guo R, et al. Associative memory extinction is accompanied by decayed plasticity at motor cortical neurons and persistent plasticity at sensory cortical neurons. Front Cell Neurosci. 2017;11(168):1–12.
180. Liu Y, et al. Activity strengths of cortical glutamatergic and GABAergic neurons are correlated with transgenerational inheritance of learning ability. Oncotarget. 2017;8(68):112401–16.

181. Zhao X, et al. Coordinated plasticity among glutamatergic and GABAergic neurons and synapses in the barrel cortex is correlated to learning efficiency. Front Cell Neurosci. 2017;11(221):1–12.
182. Ashby MC, Isaac JT. Maturation of a recurrent excitatory neocortical circuit by experience-dependent unsilencing of newly formed dendritic spines. Neuron. 2011;70(3):510–21.
183. Bourne J, Harris KM. Do thin spines learn to be mushroom spines that remember? Curr Opin Neurobiol. 2007;17(3):381–6.
184. Fifkova E, Van Harreveld A. Long-lasting morphological changes in dendritic spines of dentate granular cells following stimulation of the entorhinal area. J Neurocytol. 1977;6:211–30.
185. Fortin DA, Srivastava T, Soderling TR. Structural modulation of dendritic spines during synaptic plasticity. Neuroscientist. 2011;18(4):326–41.
186. Hongpaisan J, Alkon DL. A structural basis for enhancement of long-term associative memory in single dendritic spines regulated by PKC. Proc Natl Acad Sci U S A. 2007;104(49):19571–6.
187. Kasai H, et al. Structural dynamics of dendritic spines in memory and cognition. Trends Neurosci. 2010;33(3):121–9.
188. Kitanishi T, et al. Experience-dependent, rapid structural changes in hippocampal pyramidal cell spines. Cereb Cortex. 2009;19(11):2572–8.
189. Lendvai B, et al. Experience-dependent plasticity of dendritic spines in the developing rat barrel cortex in vivo. Nature. 2000;404:876–81.
190. Leuner B, Falduto J, Shors TJ. Associative memory formation increases the observation of dendritic spines in the hippocampus. J Neurosci. 2003;23(2):659–65.
191. Matsuzaki M, et al. Structural basis of long-term potentiation in single dendritic spines. Nature. 2004;429(6993):761–6.
192. Moczulska KE, et al. Dynamics of dendritic spines in the mouse auditory cortex during memory formation and memory recall. Proc Natl Acad Sci U S A. 2013;110(45):18315–20.
193. Murakoshi H, Yasuda R. Postsynaptic signaling during plasticity of dendritic spines. Trends Neurosci. 2012;35(2):135–43.
194. Zuo Y, Perkon I, Diamond ME. Whisking and whisker kinematics during a texture classification task. Philos Trans R Soc Lond Ser B Biol Sci. 2011;366(1581):3058–69.
195. Patterson M, Yasuda R. Signalling pathways underlying structural plasticity of dendritic spines. Br J Pharmacol. 2011;163(8):1626–38.
196. Tanaka J, et al. Protein synthesis and neurotrophin-dependent structural plasticity of single dendritic spines. Science. 2008;319(5870):1683–7.
197. Yuste R, Denk W. Dendritic spines as basic functional units of neuronal integration. Nature. 1995;375:682–4.
198. Harnett MT, et al. Synaptic amplification by dendritic spines enhances input cooperativity. Nature. 2012;491(7425):599–602.
199. Jung CK, Herms J. Structural dynamics of dendritic spines are influenced by an environmental enrichment: an in vivo imaging study. Cereb Cortex. 2014;24:377–84.
200. Schiegg A, Gerstner W, Ritz R, Leo van Hemmen J. Intracellular Ca2+ stores can account for the time course of LTP induction: a model of Ca2+ dynamics in dendritic spines. J Neurophysiol. 1995;74(3):1046–55.
201. Garin-Aguilar ME, et al. Extinction procedure induces pruning of dendritic spines in CA1 hippocampal field depending on strength of training in rats. Front Behav Neurosci. 2012;6:12.
202. Leuner B, Shors TJ. New spines, new memories. Mol Neurobiol. 2004;29(2):117–30.
203. Sanders J, et al. Elimination of dendritic spines with long-term memory is specific to active circuits. J Neurosci. 2012;32(36):12570–8.
204. Vees AM, et al. Increased number and size of dendritic spines in ipsilateral barrel field cortex following unilateral whisker trimming in postnatal rat. J Comp Neurol. 1998;400(1):110–24.
205. Stevens CF. Strengthening the synapses news. Nature. 1989;338(6215):460–1.
206. Yu J, et al. Quantal glutamate release is essential for reliable neuronal encodings in cerebral networks. PLoS One. 2011;6(9):e25219.

207. Yu J, Qian H, Wang JH. Upregulation of transmitter release probability improves a conversion of synaptic analogue signals into neuronal digital spikes. Mol Brain. 2012;5(1):26.
208. Wang J-H, et al. Both glutamatergic and Gabaergic neurons are recruited to be associative memory cells. Biophys J. 2016;110(3 Supplement):481a.
209. Wang J-H, Guo R, Wei Z. Associative memory extinction is accompanied by decays of associative memory cells and their plasticity at motor cortex but not sensory cortex. Soc Neurosci. 2017;81(09):10385.
210. Pavlov I. Conditioned reflexes: an investigation of the physiological activity of the cerebral cortex. Translated by Anrep GV. Nature. 1927;121(3052):662–4.
211. Thorndike EL. Animal intelligence: an experimental study of the associative processes in animals. Psychol Rev Monogr Suppl. 1901;2:1–109.
212. Davitz JR, et al. Conditioning of fear: a function of the delay of reinforcement. Am J Psychol. 1957;70(1):69–74.
213. Strouthes A. Desensitization and fear conditioning. Psychol Rep. 1965;17(3):787–90.
214. Weisz K, Vasenszky S, Weisz P. A method for producing a conditioned fear reflex in rats. Kiserl Orvostud. 1953;5(3):199–202.
215. Pennypacker HS, et al. An apparatus and procedure for conditioning the eye-blink reflex in the squirrel monkey. J Exp Anal Behav. 1966;9(5):601–4.
216. Theios J, Brelsford JW Jr. A Markov model for classical conditioning: application to eye-blink conditioning in rabbits. Psychol Rev. 1966;73(5):393–408.
217. Olton DS, Isaacson RL. Importance of spatial location in active avoidance tasks. J Comp Physiol Psychol. 1968;65(3):535–9.
218. Olton DS. Spatial memory. Sci Am. 1977;236(6):82–4, 89–94, 96, 98
219. Morris R. Developments of a water-maze procedure for studying spatial learning in the rat. J Neurosci Methods. 1984;11(1):47–60.
220. Wang J-H, et al. Neurons in barrel cortex turn into processing whisker and odor signals: a novel form of associative learning. Soc Neurosci. 2013;653(14):WW11.
221. Wang JH, et al. Upregulation of glutamatergic receptor-channels is associated with cross-modal reflexes encoded in barrel cortex and piriform cortex. Biophys J. 2014;106(2 Supplement):191a.
222. DeLong MR. Activity of basal ganglia neurons during movement. Brain Res. 1972;40(1):127–35.
223. DeLong MR. Putamen: activity of single units during slow and rapid arm movements. Science. 1973;179(4079):1240–2.

Chapter 2
Patterns of Learning and Memory

Abstract Learning and memory have been classified into various patterns in physiology and psychology, although cellular architectures underlying these patterns of information acquisition and storage remain largely unknown. It is critically important to reveal the correlation of memory cells and their circuits with various memory patterns in order to have a comprehensive view of learning and memory as well as develop therapeutic strategies for memory deficits. In the acquisition of information, knowledge, and experiences, learning has been classified into associative learning and nonassociative learning, in which associative learning is a major style in information acquisition. In terms of the pattern of memories to acquired signals, various classifications are assigned based on memory contents. For instance, declarative (explicit) memory or nondeclarative (implicit) memory is judged whether memories are associated with consciousness state or not. In declarative memory, episodic memory refers to the storage of events, place, time, their associated emotions, and other conception-based knowledge in relevance to specific experience, whereas semantic memory involves episodic memory relevant to generalized and summarized knowledge, theories, and views. Based on input and output, there are working memory and perceptual memory. Working memory is featured as a short-term memory from the sensory input guidance to the processing manipulation, whereas perceptual memory is featured as long-term memories to visual, auditory, and other perceptual signals. In terms of the capacity and efficiency of signal retrievals about perceptual memory, eidetic memory is used to describe that the signals learned in short time can be retrieved vividly, especially in the childhood stage. Regardless of these patterns classified for learning and memory, the purpose of studying memory formation and retrieval is to uncover cellular architectures that are suitable to interpret all of these types of memories. In this chapter, author intends to figure out a diagram constructed from these types of memories and to propose testable cellular networks for them.

Keywords Associative memory · Explicit memory · Implicit memory · Episodic memory and semantic memory

J.-H. Wang, *Associative Memory Cells: Basic Units of Memory Trace*,
https://doi.org/10.1007/978-981-13-9501-7_2

2.1 Introduction

Learning and memory are essential for cognitive events, emotional reactions, and their output through behaviors in life [1–4]. The topics about learning efficiency, memory capacity, and memory enhancement are widely interested in normal population and individuals suffering from memory deficits. Scientists in fields of physiology, psychology, pathology, pharmacology, and neuroscience have paid countless attention to reveal cellular and molecular mechanisms as bases of these topics and interests relevant to memory formation and its relevant behaviors [5]. Many patterns of learning and memory have been noticed and classified in the society for memoriology, especially in the field of psychology. Unfortunately, the cellular and molecular mechanisms underlying these patterns of learning and memory have not been clearly uncovered. Core and comprehensive diagrams for learning and memory remain to be figured out. The effort to abstract the nucleus diagram for different patterns of memories is valuable to compensate this unfortunate event and to reveal memory core [6].

In the study of learning and memory by the society for memoriology, a variety of patterns about learning and memory has been defined based on learning styles, consciousness state for memory, memory contents, memory location, memory persistence, and memory efficiency. In addition to nonassociative learning, including sensitization and habituation, the acquisition of information, knowledge, and experiences depends mainly on associative learning in that multiple signals are associatively learned and conjointly memorized. Compared with nondeclarative (implicit) memory for the storage of procedures and operations without the requirement of consciousness thoughts, declarative (explicit) memory stands for information storage with the consciousness retrieval of memory contents. Declarative memory is further classified into episodic memory for the storage of events, places, time, their associated emotions, and other conception-based knowledge relevant to specific experiences, as well as semantic memory for the storage of generalized and summarized knowledge, theories, and views from episodic memory. In terms of motion-relevant processes, memory can be defined as short-term working memory from the sensory input guidance to the processing manipulation and long-term perceptual memory for the storage of visual, auditory, and other perceptual signals. In terms of the capacity and efficiency of retrievals about perceptual memory, eidetic memory is used to describe that signals learnt in short time can be retrieved vividly, especially in childhood stage. Based on the space and time of signals, memory can be classified as spatial memory and temporal memory. In addition, memories can be classified as memories to unitary signals versus complicated signals that are organized from unitary signals and memories to exogenous versus endogenous signals. In this regard, associative memory includes complicated patterns to be realized.

Figure 2.1 presents a diagram for associative memory and its classification, in which common features for these patterns should be those in associative memory. In other words, cellular architectures for associative memory will constitute the foun-

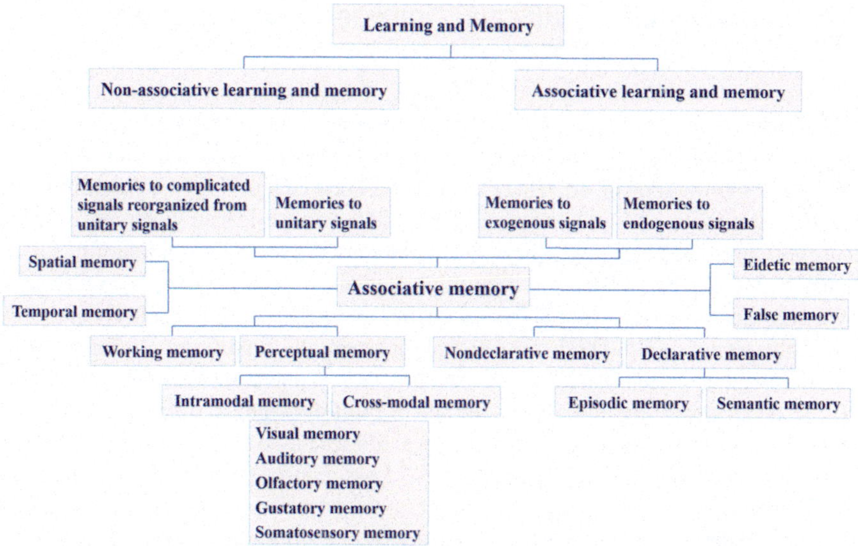

Fig. 2.1 The classification of associative learning and memory
The information acquisition can be classified into associative learning and nonassociative learning. Associative memory is classified into declarative (explicit) versus nondeclarative (implicit) memory based on the need of consciousness during memory retrieval, perceptual versus working memory based on signal sources, spatial versus temporal memory, eidetic versus false memory, memories to exogenous or endogenous signals as well as memories to unitary signals versus complicated signals reorganized from unitary signals. Perceptual memory can be intramodal versus cross-modal manner, which includes olfactory, auditory, visual, gustatory, and somatosensory memory

dation of these memory patterns. Based on the views given in Chap. 1, neural substrates for information storage are thought to be memory traces or engrams that are constituted by various cell assemblies in relevance to different memories. The efforts to correlate cellular architectures with these memory patterns should be cooperated by the society of memoriology in the fields of psychology, physiology, pharmacology, and neuroscience. Basic units in the brain are neurons, glia cells, and their connected synapses. If basic units in memory traces are these components, author intends to assume that various patterns of memory are based on cellular architectures and their circuitry styles. In this regard, sketching the simplified and common cores for associative memory and being shared by these memory patterns should be practiced. In the prediction of memory engrams and their circuits for associative memory, associative memory cells and their constituted engram circuits are selected (please refer to Chap. 5). In this chapter, author intends to summarize these patterns classified for learning and memory, to figure out the core diagrams constructed from these memory types, and to propose the testable cellular architectures for various memory patterns.

2.2 Associative Learning and Nonassociative Learning

Learning has been defined as the acquisition of new information, knowledge, and experiences that may be unitary signals or complex signals reorganized from unitary ones in individuals from external environments through their natural activities and social interactions [7–9]. There are various sorts of learning, e.g., conditioning versus habituation, simple versus complication, active learning versus inactive learning, and conscious awareness versus unconsciousness. In the learning practices, the associative acquisition of multiple signals, objects, events, or experiences by cross-modal sensory modalities or an intramodal sensory modality stands for associative learning, in which these signals are mentally linked and reinforced each other, as well as these signals are grouped together as one associative memory [4, 9, 10]. In other words, the integrative storage of multiple signals, objects, events, and experiences as well as the reciprocal retrieval of these associated signals is called as associative memory [4, 10]. Subcategories of associative learning include operant conditioning, classical conditioning, observational learning, imprinting, play, multimedia learning, episodic learning, meaningful learning, rote learning, enculturation, drug addiction, and so on [9, 11–17]. However, nonassociative learning refers to the relatively permanent changes caused by the repeated exposure to a single stimulus, such as habituation and sensitization [18, 19].

In nonassociative learning, a habituation refers to the diminishing of response strength or probability when the stimulus is repeated in a typic reflex. The habituation process is faster in response to weak or strong stimulations at high rate than those at low rate. The habituation differs from the extinction in associative process, for instance, an operant extinction shows that response declines because it is no longer followed by a reward [19]. The concept of habituation acts in direct opposition to sensitization. Sensitization stands for the progressive amplification of response strength following the repeated uses of a stimulus in defensive reflexes, such as withdrawal or escape becomes stronger after the exposure to different harmful or threatening stimulus. Sensitization is thought to underlie both adaptive and maladaptive learning processes in the organism [18]. It is noteworthy that both habituation and sensitization are featured by the changeable strength of innate reflexes without the involvement of acquiring new signals. It is with caution to adapt them as learning [6].

Associative learning of multiple signals, objects, events, or experiences makes these signals mentally linked and reinforced one another, such that these signals are grouped into an associative memory [4, 14]. In classical conditioning, the repeated pairings of a neutral or naïve stimulus with a stimulus for an innate reflex can induce a phenomenon that this neutral stimulus is able to elicit innate reflex-like response. In operant conditioning, the behavioral manipulation reinforced by reward stimulus becomes more likely to occur in this neutral stimulus, or the behavioral manipulation punished by aversive stimulus becomes less likely induced by this neutral stimulus. It remains to be studied whether stimulations of evoking innate reflexes, reward, or punishment are able to induce those reflexes by neutral stimulations after the formation of associative memory, since the retrievals of associated signals are reciprocal in human beings under realistic conditions [4, 10].

The standard protocol to induce classical conditioning is featured by repetitively pairing a neutral stimulus with an unconditioned stimulus. This unconditioned stimulus can reliably induce an innate reflex or innate response. The neutral stimulus, also called as unrelated stimulus or conditioned stimulus, does not induce this innate response though it may induce other responses. After repeated pairings, i.e., conditioning, this innate response can be induced by both unconditioned stimulus and conditioned stimulus, respectively. The response, which is originally induced by unconditioned stimulus, emerged to the conditioned stimulus has been called as conditioned reflex. For instance, in Ivan Pavlov's conditioning, the dogs are fed with meat powder that innately induces their saliva. Meat powder is thought of as unconditioned stimulus, and the salivation is unconditioned response. Pavlov rings the bell around the dogs before feeding meat powder. Initially, the bell sound as the neutral stimulus does not make dogs to salivate. However, meat powder in their mouths induces them to salivate. After numerous pairings of bell and meat powder, dogs learn that the bell signifies the food forthcoming and begins to salivate once they hear the bell. The bell becomes conditioned stimulus, and the salivation to the bell becomes conditioned response. This process is called as conditioned reflex or conditioning [20]. The classical conditioning has been observed in numerous species of animals [21].

Psychological analysis by Freud and followers is largely based on introspection method or the examination or observation of individual's mental and emotional processes. However, John Watson argues that the introspective method is too subjective and that the study of human development to directly observable behaviors should be limited [22]. In this book *Psychology as the Behaviorist Views It*, he suggests that laboratory studies should serve best to psychology in science. Watson's most famous, but controversial, experiment demonstrates how psychologists can account for the learning of emotion through classical conditioning principles. Other processes of associative learning are given below.

Observational learning is a learning process that occurs through observing the behaviors of other people. It is one of the forms of social learning. In humans, this form of learning seems to require a social model, such as a parent, sibling, friend, or teacher with surroundings without the need of reinforcement [23, 24]. Imprinting refers to associative learning during the particular life stage that is rapid and independent of behavioral consequences. In filial imprinting, young animals can form an association with another individual or an object, which is likely a parent relationship [25, 26]. Playing with inanimate objects or other animals describes a behavior without particular consequence itself, but this behavior improves performance in similar future situations [27]. Memories through playing with partners, especially in young age, may establish useful strategies and social skills for future activities [28]. For instance, children can develop emotion skills by learning to deal with the emotion of anger during playing activities. Playing may facilitate the development of thinking and language skills in children. Types of playing include sensorimotor playing through repetition, role playing, rule-based playing, construction playing, and movement playing, which are often intersecting [29]. Enculturation is the process in that people learn values and behaviors appropriate or necessary in their surrounding culture and that parents, other adults, and peers shape the individual's understanding to these values and behaviors. In contrast, acculturation refers to a situation that a person who adopts the values and societal rules of a culture different from their native one [30]. Multimedia learning refers to the learning processes participated by more than one sensory system, such as a combination of visual and auditory stimulations [31]. Transfer learning is the application of skills, knowledge, or understanding to resolve a novel problem or situation when certain conditions are met. Learning transfer is infrequent but most common when cues, primes, and guidance are given in order to clarify what it is and how it may be promoted by instructions [32]. All types of learning activities above belong to associative learning and memory through the complicated processes in comparison with conditionings.

In terms of the efficiency of associative learning, rote learning stands for a process that associative learning makes memory retrieval by learners to be exact as their reading or hearing. Major technique used for rote learning is learning by repetition. Rote learning is used in diverse areas, such as mathematics, music, and religion [12]. In contrast, meaningful learning refers to the concept that learned knowledge is fully understood to the extent relevant to others. Meaningful learning implies that there is a comprehensive knowledge of the context of the facts learned [11, 33]. In the rote learning, memories in details indicate that each of the characters in learned knowledge is imprinted by a set of memory cells in memory traces. It is noteworthy that there are other subcategories of associative learning processes in life, such as informal learning, formal learning, nonformal learning, nonformal learning with combined approaches, tangential learning, dialogic learning as well as incidental learning. These learning styles are especially seen in the educational processes [34, 35].

In summary, associative learning refers to the process that the joint acquisition of the associated signals is fulfilled by the association of new signals with an innate signal. The innate signals may be those signals genetically assigned for each of sensory modalities, as well as the unitary signals or complicated signals organized from unitary signals that have been learned. For instance, these complicated signals may be the sentences reorganized from letters and words as well as various photo images reorganized from distinct objects [10]. By the repeated pairings of innate signal and learning signal, associative memory, or integrative storage, to these associated signals can be formed. The formation of associative memory is featured by a fact that these associated signals are retrieved reciprocally by associative learning cues or automatically recalled by signal conversion among sensory modalities. There are two physiognomies for associative memory, i.e., the integrative storage and the distinguishable retrieval of associated signals [36]. On the other hand, nonassociative learning stands for the acquisition of a given sensory signal. In nonassociative learning, the repeated stimulations to a single sensory modality may lead to organism's habituation or sensitization to this sensory signal [37]. Nonassociative learning does not involve the acquisition of new experiences, concepts, and knowledge. Repetitive activations in a sensory system may lead to its functional upregulation or downregulation by the sensitization or desensitization of sensory receptors and/or cortical neurons. Based on this mechanism, nonassociative learning may be called as "review" instead of the learning. In fact, the acquisition and memory to all of the new signals entail these signals to be associated with the signals that have been resided in the brain to facilitate their joint acquisition and integrative memory [6, 10].

In processes of associative learning, multiple signals included in an environment or an object stimulate the different sensory modalities. These signals activate correspondent sensory receptors and are transmitted to sensory cortices for their integration and associative storages. For example, a fruit can be detected by the gustatory system for its sweetness, the visual system for its shape and color, the olfactory system for its aromatic odor, the auditory system for its name, and the somatosensory system for its surface sign. After the featured signals of this fruit are jointly acquired and associatively memorized, a signal induces the recalls of its associated signals or the other way around. The retrieval of these cross-modal signals can be achieved automatically through the conversion of signals among different modalities. For instance, signals in a picture can be replayed by the words and sentences in mind as well as represented by verbal language without the need of any cues. Moreover, multiple signals with identical basic characters, such as the visual images including different light intensities and colors, can be associatively detected by a single modality or one type of sensory receptors and its projected cortex. These signals are associatively acquired in an intramodal manner and primarily stored in one of the sensory cortices [4, 10].

In order to fulfill memories to associatively learned signals that are the integrative storage and reciprocal retrieval of associated signals, memory traces, engrams, or cell assemblies are proposed to be formed, which encode these associated signals and receive new synapse innervations that carry these signals taken by associative learning. This hypothesis is based on a fact that associative learning and memory are largely influenced by internal and external factors, which are presumably based on neurons, their interconnected synapses, and the arousal system that need to be examined experimentally. Currently, primary associative memory cells have been detected in the sensory cortices, which encode multiple associated signals on the basis of the synapses from innate inputs and new innervations from coactivated brain areas for the integrative storage of associated signals [36, 38–44]. Moreover, secondary associative memory cells have been detected in the prefrontal cortex, hippocampus, and motor cortex, based on the synapse innervations among them and from primary associative memory cells [45–47].

2.3 Declarative (Explicit) Memory Versus Nondeclarative (Implicit) Memory

It is a common sense that the retrieval of certain previously learned knowledge and experiences needs to pay attention with the help of consciousness and intention, whereas the retrieval of other memories to the operation skills and procedures does not need conscious and intension knowing. In this regard, two terms, declarative memory and nondeclarative memory, are issued based on whether memory retrieval is coupled with consciousness and intention after memory formation [48, 49]. For instance, declarative (explicit) memory involves conscious recollection, compared with nondeclarative (implicit) memory which involves an unconscious and nonintentional form of memory. Remembering the specific driving lesson is an example of declarative memory, whereas improving the driving skill during the lesson is an example of implicit memory.

2.3.1 Declarative Memory

Another name as explicit memory refers to the process that after memory formation, memory retrievals with conscious and intention components to represent and recollect previously learned knowledge, experiences, and information [48], such as remembering the time of a meeting and recollecting events during that meeting from years ago. Declarative memory depends upon conceptually driven and top-down sequential processing. Subjects reorganize the data through the association of one object with previously relevant stimulations or experiences in a logical sequence, and then they are jointly memorized. The subsequent retrieval of this object and relevant data is largely influenced by the way how the information, relevant to this object and the data, has been originally processed. The deep processing by an individual who has thought the meaning or shape of the information improves subsequent retrieval. In order to create declarative memory, individuals have to do something with his or her experiences and knowledge previously learned by thinking about them, talking about them, writing them down, studying them, and so on. The more the individual practices, the better the individual will remember. As this individual has played an active role in the association of processing explicit information with endogenous signals in his or her mind, the endogenous cues that have been used in associative processing can also be used to initiate the spontaneous retrieval of the information. For example, an individual has talked about certain experiences; the words that have been used will help him or her to remember these experiences in later time. In summary, explicit memory can be thought as the memory of the information including relevant subject and contents by conscious and intention organization, in which the well retrieval of the entire information is helped by more rehearsal and practices.

In terms of neural correlates involved in declarative memory, several neural structures have been proposed to be relevant to memory traces for declarative memory, especially the prefrontal cortex, amygdala, hippocampus, and rhinal cortex in the temporal lobe. Nuclei in the thalamus are also included, since interconnections between the prefrontal cortex and temporal cortex are mediated through the thalamus. These brain structures to constitute neural circuits for declarative memory receive axon and synapse inputs from the neocortex as well as the brain stem system including nuclei that release acetylcholine, serotonin, and noradrenaline [50]. It remains far away to know about cellular architectures for declarative memory. As declarative memory is associative in nature, cellular assemblies in memory traces or engrams with the associative function may be a candidate of neuronal substrates for explicit memory [6].

There are two subtypes of declarative memory, episodic memory and semantic memory [51–58]. Episodic memory, also called autobiographical memory, refers to memory and retrieval to single events and experiences in individual's life. Episodic memory is necessary for time traveling through remembering individual's past and imagining his or her future and is considered as a unique human quality that depends on maturation and has not been found in babies and young children. Semantic memory is not factual but consists of all of explicit memories by summary and generalization, such as knowledge organized from historical events and figures, capability to recognize friends and associates as well as information learnt about specialized vocabularies, reading, writing, and mathematics in schools.

The neural basis of episodic and semantic memories remains not known though some scientists suggest that episodic memory may depend on the right hemisphere and semantic memory on the left hemisphere. As knowledge and information in semantic memory are reorganized from single events and experiences in episodic memory, cell units in engrams for episodic memory may be integrated with convergent inputs and outputs in cellular architectures for semantic memory [6].

2.3.2 *Nondeclarative Memory*

Implicit memory refers to a process that the retrieval of previous experiences, such as operations and procedures, helps in performing a specific task without consciously knowing these experiences [48]. Nondeclarative memory appears priming, in which individuals show improved performance on tasks that they have subconsciously prepared. Nondeclarative memory may lead to an illusion-of-truth effect, in which individuals prefer to ensure the information that they have heard as true statement, regardless of their veracity. People apply implicit memory every day in the form of procedures that allow them to remember how to tie their shoes or to ride bicycles without consciously thinking about such activities. In the study of nondeclarative memory, scientists suggest that implicit memory operates through the different psychological processes than explicit memory.

Illusion-of-truth effect describes a phenomenon that a person is more likely to believe a familiar statement than an unfamiliar one. In an experiment, participants are guided to read 60 plausible statements every 2 weeks. Subsequently, they are asked to rate these statements based on their legality. Parts of these statements are presented more than once in different periods, in which some statements are true and other statements are false. In most cases, these participants prefer to rate the statements that they have heard as true statements, even though they do not remember clearly when or where these statements have been heard as well as do not mind whether these statements are actual validity. Because illusion-of-truth effect occurs without consciousness and explicit knowledge, this effect is thought of as the direct outcome of implicit memory. In this test, a lot of participants prefer to rate those sentences that have been previously heard as true, even when they are told that these sentences are false [59]. The illusion-of-truth effect in nondeclarative memory shows potential danger as it can lead to unconscious decisions about the statement's veracity. This term and experiment include the complicated situations, i.e., cognitive processes consist of memory plus judgment in relevance to numerous memory contents. In fact, memories themselves are based on learnt signals without any preference regardless they are true or false. As long as the learnt signals activate cortical neurons through their specific afferent pathways to recruit them to be memory cells, these memory cells can be activated by cues in memory retrievals, and those memory cells historically recruited may be reactivated before this reading session and are reactivated many times during rehearsal reading. Based on a rule of being active together, connecting together, and strengthening together, activity-dependent positive cycle in the recruitment and refinement of associative memory cells [4, 6, 10] makes those memory cells historically recruited to be more active, such that their encoded storage of statements is retrieved preferentially.

Another form of nondeclarative memory used every day is called procedural memory. Procedural memory allows us to perform some actions, such as writing words or riding a bicycle, without consciously thinking about it. In two groups of people, both amnesic patients with heavily impaired short-term memory and healthy subjects are asked several times to solve the Tower of Hanoi puzzle, i.e., a complex problem-solving game that requires 31 steps to complete. Group one of amnesic patients demonstrates the improvement over time similar to group two of healthy people, even though some participants claim that they do not remember having seen this puzzle before. This result suggests that procedural memory belongs to implicit memory without the need of consciousness, or, completely independent of explicit memory [60]. In another study, two groups of people are given flavored carbonated drink. Participants in group one is later exposed to motion sickness. These participants develop the taste aversion against this carbonated drink, even though they are told that this drink does not cause motion sickness. This result indicates that the implicit procedural memory subconsciously associates with the sickness and the drink flavor [61]. It is argued whether implicit attitudes are attributed to the category of implicit memory or they are a relevant but different phenomenon. In some ways, implicit attitudes resemble procedural memory because they rely on an implicit unconscious piece of knowledge that has been learned [62, 63].

In terms of evidences for the separation of implicit and explicit memory, researchers suggest that implicit memory is obviously different from explicit memory and that they operate different processes in the brain. Recent effort has been directed toward studying their difference, especially studying amnesic patients and priming effect [63]. For instance, amnesic patients demonstrate unimpaired capability to learn the tasks and procedures that do not rely on explicit memory but an impaired capability in long-term verbal memory [60]. This result implies that the mechanism for long-term declarative memory does not work for implicit memory. Furthermore, the results from the study of the priming in amnesic patients indicate that the intact implicit memory is associated with the impaired explicit memory. Participants in control group and amnesic patient group demonstrate similar improvements in word completion by priming under the condition that they have no memory to be involved in the previous test [64]. The production of the priming without the involvement of explicit memory indicates that explicit memory and priming use different processes in the brain. In another study of priming words through an auditory test and then visual test, the influence of the priming on memory is examined. In the result, little decline in the priming effect is seen in patients who are examined by explicit memory, in which the patients are simply asked whether they recognize the word heard in the first part of the experiment. Subsequently, on the word identification test of implicit memory, the priming effect is obviously impaired by alternations in modality from the studying function to the testing function [65].

In the investigation of depth-of-processing effect, participants are given a list of words and asked to engage with these words. Some of these words are required to be learnt superficially, such as counting the number of letters in each given word. In one set of words, participants are required to perform tasks that they use elaborative process, such as answering questions about a word's meaning. They are subsequently assessed in their capability to recognize whether they have seen these words in the experiment. As the depth-of-processing effect assists explicit memory to the words, participants should demonstrate better memories to those words with elaborative process on these studies. On the other hand, when their implicit memory is examined by flashing words on a screen, participants are asked to identify them. There is no difference for participants to identify words flashed on screen that have been superficially learned and elaboratively processed. In other words, the priming effect is quite similar for the words with elaborative process versus the words with superficial learning. Therefore, implicit memory is not dependent upon depth-of-processing effect as explicit memory is [66]. In fact, these words have been learned and memorized in explicit way; before this experimental task, participants are asked to review them either superficial or elaborative processing. The retrieval of these words needs sufficient time when cues are given. Different ways for their retrieval may lead to different results.

Another example comes from the test of explicit recognition. The efforts to intervene memories to a list of words influence people's capability to recognize words significantly. This intervention does not influence people's implicit memories to the words [66]. In the tests of word completion, there appears no correlation statistically between people's capability to explicitly remember a list of words and their capability to subconsciously apply the priming effect to assist their performance in identifying words that have previously seen [67]. The results above firmly indicate that implicit memory is not only present but also present in itself identity and functions that differ significantly from explicit memory.

In summary, the main difference between declarative memory and nondeclarative memory is whether retrievals to memorized signals are associated with consciousness and intention. This classification ignores the fact in that the fulfillment of nondeclarative memory in skills and procedures has been done more practices than that of declarative memory in knowledge and experiences. In common sense, there appear to be more practicing, more familiar and easier to be recalled in memorized contents. Likely to riding bicycles without the requirement of consciousness and intention, teaching the familiar knowledge by experienced teachers with uncounted times does not need for them to look the book and to be consciously intension. On the other hand, likely to learning knowledge, the initial learning and retrieval of skills in riding bicycles need consciousness and attention for further practices. It seems to us that there is a conversion or transferring of declarative memory into nondeclarative memory through practices and rehearsals [6]. In terms of cellular architectures, the repeated activations of memory cells that store knowledge and procedures lead to the activity-dependent positive cycle in the recruitment and refinement of associative memory cells [4, 6, 10] for this transferring from declarative memory to nondeclarative memory. Moreover, the activation of associative memory cells without the enriched number during the period of initial learning needs the help from the arousal system for the wakefulness and consciousness. In brief, the formation of explicit memory and implicit memory may use similar cellular mechanisms though their retrievals appear different in aspects of consciousness and intention.

2.4 Episodic Memory Versus Semantic Memory

According to the natures of memory contents, declarative memory is classified into episodic memory and semantic memory [51–58]. Episodic memory stands for memories to events, times, places, their associated emotions, and other conception-based knowledge in relation to an experience. Semantic memory refers to memories to meaning, understanding, and other concept-based knowledge generalized to multiple facts. The conscious retrieval of factual information and general knowledge about the world is generally thought to be independent of context and personal relevance. Semantic and episodic memory constitutes declarative memory, one of two major divisions in memory, i.e., explicit and implicit memory [56, 57]. In spite of this classification of episodic and semantic memory in the category of declarative memory, such two types of declarative memory are related to each other under some conditions [55, 56, 58]. Their distinction and association are presented here in order to figure out their potential mechanisms.

In terms of the relationship between episodic memory and semantic memory, their distinctions are obvious [52, 59]. Episodic memory is considered as a one-shot learning mechanism, i.e., remember an affair through one exposure to an episode. On the other hand, semantic memory is thought of as multiple exposures to each of the references and is updated on each exposure. Furthermore, episodic memory can be understood as storing each of items in a map that links items together in semantic memory. For example, semantic memory tells what a bicycle looks like and what its functions are, while episodic memories in all new experiences about this bicycle modify a single semantic retrieval of bicycle. It is noteworthy that episodic memory can be converted into semantic memory over time [68]. In this conversion, most of the episodic memories about a particular event is generalized, and then the context of the specific event is lost, i.e., semantic memory. One modification of this view is that episodic memory, which is often recalled, is remembered as a kind of monologue. If someone tells and retells one story repeatedly, he or she may feel that the event may no longer be remembered, but a pre-written story is emerged in his or her recall, such as personal semantics [55].

2.4.1 Episodic Memory

This term is given based on memories to spatial and temporal contents in relevance to each of events, experiences, and associated emotional reactions as well as concept-based knowledge in a single affair. The features of episodic memory include that memory contents are individual affairs and facts in specific location and time, memory contents are related to properties of single object and event, its formation can be fulfilled through an exposure as well as episodic memory to an affair, or a fact in multiple exposures can be generalized and converted into semantic memory [55, 56, 58, 68]. Episodic memory does inform semantic knowledge, and episodic memory is reliant upon semantic knowledge. Some people do not believe that all episodic memories will inevitably distill away into semantic memory.

A few interesting phenomena are pointed here. There is a gender difference in episodic memory behaviors. Based on brain activations during episodic memory retrieval, women tend to outperform men on episodic memory tasks. Another one is age differences in episodic memory behaviors. The activation of specific brain areas, mostly the hippocampus, may lead to episodic memory retrieval differently in young and older people [69]. Older people tend to activate both left and right hippocampus, while young people activate only the left one [70]. Episodic memories are usually associated with emotional reactions. Although the relationship between emotion and memory is complicated, the emotion tends generally to increase a likelihood that an event will be remembered easily and vividly, in which flashbulb is an example [71].

In terms of neural substrates for episodic memory, many studies have been done by neural lesions, imaging, or pharmacology. The formation of new episodic memories requires the medial temporal lobe, a structure including the hippocampus and limbic system [72–74]. Without the medial temporal lobe, individuals are able to form new procedural memories but not remember events that have happened [75]. The prefrontal cortex in the left hemisphere is involved in the formation of new episodic memories. Patients with damaged prefrontal cortex can learn new information in a disordered way. They show normal recognition to an object they have seen in the past but fail to recollect when or where it has been viewed [76, 77]. Some researchers believe that the prefrontal cortex may participate in organizing information for more efficient storage, indicating its role in the executive function. Others believe that the prefrontal cortex works for semantic strategies which enhance the encoding, such as thinking about the meaning of the studied materials or rehearsing them during working memory [78–80]. It is still debated about how long episodic memories to a particular event can be maintained in the hippocampus. In this discussion, scientists think that episodic memories always depend upon the hippocampus. Some scientists prefer to trust episodic memories are maintained in the hippocampus for the short period, instead, they believe that episodic memories are consolidated to the neocortex [81]. The short-term presence of episodic memories in the hippocampus is granted by data that hippocampal neurogenesis in the adult can facilitate the erasing of memories to old signals as well as increasing the efficacy of memories to new signals [82].

It has been found that episodic memory is modulated by some neurotransmitter systems. Long-term visual episodic memory in healthy subjects and brain injury patients is upregulated with acetylcholine, serotonin, dopamine, and glutamate [83–88]. For instance, the elevation of synaptic acetylcholine by the administration of acetylcholinesterase inhibitor donepezil is able to specifically enhance episodic memory. This result has been granted by other investigations in transgenic mice, in which the deletion or knockout of alpha-7 subunit of nicotinic receptors impairs episodic memory [89, 90]. In elder individuals, there is loss of dopamine neurons as part of normal aging, which has been detected by using pharmacologically functional MRI in elder population. The dopamine precursor levodopa leads to a dose-dependent (inverted U-shaped) persistent benefit to episodic memory by seeing images of scenes, independent of whether encoding-relevant hippocampal fMRI activity is weak or strong (U-shaped dose-response relationship). This result implies that dopamine involves a consolidation of episodic memory in a narrow dosage [91].

In terms of the pathology of episodic memory, patients who have suffered from an episode of transient global amnesia manifest memory deficits, especially problems in the retrieval of episodic memories and recent semantic information with the capacity of reliving. These deficits may not result from the deterioration of memory by itself but rather from difficulties in accessing memories [92, 93]. Studies suggest that there may be selective damage to the limbic-prefrontal episodic memory system in some people with autism and that autistic deficits include the episodic or self-conscious memory of personally experienced events [94–96]. Amnesia is often given to patients with deficits in episodic memory, especially Alzheimer's disease that tends to damage the hippocampus before other brain areas with the dominant impairment of episodic memory. Alzheimer's patients are often classed as amnesiacs [97, 98]. In a case report, a patient with amnesic syndrome at 10 years old demonstrates the impaired episodic memory for both verbal and visual materials. The deficit has also been seen in a wide variety of semantic memory tasks including reading vocabulary and verbal fluency tests, semantic classification, and lexical decision tasks as well as verbal intelligence test. On the other hand, this patient shows normal learning and retention of two procedural tasks. This study is inconsistent with the view that the amnesic syndrome represents a selective defect of episodic memory that leaves semantic memory relatively unaffected [99]. These studies have not taken care about the time sequence in the change of episodic and semantic memory in amnesic patients, for instance, an initial impairment of episodic memory and subsequent impairment of semantic memory in Alzheimer's disease.

2.4.2 Semantic Memory

This term is defined based on memories to meanings, understandings, principles, and generalized knowledge as well as the conscious retrievals of factual information and general knowledge about the world [56, 58]. Semantic memory includes generalized knowledge and principle that do not involve the memory of specific events. For example, someone can answer a question like "Are wrenches pets or tools?" without remembering any specific event in which you learned that wrenches are tools. Semantic and episodic memory constitutes the category of declarative (explicit) memory. The counterpart to declarative memory is implicit memory in relevance to procedures [48].

The concept of semantic memory was introduced by Endel Tulving [57], who intended to distinguish episodic memory from another memory for general knowledge what he termed semantic memory. His idea was influenced by Reiff and Scheers who issued two primary forms of memory, i.e., remembrances (memories to experience) and memoria (memories without experiences). Semantic memory reflects knowledge about the world with generic information that is acquired across various contexts and is used in different situations. In other words, semantic memory is the sum of all knowledge, such as principle, math formula, regulation, and facts. Different forms of episodic memory refer to unique and concrete personal experiences; semantic memory refers to general facts and meanings for sharing among people. Tulving's suggestion about the distinction between semantic and episodic memory has been widely accepted because his theory allows separating the conceptualization of knowledge in the world from single activities. Tulving discusses these separate systems of conceptualization of episodic and semantic memory in his book entitled *Elements of Episodic Memory* [100]. Both episodic and semantic memories differ in several following factors, such as the characteristics of their operations, the kind of information they process as well as their application to the real world and the memory laboratory. Before Tulving's proposal, the field of human memory has been ignored by experimental psychologists. Numerous experiments have been conducted to validate Tulving's hypothesized distinction of episodic and semantic memory [48].

In an experiment [101], four groups of college students are selected to conduct the test of word memory under the hypnotized condition. It is noteworthy that these students demonstrate various levels of susceptibility to hypnosis. They are asked to study a list of 16 common words by a multi-trial free-recall method under the hypnosis. After these students can recall this list of words twice perfectly in succession, they are notified that they would not remember any of these learned words. Although this notice has been given by the experimenters, students can recall the learned words and recall the words from the list as well. After these subjects were awakened, this experiment task is measured in the following stages. In stage one, the quantity of words that students are able to recall is applied to merit the performance for the episodic task of free information retrieval. Most of the participants can remember the list of learned words. In stage two, the task for semantic memory is assessed. Each student is required to experience the semantic free association test. In this test, some stimulus words are showed to students for eliciting the learned words. After these two stages of experiments are done, the levels of students' susceptibility to hypnosis are estimated, in which their scores are determined on the Stanford Hypnotic Susceptibility Scale. These participants are grouped based on the level of susceptibility to hypnosis. As results, the probabilities of semantic free association test are relatively identical across different groups of hypnosis susceptibility. However, the probabilities of episodic free recall are significantly different across these groups. There appears an inverse relationship between the percentage of successful recall and the levels of the hypnotizability in subjects. The subjects in the high levels of susceptibility group recalled almost nothing, whereas the medium and low groups recalled 86% of the learned words. This study provides evidence for granting the distinction between episodic and semantic memory hypothesized by Tulving. Moreover, this study indicates that conscious state is critical to influence the formation of episodic memory.

In Jacob and Dallas experiment (1981), participants are presented with 60 words once at a time and are asked different questions. Some questions cause participants to pay attention to the visual appearance of these words, such as are these words written by bold letters? Other questions result in participants to pay attention to the auditory sound of the words, such as do these words rhyme with ball? Certain questions induce participants to pay attention to the meaning of the words, such as do these words refer to the form of communication? In these questions, 50% answered no, and other 50% answered yes. In two phases of this experiment, 60 old words seen in stage one and other 20 new words are shown to the participants once at a time. Subsequently, participants are asked to perform one of the following two tasks:

1. Semantic memory is tested by perceptual identification task. In this task, participants are showed the words flashed on a video-screen for 35 ms. Then, they are asked to speak what these words are.
2. Episodic memory is tested by episodic recognition task. In this task, these participants are showed each one the words and have to ensure whether they have looked the words in the previous phase of the experiment.

Through data analysis, the percentages of the correction in semantic memory (perceptual identification task) are not changed about the encoding conditions of appearance, sound, or meaning of these words. However, the percentages of episodic memory (episodic recognition task) increase in the encoding conditions from the appearance to the sound and the meaning of these words. These effects are also greater for the answer of "yes" encoding words than the answer of "no" encoding words. This result indicates strong distinction in the performance of episodic memory and semantic memory, i.e., supporting Tulving's hypothesis [52, 68]. It is noteworthy that some words learned in these tasks may have known by participants, so that the experiments here may not be purely learning and memory processes and be cognitive processes involved.

In terms of mechanisms underlying semantic memory, various theoretical models have been proposed. In a simplified model, the network consists of a set of link-connected nodes. These nodes represent concepts, words, and perceptual features or nothing at all. These links may be weighted, i.e., some are stronger than others. These links may be equivalent in length such that some links take longer to traverse than others. All these features of networks have been employed in models of semantic memory. In a semantic network [102], each node can be interpreted to represent specific concept, word, or feature. Semantic networks generally do not employ the distributed representations for concepts described in a neural network. Semantic network is featured as its links almost directed. These links only point to one direction from the base to the target. These links come in many distinct types, in which each standing for a particular relationship can hold between any two nodes. The processing in this semantic network often works in a form of spreading activation. Neural networks for semantic memory have been widely applied in the models of discourse, logical comprehension, and artificial intelligence [103]. In these models, the nodes of semantic networks represent words or word stems, whereas those links correspond to syntactic relations between them, such as the semantic network for a computational implementation in knowledge representation [104].

Another model for semantic memory is associative model. In this model, the association in memories refers to the relationship between two signals. As the fundamental concept in psychology, the associations at the different levels of mental representation are essential to memory and cognition. A subset of associations among the collected items in memory is presented as the links among network nodes. Each of these nodes presents to a particular thing in memory. Neural networks for semantic memory may be featured as associative models of cognitive processes. In the presentation of associations by the nodes as N-by-N matrix, N denotes the quantity of things being memorized. Each cell in this matrix relates to the strength of associations between items in columns and rows [105, 106]. Associative learning is presumably based on a Hebbian process. Two items in memory are coactive, the association between these two items becomes stronger, and these two items are to activate each other [6]. A standard model of associative memory is called as the Search of Associative Memory (SAM) model [107]. Although the SAM is originally designed to be a model for episodic memory, its mechanism is sufficient to grant some representations in semantic memory. When two items fall into a working memory buffer simultaneously, the weight of their links is enhanced. The items that are more coactive will be more closely linked. In addition, the items in the SAM are linked with particular context. The strength of their associations is influenced by a process of how long these items are maintained in the given context. In the SAM, the memories may be composed of a group of associated items in memory as well as the associations between items and contexts. The simultaneous presence of a set of items with a context is very likely to result in some subsets of items being memory. The strength in the interaction among these items by virtue of their shared context or their coexistence indicates the items in relevance of semantic memory [108].

Neural substrates for semantic memory are largely unknown, although certain brain areas are believed to be involved in semantic memory [56, 109, 110]. Many scientists and clinicians assume that semantic memory uses the same brain system for episodic memory, such as the hippocampus and medial temporal lobes. In this system, the hippocampal formation encodes memories or makes this system to be possible for memories. The cerebral cortex stores those memories after the initial encoding processes are completed. New evidences have been shown to grant more precise interpretation of this hypothesis. The hippocampal formation includes the hippocampus as well as the entorhinal cortex and the perirhinal cortex (parahippocampal cortices). The amnesic patients featured by hippocampal damage but spared parahippocampal cortex show certain degree of intact semantic memory in spite of loss of episodic memory in complete manner. This result indicates that the encoding of semantic memory is not entirely based on its physiology in the hippocampus [56, 109, 110]. Other scientists assume that the hippocampus participates in episodic memory and spatial cognition by questioning the location of semantic memory. In addition, researchers assume that semantic memory resides in the temporal neocortex. However, researchers believe that the knowledge in semantic memory is broadly distributed across all of the brain regions. To support this latter view, researchers holding a view about distributed semantic memory believe that the knowledge of the dog's sound resides in the auditory cortex, while the capability to recognize and imagine the visual features of a dog resides in the visual cortex. Perhaps all these representations are indexed by the left temporal pole that is particularly vulnerable to damage in semantic dementia [68, 93, 111].

In addition to the hippocampus, the left inferior prefrontal cortex and the left posterior temporal areas are involved in semantic memory. Temporal lobe damage affecting the lateral and medial cortices leads to an impairment of semantic memory. Damage to different areas of the brain affects semantic memory differently [110, 112]. Brain scan shows that the activity of the left hippocampal areas increases in semantic retrieval as well as the right middle frontal gyrus and the right inferior temporal gyrus show an increase in activity. The damage to brain areas involved in semantic memory results in different effects, depending on the area and type of damage. For instance, category-specific impairments can occur where patients have different knowledge deficits for one semantic category over another, depending on location and type of damage [113, 114]. This category-specific impairment indicates that knowledge may rely differentially on sensory and motor properties encoded in separate areas [115, 116]. The different diseases and disorders can affect the biological network of semantic memory. A variety of studies has been done in the attempts to determine the influence of different factors such as herpes simplex virus encephalitis and semantic dementia on various aspects of semantic memory. For example, the deficits in semantic memory induced by herpes simplex virus encephalitis appear to be more category-specific impairments. Semantic dementia seems to cause more generalized semantic impairments [113]. Other neurological diseases also influence semantic memory. For instance, patients suffering from Alzheimer's disease show impairments in recognizing, naming, or describing objects. These features are presumably caused by the injury of in the degradation of semantic knowledge [117]. In fact, the distinct stages of neurological disease and psychological disorders are associated with semantic and episodic memories differently, which depend upon the impairment in the patterns of memory traces consisting of associative memory cells (please refer to Chap. 9). For instance, patchy damages in groups of associative memory cells may lead to impairments in episodic memory and category-specific semantic memory in the early stage, while widespread damage in numerous groups of associative memory cells will result in a more generalized semantic impairment [6].

In diseases relevant to semantic memory, category-specific impairments include widespread, patchy, or localized damage. There are about four categories for specific semantic impairments. Perceptual/functional features, topographic organization, informativeness, and intercorrelations are four aspects of the decreased function in deficits of semantic memory [118, 119]. Alzheimer's disease is thought of as deficits in semantic memory featured by category-specific impairments. Patients demonstrate the incorrection in recognizing, naming, and describing objects previously learnt. Semantic dementia is featured by language disorder. Patients show impairments in understanding and recognizing verbal language. The deterioration includes the inabilities to generate familiar words, to recognize visual images, and to name objects. The data indicate that the temporal lobe of cerebral cortices may take in charge for category-specific impairment of semantic memory disorders. Moreover, modality-specific impairment is another feature of semantic memory disorders [120]. In terms of semantic memory about modalities, different sensorimotor pathways and cortices influence the representation of semantic memories. Modality-specific deficits in semantic memory can be classified into the separate subsystems based on sensory input modalities. These input modalities include olfactory, gustatory, visual, auditory, and tactile sensations. Modality-specific deficits can also be classified into the subsystems based on the types of associated signals, such as visual signal versus verbal signal as well as perceptual signal versus functional signal [120]. These modality-specific deficits can account for category-specific impairments in semantic memory disorders. The damage to visual semantic memory primarily impairs knowledge of living things. The damage to the function of semantic memory primarily impairs knowledge of nonliving things.

In summary about present and future research, semantic memory has been taken back the interest in recent 20 years. The interest is largely due to the emergence of new techniques, especially the functional neuroimaging that mainly includes positron emission tomography (PET) and functional magnetic resonance imaging (fMRI). By using functional neuroimaging, certain key mysteries about the characters of semantic memory are becoming clear. Instead of a view that any of the brain areas is involved in the dedicated and privileged retrievals of all sorts of semantic knowledge, semantic memory is based on the information collection from different systems that are functionally and anatomically independent. Each of attribution-specific systems is linked to a sensorimotor modality and even more specific feature within the single modality. In addition, the studies from neuroimaging indicate differences between processes in semantic memory and the processes of sensorimotor encodings. It has been suggested that in early stage of development, semantic memory, like perception, is featured by the subtypes in the color, size, form, and motion of visual signals. To the distributed functions of brain regions and these subtypes, the left or bilateral ventral temporal cortices may be relevant to retrieving the knowledge of color and form. The left lateral temporal cortex works for the knowledge of motion. The parietal cortex encodes the knowledge of size [121, 122]. Studies by using neuroimaging indicate the large distributed network for the information retrieval of semantic memory, in which the signals are organized minimally by attribution and additionally by category. These networks include the ventral temporal cortices encoding the form and color knowledge, the lateral temporal cortex encoding the motion knowledge, the parietal cortex encoding the size knowledge, and the premotor cortex encoding the manipulation knowledge. Other regions, such as more anterior areas of the temporal cortex, may work for the representation of verbal conceptual knowledge in the category-organized fashion [121].

In general, episodic memory is featured to be the memories to events, times, places and conception-based knowledge related to an experience. Semantic memory stands for memories to meaning, understanding, and concept-based knowledge generalized to multi-facts. The conscious retrievals of factual information and general knowledge about the world are presumably independent of context and personal relevance [56, 57]. In spite of this classification of episodic and semantic memory, these two types of declarative memory are related to each other under certain conditions [55, 56, 58]. The differences between the memory of spatial and temporal features for specific objects and the memory of generalized knowledge from multiple objects seem to come from the abstraction and generalization through repeated practices, i.e., a feature about more practices, more summary, and more potential conversion of episodic memory into semantic memory.

2.5 Working Memory and Perceptual Memory

Memory processes can also be classified as short-term working memory from the sensory input guidance to the processing manipulation [123–125] and long-term perceptual memory that includes the storage of visual, auditory, olfactory, gustatory, and somatosensory signals [126, 127]. Perceptual memory is defined as a long-term memory of subject's perceptual experiences, i.e., memory for people faces, voices, appearance, and so on, in which these signals are inputted from sight, sound, smell, taste, and tactile sensations. Working memory functions as the buffer of cognitive memory. This buffer has a limited capacity for the transient holding, processing, and manipulation of information in the short term and plays critical roles in the execution of reasoning as well as the guidance of decision-making and other behaviors. Both perceptual memory and working memory are thought of as clear difference, though there is certain interaction between perceptual memory and working memory [128–130].

2.5.1 Working Memory

This term is a theoretical construct in cognitive psychology that refers to the structures and processes used for temporarily storing and manipulating information. Many theoretical models are proposed for the structure of working memory. Some specific areas in the brain are presumably to be relevant to working memory. Most scientists have believed that the frontal cortex, the anterior cingulate, the parietal cortex, and the basal ganglia are critically involved in working memory. The data about neural substrates for working memory originate from the experimental studies in the lesion of brain areas in animals and in neuroimaging of human brains. Many applications of working memory are based on the use of working memory capacity to explain the intelligence and other cognitive capabilities, to understand autism and Alzheimer's disease, to improve teaching approaches, and to create artificial intelligence simulating the human brain [125]. The term is firstly used in the context of theories that likened the mind to a computer. Working memory also refers to short-term memory, primary memory, immediate memory, operant memory, or provisional memory [123]. Short-term memory refers to an ability to remember information over a brief period of time (in the order of seconds). Most theorists apply working memory to replace or include the older concept of short-term memory, in which they strongly emphasize active information manipulation instead of passive maintenance.

Many models have been proposed regarding how working memory operates anatomically and cognitively. Three models have been widely accepted. Baddeley has firstly introduced the multicomponent model for working memory. Two slave systems are responsible for short-term maintenance of information, and one central executive is responsible for supervising information integration and for coordinating slave systems. One of the slave systems stores phonological signals, i.e., sounds in language, and prevents the decay of these signals by repeatedly articulating their contents or refreshing these signals in a rehearsal circuit. Another slave system memorizes visual and spatial signals. This visuospatial sketch pad can be routinely used to construct and manipulate visual pictures as well as to present mental charts. This visuospatial sketch pad is further separated into a visual subsystem that process shape, color, and texture, as well as a spatial subsystem that encodes the location. The function of the central executive system is used to direct the attention in relevance to signals learned, to suppress irrelevant signals and inappropriate intervenes as well as to coordinate cognitive activities if more than one task should be treated simultaneously. Subsequently, this model is extended by introducing the fourth component or episodic buffer. This episodic buffer is used to hold the representations that integrate signals from phonological, visual, and spatial contents and to process signals that have not been covered by slave systems, such as semantic signals and musical signals. This fourth component is thought to be episodic in nature, since it is presumably to bind the signals into unitary episodic retrievals. This episodic buffer is similar to Tulving's concept about episodic memory; however, it is a temporary store [123].

Another model in working memory refers to a subset of the representations in long-term memory. Working memory is organized in two embedded levels. The first level consists of long-term memory representations that are activated without the limitation. The second level is called as the focus of attention process regarded to be the capacity limitation and holds up to four times of the activated representations [131, 132]. Subsequently, the third component is added. This component is a narrow focus of attention process that holds one chunk at a time. With the embedding of one-element focus in four-element focuses, this component serves to select a single chunk for its processing. For instance, an individual can hold four digits in mind simultaneously in Cowan's focus of attention. Somebody expects to perform some events on each of these digits, e.g., adding number two with each digit. As most individuals are unable to perform several mathematical processes simultaneously, the separate encoding to each of these digits is required. This attention component selects one of these digits for dealing with and then shifts attentional focus to the next digit, until all of these digits have been dealt with [133].

Most adult people are able to repeat seven digits in the correct order. Certain individuals have demonstrated impressive enlargements. This achievement is possibly fulfilled by the extensive trainings on the encoding strategy, by which the digits in a list are grouped in three to five as well as these groups are encoded as the single unit. People must be able to recognize groups as some known codes. For instance, an individual uses his extensive knowledge of racing times from the history of sports. Furthermore, several units can combine into the higher-order unit to form a hierarchy of chunks. A small quantity of chunks at the highest level of the hierarchy should be retained in working memory. These chunks are unpacked again during the retrieval. The chunks in working memory act as retrieval cues that direct to the digits. It is noteworthy that practicing memory skills does not upregulate working memory capacity. This point can be shown by using different materials. People who can recall 80 digits are not exceptional when they come to represent the words. However, people are able to use memory skills in most daily tasks. The tasks, such as reading about more than seven chunks, need to be preserved in memory with a capacity of seven chunks. Working memory will be full after a few groups of words are shown. People will not be able to understand those complicated relations among those thoughts that are presented in a novel or a scientific story. Eventually, people achieve this task by memorizing most of the contents that they have read in long-term memory and by gathering them in the retrieval approaches. People require to hold only a few concepts in working memory. These concepts serve as cues to retrieve everything associated with them by retrievals, i.e., long-term working memory [134].

In general, working memory is considered to have limited capacity. The earliest quantification of capacity limitation associated to short-term memory is magical number seven [135]. The memory span of young adults is around seven elements, or chunks, regardless whether these elements are digits, letters, words, or other units. Studies show that the span depends upon the category of elements, such as seven chunks for digits, six chunks for letters, and five chunks for words, as well as upon the characters of elements within a category. In general, the memory span is lower for long words than short words. The memory span for verbal contents, e.g., letters, words, and digits, depends upon the time to speak contents clearly as well as upon the lexical status of the contents. In other words, the memory span is influenced by whether the contents are words known by people or not. In addition, other factors may influence people's memory span. It is hard to clearly tell the capacity of short-term or working memory through the number of chunks [136, 137]. In addition, the capacity of working memory seems to be regulated by norepinephrine in the locus coeruleus and dopamine in substantia nigra of the brain stem [124, 138] as well as other psychological states [139].

Working memory capacity can be tested by a variety of tasks. A common measure is a dual-task paradigm by combining memory span measure with simultaneous processing task. In memory span about reading, people are asked to read a number of sentences and remember the last word of each sentence. At the end of the list of sentences, participants repeat back words in their correct order. The rate of the correct order is merited. In addition, other tasks without this dual-task nature have been shown to be good measures of working memory capacity [124]. The measurements of working memory capacity are influenced by the performances in complicated cognitive events, e.g., understandings, reading comprehensions, problem-solving, and intelligence quotient. With the involvement of cognitive tasks, working memory capacity reflects the efficiency of executive functions, particularly the capability to preserve a few task-related representations in the front of disturbing irrelevant signals. These tasks appear to manifest individual differences in the capability to pay attention, especially when other activities serve to seize attention [140]. The influence by cognitive tasks appears to be a function of the frontal lobe in the cerebral brain [136, 137].

In terms of the location for working memory in the brain, the first insight into neural substrates for working memory comes from animal research [141]. Electrical activities are recorded at the neurons in the prefrontal cortex of monkeys, while they are doing the delayed matching task [142]. Each monkey observes how the experimenter puts a small amount of food under one of two cups that have identical appearance. A shutter that screens off the two cups from the monkey's view is subsequently lowered for changeable delay periods. Following the delay, this shutter is raised. In the meantime, the monkey is permitted to take the food underneath one of two cups. The successful retrieval of the food in the first attempt requires food location to be held in memory over the delay period. The animals can achieve something after certain trainings in the task. The neurons in the prefrontal cortex produce spikes mostly during those delay periods. This result indicates that prefrontal cortical neurons work for encoding food location while the shutter is down. Further studies have also shown these delay-active neurons in the posterior parietal cortex, the caudate, the thalamus, and the globus pallidus [141]. Of course, the delay-active neurons may be relevant to monkey's appetite to obtain the food and emotional reaction to hate this shutter and earnestly hope the shutter lifting, which may be memory-driven processes.

The localization of human brain functions can be well achieved by the advanced brain imaging approaches, such as positron emission tomography and functional magnetic resonance imaging [143]. Studies have confirmed that the prefrontal cortex is more active during working memory. Many arguments are focused on the different roles of the ventrolateral area and the dorsolateral area of the prefrontal cortex in working memory. In a view, the dorsolateral area works for spatial working memory, whereas the ventrolateral area works for nonspatial working memory. Another view about their functional differences suggests that the ventrolateral area works for the pure maintenance of memory contents, while the dorsolateral area is likely relevant to tasks that encode memorized materials. Although the debate is not stopped, most of the evidences grant functional differences from these areas of the prefrontal cortex in working memory [144]. Neuroimaging also shows that the working memory is not limited to the prefrontal cortex. The activated regions during the execution of working memory tasks scatter over a large portion of cerebral cortices. These data indicate that working memory is not a simple memory process but a mixture of memory, cognition, and emotion since these experiments are tasks of working memory. In spite of this wide scattering, there is the location dominance for working memory. More right-hemisphere areas are recruited for spatial working memory, while more left-hemisphere areas are recruited for verbal and object working memory. In verbal working memory, brain activations can be broken down into a component that reflects working memory maintenance in the left posterior parietal cortex and another component that reflects subvocal rehearsal in the left frontal cortex. An emerging consensus comes out that most of the working memory tasks recruit the network of the prefrontal cortex and the parietal areas. For instance, the connectivity between these areas increases during a working memory task. These regions are required for working memory and not just inadvertently activated during working memory tasks. This requirement is tested by temporarily blocking these regions via transcranial magnetic stimulation to produce impairments in task performance [145].

Although the locations for execution of working memory have been studied in detail, less is known about how the brain achieves the short-term maintenance of guided signals and the goal-directed manipulations [146]. As discussed previously, certain neurons with the persistent firing of sequential spikes during the delay period of working memory indicate that these neurons in the brain work for maintaining representations active without external inputs. If the working memory task demands maintaining more than one element of information, maintaining representations active is not sufficient. In addition, the components and features of each element must be integrated to prevent them from mixing up. To remember a red triangle and a green square at the same time, an individual must make sure that red is bound to triangle and green is bound to square. To fulfill these bindings, those neurons that encode the features of the same element should produce spikes in synchrony, whereas the neurons that encode the features about different elements should produce spikes nonsynchronous. For example, the neurons that encode redness produce spikes in synchrony with neurons that encode triangular shape but out of synchrony with those neurons that encode square shape. This feature for neurons to bind color and shape is similar to the encoding of associative memory cells [6]. No direct evidence shows that the working memory applies this binding mechanism and other mechanisms being proposed. There is an assumption that the synchronous activity of the neurons during working memory oscillates in theta frequencies (4–8 Hz). The power of theta frequency in the EEG increases with working memory. The oscillations in the theta band over different regions of the skull become more coordinated when individuals try to remember the binding between two components of information [146–149]. Cellular architectures, especially in morphology, for the synchrony or nonsynchrony of firing sequential spikes among neurons that encode the different features of information for working memory, remain to be positioned far away from our profiles of comprehensive diagram in memory formation.

2.5.2 *Perceptual Memory*

This term stands for long-term memories to subject's perceptual experiences, i.e., memory for people faces, voices, appearance, and so on, in which these signals are inputted from sight, sound, smell, taste, and tactile sensations [128–130]. Based on the routes of information inputted from different modalities, perceptual memory is further sorted as visual memory, verbal memory, spatial memory, olfactory memory, and so on [127, 150, 151]. Perceptual memory is improved by cholinergic system in the central nervous system, in which perceptual memory changed the manipulations of this system by surgical lesion, pharmacological approach, and genetic deletion [152].

Visual memory is a part of memory that maintains some characters of subject's sensations relevant to visual experiences. After visual memory formation, the subjects are able to consign the memorized things that are featured as similar objects, places, animals, or people in sort of a mental image for subsequent retrievals. This ability is called sometimes as a "mind's eye," by which individuals can retrieve memory from a mental image of the original objects, places, animals, or persons. With a sort of mental reorganized images, visual memory appears not to be an episodic memory but more likely to be a semantic memory in nature [153–156]. Eidetic imagery [157] has been thought of as actual visual memory through comparing the similarity between imagery and actual picture, though this visual memory is initially used to test the hallucination in schizophrenia [158]. In the study of visual memory, the experimenter shows participants an image for 30 s. After this image is taken out, participants are asked whether they can see anything. For example, children in elementary school are demonstrated with an illustration of *Alice's Adventures in Wonderland*. After this illustration is removed, some children are able to recall and describe this illustration with a vivid accuracy in the image they have seen [159]. Eidetic imagery seems to be more effective in children than adults who do not describe similar experience. This difference is attributed to a lack of verbal and conceptual systems in children, compared to adults [160], or mutual synapse innervations among different sensory modalities are not well developed in young ages [6]. It is noteworthy that a few kinds of perceptual memory, such as visual memory, photographic memory, and iconic memory, are related to eidetic imagery.

Spatial memory is a part of memory responsible for storing information about organism's environment and spatial orientation [161, 162]. Spatial memory will be formed after an organism gathers and processes sensory information about its surroundings, especially by vision and proprioception. Human's spatial memory is featured as navigation around a familiar city; similarly rat's or mouse's spatial memory is needed to learn the location of food at the end of a multiple-arm maze or the location of a plate in water maze. As the hippocampal lesion is associated with the impairment of spatial memory, the hippocampus has been thought as the location for spatial memory [163]. Moreover, some hippocampal principal neurons are featured by showing location-specific firing, also called place cells [162, 164]. The link of place cell activity to hippocampal memory function has been tested, in which the recording activities from single neurons or ensembles of neurons are done during spatial memory. The insights into the cellular mechanisms of spatial memory based on place cells are beginning to emerge [165]. The cellular mechanisms underlying spatial memory has been proposed to be the function of place cells in the hippocampus [166]. Visuospatial memory can be considered to be a subcategory of visual memory because it relies on cognitive map.

Olfactory memory refers to the storage and retrieval of odor signals, one type of perceptual memory. Many studies have shown that olfactory memory has various features of common memories including persistence and high resistance to interference [167–169]. After the birth, the alternations in the olfactory system, such as the olfactory bulb and cortices, are critically important to influence the maternal behaviors. Olfactory cues in mammalians play a critical role in the coordination of mother-infant links and in the subsequent development of the offspring normally. For instance, breast odors from mothers have individual features, which constitute a foundation for offsprings to recognize their mothers. In the entire evolutionary process, the olfactory system has become more and more important for various purposes including the survival of different species as well as the development of communication between mothers and infants or between males and females. To human beings and primates, the aspects of survival and communication remain to be functionally important. The enriched odor exposures increase the number of newborn neurons in the adult olfactory bulb and improve olfactory memory [170]. On the other hand, individuals suffering from brain degenerative diseases, such as Alzheimer's disease and other types of dementia, also show deficits in olfactory memory. These individuals show the deteriorated capability to distinguish odors when their neurodegeneration worsens, though it is not known whether the neurodegeneration impairs olfactory capability or the other way around. In this regard, the reports show that olfactory memory deficits can be used to help identifying some types of mental disorders. For example, major depression, as one type of psychiatric disorder, possesses its own featured pattern in olfactory deficiency [126, 171, 172].

By using neuroimaging, researchers have found that the medial prefrontal cortex is activated during various tasks of olfactory memory. Medial prefrontal cortical neurons receive extensive synapse innervations from olfactory afferent pathways. As the primary areas of encoding olfactory signals, these neurons are activated immediately after offspring birth. Although the main or accessory olfactory systems have no functional specificity in the development of maternal behaviors, the main olfactory system is influenced when the individual odor discrimination to offspring is needed. Once exposure to offspring odors, this main olfactory system experiences significant change. The changes in synaptic circuitry contribute to the level of maternal responsiveness and memorization to these odors. In addition, the fetus may be able to identify chemical cues after its intrauterine environments are formed. Olfactory learning and memory are present during the intrauterine period, since newborn infants demonstrate positive responses to the odor of their own amniotic fluid. Newborn infants also respond to olfactory cues linked to the odors from their maternal breast. For example, infants show favorable recognition and reaction to smells secreted from their mother's breasts, though they may be attracted to breast odors from unfamiliar nursing females. This unique scent of the mother to the infant is called as maternal olfactory signature. Besides mother's breasts can be a source of the unique olfactory cue, infants are also able to identify and react to scents from their mother underarm in familiarity and preference manners [173, 174]. The formation of these olfactory signatures from intrauterine scent and breast scents is based on associative learning and memory. This phenomenon about olfactory signature is also seen in rodents and mammalians [168].

2.6 Eidetic Memory and False Memory

According to the accuracy and the efficiency of memory formation and retrieval, declarative memory in the category of associative memory can be classified into eidetic memory and false memory in psychology. Eidetic memory, also called as total recall, photographic memory, or eidetic imagery, refers to a situation that individuals are able to recall images, sound, or objects with extreme accuracy and abundant volume or the vivid retrieval of their memories [175]. On the other hand, false memory stands for memories to events or knowledge that have never been happened or the distortion of memories to events or knowledge that have occurred as determined by externally validated facts [176, 177]. In addition to the difference of the accuracy of memory contents and the efficiency of memory retrieval, the speed of occurrence is different in these two types of memory. Eidetic memory is examined immediately after images, sound, or objects are given, and false memory is usually examined with the delay. In this regard, false memory is influenced by many facts, such as time, goals, and other cognitive states, beyond memory mechanisms.

2.6.1 Eidetic Memory

This type of memory is featured by the vivid recall of visual images and word language with extraordinary detail [157, 158]. Experts in the field of memoriology often use picture elicitation method to detect its retrieval. Eidetic memory used in children is typified by their capability to learn a picture lasting for 30 s and to preserve a nearly perfect recall to that picture in short period after the image has been taken away. Children in the test declare to have imaginary about this picture on a blank background as vividly perfect in detail as its presence there [160, 175]. Although many adult people demonstrate exceptional memory capabilities, there is no data showing whether truly eidetic memory can be maintained in adulthood [160]. As we have known, numerous famous artists and composers are usually thought to possess the ability of eidetic memory. However, it is likely that their exceptional memory to professional skills and knowledge may be based on repeated trainings in their respective fields. In other words, they have devoted large amount of their time and energy to improve their capability. Based on a rule about activity-dependent positive cycle in the recruitment and refinement of associative memory cells [6, 10], the focus and attention to individual arts and knowledge with repeated practices improve relevant memories through the recruitment of more memory cells and the upregulation of their encoding capability to facilitate memory retrievals, such that these artists and composers show the amazing capabilities in their professions.

Although individuals with eidetic memory recall visual information very precisely, their eidetic memory is not limited to visual recall, and they can recall other aspects of given objects in sensory information, such as auditory, tactile, gustatory, olfactory, and other dimensions [160, 175]. The discussion about eidetic memory tends to shift toward eidetic imagery which is basically the portion of eidetic memory that is visual in nature. Like other memories, the intensity of the recall may be influenced by several factors, such as the time and frequency of exposure to the stimulus, the conscious observation, the relevance to person, and so on. Individuals who show good memory declare to have eidetic memory, but there is difference in the method of information processing. People with skillful memory often apply mnemonic device to retain information, such as the division of an idea into enumerable elements. People with eidetic memory remember very specific details, such as the locations of a person standing and somethings a person is wearing. It can occur that certain individuals may experience sporadic eidetic memory, in which they may describe some number of memories in very close detail. The sporadic eidetic memories are not triggered consciously in most cases [175].

It is noteworthy that some autistic patients display extraordinary memory. For instance, autistic savants are socially rare but express certain specific signs of remarkable memory. However, most autistic patients by diagnosis do not have eidetic memory. Synesthesia is featured by potentiated auditory memory but only for signals that initiate synesthetic reaction. Certain synesthetes possess more than normal perfection in color sense. They are able to match color shades virtually perfect after extended periods without synesthetic reaction. Recent controversies about eidetic memory come from the excessive application of eidetic memory to those individuals who show exceptional memory skill. The presence of exceptional memory skills has been well documented. Extraordinary memory skills may result from the combinations of innate capability, learnt tactics, and enriched knowledge foundation. Technically, although eidetic memory refers to memory for sensory events as accurate as if individuals are still viewing or hearing original objects or events, almost all declarations of eidetic memory are well beyond this narrow definition. There may be a few people who are capable of a limited amount of eidetic recall [160, 175]. The documented eidetic capabilities, however, appear to be far more circumscribed and far less common than popularly imagined.

The mechanisms underlying eidetic memory have never been known. If we believe the accomplishment of memory by memory cells that memorize unitary signals in images, verbal, and other sensory dimensions, the vivid recalls of these integrated signals may be based on the efficient mutual synapse innervations and the fully activation of all of associative memory cells, especially primary associative memory cells in the sensory cortices [6].

2.6.2 False Memory

This term is used to describe memories to events or knowledge that have not been occurred or memory distortion in events or knowledge that do occur as determined by externally corroborated facts [176, 177]. It is a common experience that human memory may be unreliable to some degree by failing to remember or by remembering incorrectly. Our sense to identities is linked to our memory that can be disturbed by some challenges, e.g., amnesia, Alzheimer's disease, and post-traumatic stress disorder with dramatic loss of memory. Memory is a complicated process and only partly understood. The qualities of memories do not provide a reliable way to determine accuracy. A vivid and detailed memory may be mistaken, and a vague memory may be correct. In other words, the continuity of memory is no guarantee of truth, and the disruption of memory is no guarantee of falsity. In addition, retrieved memory contents are believed to be a reconstructed phenomenon and can usually be strongly influenced by numerous situations, expectations, emotions, beliefs to others, inappropriate interpretation, or desired outcome [6, 177, 178].

Although false memories are affected by various factors, the emotion has special impact for theoretical and practical reasons. Emotion's influences on false memory depend upon whether it is embedded in contents to be remembered events or in person's moods, where the mood is an aspect of the context of the encoded events. In the experiments of spontaneous and implanted false memories, the negatively valenced content provokes distortion, but negatively valenced moods protect against any distortion. The studies about the correlation between moods and false memory indicate that long-lasting negative natural moods, such as major depression, initiate false memory. Content-context disassociation in false memory has been tried to be explained by current opponent-process models for false memory, such as fuzzy-trace theory. Variations in emotion contents mainly influence memories to the essence of the processes. Variations in emotion contexts mainly influence memories to the verbatim form of processes. How these effects are modulated by variations in memory tests and in arousal states remains to be examined [179, 180].

In terms of the foundation of false memory, a hypnosis state has been used to explain the formation of false memory. It has been known that the hypnosis technique can induce dream-like imagination and raise the subjective certainty to fantasy. For example, certain false memories are developed by the rehearsal of an activity that has been ensured as fantasy. After repetitive thinking and imagining this activity, individual may start to remember this event as if it has actually happened. By cuing or questioning, this individual may confidently represent this activity that is only previously imagined to be familiar. In fact, this false memory is formed by the retrieval of the stored endogenous signals that come from imagining [6]. As we have known, the conversion of short-term memory into long-term memory is based on the rehearsal mechanism for memory consolidation. Similarly, the rehearsal of those incorrect signals induces the formation of long-term false memory. This rehearsal mechanism is suitable for both real memory and implanted false memory [176]. In addition, memory includes reconstructions to various events, not simply recall. A lot of people experience the learning and memory process in that a childhood memory actually happens to a sibling. For example, a child remembers the debate between daddy and mommy, in which daddy takes a sharp knife and has an angry mood toward mommy who looks crying and scared. This memory may be partially correct. In fact, this child only has half-remembering fragments that he or she is guided to interpret and remember correctly because daddy has been violent. Results from this indicate that memories can be implanted. However, we have evidence about implanted traumatic memories and their relevant effects, such as post-traumatic stress disorder and dissociative identity disorder, as these studies are involved in ethical issue [180].

In clinical, there is a false memory syndrome. This term describes a state of mind, in which patients possess a high quantity of memories with highly vivid recall but belonging to false memory. These false memories are usually related to abuse events during their childhood period. In studies with this situation, victims have confessed to make completely made-up stories. In false memory in relevance to child abuse, the alleged sufferers experience the detachment. This disassociation leads to the repression of traumatic memory until later life when these memories resurface either naturally or with professional aid. Many advocates of false memory syndrome argue against the methods of memory recovery, declaring that therapists and psychiatrists accidentally implant false memories [181]. As the impressive memory in this syndrome suffers, numerous associative memory cells may be recruited when the event occurs. With long-waiting period to later life, their accidental coactivations with other groups of primary and secondary associative memory cells irrelevant to this event may induce their mutual synapse innervations and recruit advanced associative memory cells that associate this event and other endogenous signals, i.e., the formation of false memory [6]. The long-tasting time from the impressive event to its recall gives a chance for the interference of other events to implant false memory.

2.7 Other Types of Memory

According to the sources of information and signals, memory can be classified into memories to exogenous signals as well as memories to endogenous information and signals. Memories to those exogenous signals, similar to perceptual memory, are featured by the sources of learned signals being from different sensory systems, such as visual and auditory systems. Memories to the endogenous signals are featured by the sources of memorized signals being from cognitive processes and emotional reactions, such as logical reasoning, associative thinking, imagination, computation, judgment, decision-making, pleasure, anger, and other cognitive events [4, 6, 10]. In addition, based on the complex of information and signals, memory can be classified into memories to unitary signals and memories to complicated signals that are reorganized from unitary signals [6]. Furthermore, memory can be classified into spatial memory and temporal memory. Spatial memory stands for memories to the information about object's location and orientation. Temporal memory refers to memories to the signals with time points and sequences, for instance, imaging motion processes and verbal presentation orders [6]. Finally, the associated signals during associative learning can come from an intramodal sensory system and/or cross-modal sensory systems. Therefore, associative memory can be classified into memories to intramodal signals and memories to cross-modal signals [4, 10] (Fig. 2.1).

2.8 Learning and Memory in Relevance to Ages

Regardless declarative memory or nondeclarative memory, learning efficiency and memory accuracy are largely influenced by the age, in which both of them appear to a bell-shape, or the increment up to young adults and the decay with aging [6, 182, 183]. In addition, unitary signals and their organized contents (topographic organization) to be learnt and memorized in manners of intramodal and cross-modal are also different during postnatal development [6, 47, 173]. In this regard, learning includes the acquisition and storage of unitary signals as well as the acquired memories to the complicated signals that are reorganized from unitary signals.

In the early stage of postnatal development, newborns or infants start to possess sensory learning based on their experience through various sensory systems, such as visual, auditory, olfactory, gustatory, and tactile sensations [184]. In sensory learning, the acquisition and memory of signals from various sensory systems occur in sensory cortices [185]. For instance, the repeated trainings can improve those perceptual capabilities. This improvement offers important insight into the physiological basis of learning in the cerebral cortex. The quick time course and facilitation improved by perceptual capabilities suggest that the learning is an integration process of normal perception. Perceptual learning can occur in a variety of areas and involve multiple mechanisms, especially the task-specific suppression of signals that interfere with the performances [186]. The contents learnt from their parents or their neighbors include unitary signals, such as a sound with single frequency or tone, a simple image (e.g., blue sky), and a simple taste (e.g., milk), as well as complicated signals, such as image complexes with multiple objects and colors, which are organized from unitary signals in different ways. Each unitary signal is detected by sensory receptors for single feature in one sensory modality and is transmitted to their innate sensory neurons in the sensory cortex by specific afferent nerves, which strengthen the encoding of sensory signals and synapse interconnections in given pathway. A group of sensory cortical neurons fulfills the encoding in relevance to a unitary signal. However, the complicated signals in the learning need be integrated by sensory cortical neurons to form complete neural signals correspondent to the complicated exogenous signals since these complicated signals in sensory stimulations have been dissected into the unitary signals by sensory receptors up to their sensory cortical neurons [6]. For instance, a complicated sound is dissected to the different frequencies and detected by hair cells on the cochlea base membrane. A complicated image is dissected to different photon beams with their colors and light intensities and detected by cone cells in the retina. One way for their integrations by sensory cortical neurons is intramodal, by which the coactivations of these sensory cortical neurons by unitary signals in complicated signals lead to their mutual synapse innervations for the integration of unitary signals into complicated signals in the neuron form or neural integrated signal. Another way for their integration is by cross-modal sensory cortical neurons, by which the coactivations of sensory cortical neurons among respective modalities by cross-modal learned signals lead to mutual synapse innervations among sensory cortical neurons in these modalities for the integration of signals from different modalities. For instance, learning in association of visual and auditory signals leads to mutual synapse innervations among visual and auditory cortical neurons. By more associations, mutual synapse innervation and signal integrations will become more enriched. These mutual synapse innervations will constitute the foundation for the recruitment of associative memory cells [6].

After growing to the age when the systemic education is given by parents and schools, children experience more cross-modal associative learning, i.e., cross-modal associative learning gradually moves forward from unitary to complicated signals, especially the association of auditory signals with visual signals, smell signals, taste signals, and tactile signals for their definitions, compared with intramodal associative learning. In the associative learning of these complicated signals reorganized from unitary signals, mutual synapse innervations that lead to the recruitment of associative memory cells become more and more enriched. In the subsequent learning, the strengthening of these mutual innervated synapses through the conversion of inactive synapses into active synapses or silent synapses into functional synapses leads to groups of associative memory cells to be activated for memory formation and memory retrieval [6].

In terms of the mechanisms underlying age-relevant memory efficiency in the formation and retrieval of associated signals, the development in the thickness of cerebral cortices and the apical dendrite length of cortical neurons are likely involved [187]. Although cellular architectures for age-related memory efficiency are not clear, the increased thickness of cerebral cortices during postnatal development, especially in the first 2 years, may grant the axon projection for long distance, the dendritic arborization for enrichment and receptive field, as well as more synapse formations. With these supports, axon prolongation and mutual synapse innervations for the union of ready-recruited neurons through doing activity together, connecting together, and strengthening together can be executed in cerebral cortices [6].

Associative memory cells have stored various unitary signals, in which the capacity of associative memory and the number of associative memory cells are variable among individuals based on the experiences and repetitions during associative learning. In this regard, memory strength and maintenance at different levels will influence their subsequent learning and cognition, i.e., interaction among learning, memory, and cognition in a forward manner for experiences influencing events as well as in a feedback manner for new experiences influencing the strength and the maintenance of previous experiences. In this regard, the protocols designed in psychology for various learnings and cognitions (e.g., recognition and identification to new objects) are influenced by previous experiences and memories. The reorganization of associative memory cells based on their recruitment and refinement in advanced learning constitutes the foundation of memory rebalance and memory reediting.

References

1. Kandel ER, Dudai Y, Mayford MR. The molecular and systems biology of memory. Cell. 2014;157(1):163–86.
2. McGaugh JL. Memory – a century of consolidation. Science. 2000;287(5451):248–51.
3. Tonegawa S, et al. Memory engram cells have come of age. Neuron. 2015;87(5):918–31.
4. Wang JH, Cui S. Associative memory cells: formation, function and perspective. F1000Res. 2017;6:283.
5. Squire LR, Knowlton B, Musen G. The structure and organization of memory. Annu Rev Psychol. 1993;44:453–95.
6. Wang J-H. Searching basic units of memory traces: associative memory cells. F1000Res. 2019;8(457):1–28.
7. Byrne JH. Learning and memory: basic mechanisms. In: Square LR, et al., editors. Fundament neuroscience. 2nd ed. Amsterdam: Academic; 2003. p. 1275–98.
8. Kandel ER. Nerve cells and behavior. In: Kandel ER, Schwartz JH, Jessell TM, editors. Principles of neural science. 4th ed. New York: McGraw-Hill; 2000. p. 19–35.

9. Wasserman EA, Miller RR. What's elementary about associative learning? Annu Rev Psychol. 1997;48:573–607.
10. Wang JH, Cui S. Associative memory cells and their working principle in the brain. F1000Res. 2018;7:108.
11. Cadorin L, et al. An integrative review of the characteristics of meaningful learning in healthcare professionals to enlighten educational practices in health care. Nurs Open. 2014;1(1):3–14.
12. Cake MA. Deep dissection: motivating students beyond rote learning in veterinary anatomy. J Vet Med Educ. 2006;33(2):266–71.
13. Gewirtz JL. Conditional responding as a paradigm for observational, imitative learning and vicarious-reinforcement. Adv Child Dev Behav. 1971;6:273–304.
14. Fanselow MS, Poulos AM. The neuroscience of mammalian associative learning. Annu Rev Psychol. 2005;56:207–34.
15. Frenois F, Le Moine C, Cador M. The motivational component of withdrawal in opiate addiction: role of associative learning and aversive memory in opiate addiction from a behavioral, anatomical and functional perspective. Rev Neurosci. 2005;16(3):255–76.
16. Letzkus JJ, Wolff SB, Luthi A. Disinhibition, a circuit mechanism for associative learning and memory. Neuron. 2015;88(2):264–76.
17. Suzuki WA. Associative learning signals in the brain. Prog Brain Res. 2008;169:305–20.
18. Byrne JH, Eskin A, Scholz KP. Neuronal mechanisms contributing to long-term sensitization in Aplysia. J Physiol Paris. 1988;83(3):141–7.
19. Wood DC. Habituation in Stentor: produced by mechanoreceptor channel modification. J Neurosci. 1988;8(7):2254–8.
20. Pavlov I. Conditioned reflexes: an investigation of the physiological activity of the cerebral cortex. Translated by Anrep GV. Nature. 1927;121(3052):662–4.
21. Baars BJ, Gage NM. In: Baars BJ, Gage NM, editors. Cognition, brain, and consciousness: introduction to cognitive neuroscience. London: Elsevier Ltd; 2007.
22. Watson J. Psychology as the behaviorist views it: the behaviorist manifesto. Baltimore: John Hopkins University; 1913.
23. Carcea I, Froemke RC. Biological mechanisms for observational learning. Curr Opin Neurobiol. 2019;54:178–85.
24. Zentall TR. Perspectives on observational learning in animals. J Comp Psychol. 2012;126(2):114–28.
25. Bolhuis JJ, Honey RC. Imprinting, learning and development: from behaviour to brain and back. Trends Neurosci. 1998;21(7):306–11.
26. Horn G. Imprinting, learning, and memory. Behav Neurosci. 1986;100(6):825–32.
27. Hedges JH, et al. Play, attention, and learning: how do play and timing shape the development of attention and influence classroom learning? Ann N Y Acad Sci. 2013;1292:1–20.
28. Hill HM, Dietrich S, Cappiello B. Learning to play: a review and theoretical investigation of the developmental mechanisms and functions of cetacean play. Learn Behav. 2017;45(4):335–54.
29. Cardoso-Leite P, Bavelier D. Video game play, attention, and learning: how to shape the development of attention and influence learning? Curr Opin Neurol. 2014;27(2):185–91.
30. Pearce MT. Statistical learning and probabilistic prediction in music cognition: mechanisms of stylistic enculturation. Ann N Y Acad Sci. 2018;1423:378–95.
31. Grunwald T, Corsbie-Massay C. Guidelines for cognitively efficient multimedia learning tools: educational strategies, cognitive load, and interface design. Acad Med. 2006;81(3):213–23.
32. Eiriksdottir E, Catrambone R. Procedural instructions, principles, and examples: how to structure instructions for procedural tasks to enhance performance, learning, and transfer. Hum Factors. 2011;53(6):749–70.
33. Agra G, et al. Analysis of the concept of meaningful learning in light of the Ausubel's theory. Rev Bras Enferm. 2019;72(1):248–55.

34. Giannini AJ, Giannini JN, Condon M. Use of tangential visual symbols to increase the long-term learning process: applications of linkage in teaching pharmacological principles of addiction. J Clin Pharmacol. 2000;40(7):708–12.
35. McLaughlin B. "Intentional" and "Incidental" learning in human subjects: the role of instructions to learn and motivation. Psychol Bull. 1965;63:359–76.
36. Wang D, et al. Neurons in the barrel cortex turn into processing whisker and odor signals: a cellular mechanism for the storage and retrieval of associative signals. Front Cell Neurosci. 2015;9:320.
37. Byrne JH. Cellular analysis of associative learning. Physiol Rev. 1987;67(2):329–439.
38. Feng J, et al. Barrel cortical neuron integrates triple associated signals for their memory through receiving epigenetic-mediated new synapse innervations. Cereb Cortex. 2017;27(12):5858–71.
39. Gao Z, et al. Associations of unilateral whisker and olfactory signals induce synapse formation and memory cell recruitment in bilateral barrel cortices: cellular mechanism for unilateral training toward bilateral memory. Front Cell Neurosci. 2016;10(285):1–16.
40. Guo R, et al. Associative memory extinction is accompanied by decayed plasticity at motor cortical neurons and persistent plasticity at sensory cortical neurons. Front Cell Neurosci. 2017;11(168):1–12.
41. Liu Y, et al. Piriform cortical glutamatergic and GABAergic neurons express coordinated plasticity for whisker-induced odor recall. Oncotarget. 2017;8(56):95719–40.
42. Wang JH, et al. Upregulation of glutamatergic receptor-channels is associated with cross-modal reflexes encoded in barrel cortex and piriform cortex. Biophys J. 2014;106(supplement 2):191a.
43. Wang J-H, et al. Both glutamatergic and gabaergic neurons are recruited to be associative memory cells. Biophys J. 2016;110(supplement 3):481a.
44. Yan F, et al. Coordinated plasticity between barrel cortical glutamatergic and GABAergic neurons during associative memory. Neural Plast. 2016;2016(ID5648390):1–20.
45. Wang J-H, Guo R, Wei Z. Associative memory extinction is accompanied by decays of associative memory cells and their plasticity at motor cortex but not sensory cortex. Soc Neurosci. 2017;81(09):10385.
46. Wang J-H, et al. Prefrontal cortical neurons are recruited as secondary associative memory cells for associative memory and cognition. Biophys J. 2018;114(3):155a.
47. Wang JH, et al. Secondary associative memory cells and their plasticity in the prefrontal cortex. Biophys J. 2019;116(3):427a.
48. Squire LR. Declarative and nondeclarative memory: multiple brain systems supporting learning and memory. J Cogn Neurosci. 1992;4(3):232–43.
49. Squire LR, Zola SM. Structure and function of declarative and nondeclarative memory systems. Proc Natl Acad Sci U S A. 1996;93(24):13515–22.
50. Gais S, Born J. Declarative memory consolidation: mechanisms acting during human sleep. Learn Mem. 2004;11(6):679–85.
51. Garcia-Lazaro HG, et al. Neuroanatomy of episodic and semantic memory in humans: a brief review of neuroimaging studies. Neurol India. 2012;60(6):613–7.
52. Horner MD. Psychobiological evidence for the distinction between episodic and semantic memory. Neuropsychol Rev. 1990;1(4):281–321.
53. Moscovitch M, et al. The cognitive neuroscience of remote episodic, semantic and spatial memory. Curr Opin Neurobiol. 2006;16(2):179–90.
54. Platel H. Functional neuroimaging of semantic and episodic musical memory. Ann N Y Acad Sci. 2005;1060:136–47.
55. Renoult L, et al. Personal semantics: at the crossroads of semantic and episodic memory. Trends Cogn Sci. 2012;16(11):550–8.
56. Squire LR, Zola SM. Episodic memory, semantic memory, and amnesia. Hippocampus. 1998;8(3):205–11.
57. Tulving E. Multiple memory systems and consciousness. Hum Neurobiol. 1987;6(2):67–80.

58. Wood F. Focal and diffuse memory activation assessed by localized indicators of CNS metabolism: the semantic-episodic memory distinction. Hum Neurobiol. 1987;6(2):141–51.
59. Begg IM, Nicholson S. Semantic and episodic relations are experimentally dissociable and stochastically independent. Can J Exp Psychol. 1994;48(3):399–417.
60. Cohen NJ, et al. Different memory systems underlying acquisition of procedural and declarative knowledge. Ann N Y Acad Sci. 1985;444:54–71.
61. Arwas S, Rolnick A, Lubow RE. Conditioned taste aversion in humans using motion-induced sickness as the US. Behav Res Ther. 1989;27(3):295–301.
62. Roediger HL 3rd. Implicit memory. Retention without remembering. Am Psychol. 1990;45(9):1043–56.
63. Roediger HL 3rd, McDermott KB. Remembering what we learn. Cerebrum. 2018;pii:cer-08-18.
64. Brooks DN, Baddeley AD. What can amnesic patients learn? Neuropsychologia. 1976;14(1):111–22.
65. Jacoby LL, Dallas M. On the relationship between autobiographical memory and perceptual learning. J Exp Psychol Gen. 1981;110(3):306–40.
66. Graf P, Schacter DL. Implicit and explicit memory for new associations in normal and amnesic subjects. J Exp Psychol Learn Mem Cogn. 1985;11(3):501–18.
67. Challis BH, et al. Perceptual and conceptual cueing in implicit and explicit retrieval. Memory. 1993;1(2):127–51.
68. Greenberg DL, Verfaellie M. Interdependence of episodic and semantic memory: evidence from neuropsychology. J Int Neuropsychol Soc. 2010;16(5):748–53.
69. Maguire EA, Frith CD. Aging affects the engagement of the hippocampus during autobiographical memory retrieval. Brain. 2003;126(Pt 7):1511–23.
70. Maguire EA, Frith CD. Lateral asymmetry in the hippocampal response to the remoteness of autobiographical memories. J Neurosci. 2003;23(12):5302–7.
71. Diamond DM, et al. The temporal dynamics model of emotional memory processing: a synthesis on the neurobiological basis of stress-induced amnesia, flashbulb and traumatic memories, and the Yerkes-Dodson law. Neural Plast. 2007;2007:60803.
72. Nyberg L, et al. Activation of medial temporal structures during episodic memory retrieval. Nature. 1996;380(6576):715–7.
73. Nyberg L, Cabeza R, Tulving E. Asymmetric frontal activation during episodic memory: what kind of specificity? Trends Cogn Sci. 1998;2(11):419–20.
74. Tulving E, Markowitsch HJ. Episodic and declarative memory: role of the hippocampus. Hippocampus. 1998;8(3):198–204.
75. Buzsaki G. Theta rhythm of navigation: link between path integration and landmark navigation, episodic and semantic memory. Hippocampus. 2005;15(7):827–40.
76. Janowsky JS, et al. Cognitive impairment following frontal lobe damage and its relevance to human amnesia. Behav Neurosci. 1989;103(3):548–60.
77. Janowsky JS, Shimamura AP, Squire LR. Source memory impairment in patients with frontal lobe lesions. Neuropsychologia. 1989;27(8):1043–56.
78. Gabrieli JD, Poldrack RA, Desmond JE. The role of left prefrontal cortex in language and memory. Proc Natl Acad Sci U S A. 1998;95(3):906–13.
79. McIntosh AR, et al. Differential functional connectivity of prefrontal and medial temporal cortices during episodic memory retrieval. Hum Brain Mapp. 1997;5(4):323–7.
80. Wagner AD, et al. Prefrontal cortex and recognition memory. Functional-MRI evidence for context-dependent retrieval processes. Brain. 1998;121(Pt 10):1985–2002.
81. Wheeler MA, Stuss DT, Tulving E. Frontal lobe damage produces episodic memory impairment. J Int Neuropsychol Soc. 1995;1(6):525–36.
82. Deisseroth K, et al. Excitation-neurogenesis coupling in adult neural stem/progenitor cells. Neuron. 2004;42(4):535–52.
83. Croxson PL, et al. Acetylcholine facilitates recovery of episodic memory after brain damage. J Neurosci. 2012;32(40):13787–95.

84. Danion JM, et al. Pharmacology of human memory and cognition: illustrations from the effects of benzodiazepines and cholinergic drugs. J Psychopharmacol. 1993;7(4):371–7.
85. Grogan J, et al. Dopamine and consolidation of episodic memory: timing is everything. J Cogn Neurosci. 2015;27(10):2035–50.
86. Huang M, et al. 5-HT1A parital agonism and 5-HT7 antagonism restore episodic memory in subchronic phencyclidine-treated mice: role of brain glutamate, dopamine, acetylcholine and GABA. Psychopharmacology. 2018;235(10):2795–808.
87. O'Halloran KD. Blast from the past! Phrenic motor memory of antecedent episodic hypercapnia is serotonin dependent: relevance to respiratory rehabilitation and sleep-disordered breathing? Exp Physiol. 2016;101(2):258–9.
88. Papenberg G, et al. Dopamine and glutamate receptor genes interactively influence episodic memory in old age. Neurobiol Aging. 2014;35(5):1213 e3–8.
89. Dempster EL, et al. Episodic memory performance predicted by the 2bp deletion in exon 6 of the "alpha 7-like" nicotinic receptor subunit gene. Am J Psychiatry. 2006;163(10):1832–4.
90. Fernandes C, et al. Performance deficit of alpha7 nicotinic receptor knockout mice in a delayed matching-to-place task suggests a mild impairment of working/episodic-like memory. Genes Brain Behav. 2006;5(6):433–40.
91. Chowdhury R, et al. Dopamine modulates episodic memory persistence in old age. J Neurosci. 2012;32(41):14193–204.
92. Eustache F, et al. Episodic memory in transient global amnesia: encoding, storage, or retrieval deficit? J Neurol Neurosurg Psychiatry. 1999;66(2):148–54.
93. Guillery-Girard B, et al. Long-term memory following transient global amnesia: an investigation of episodic and semantic memory. Acta Neurol Scand. 2006;114(5):329–33.
94. Cooper RA, et al. Reduced hippocampal functional connectivity during episodic memory retrieval in autism. Cereb Cortex. 2017;27(2):888–902.
95. Gaigg SB, Bowler DM, Gardiner JM. Episodic but not semantic order memory difficulties in autism spectrum disorder: evidence from the Historical Figures Task. Memory. 2014;22(6):669–78.
96. Wojcik DZ, Moulin CJ, Souchay C. Metamemory in children with autism: exploring "feeling-of-knowing" in episodic and semantic memory. Neuropsychology. 2013;27(1):19–27.
97. Gron G, Riepe MW. Neural basis for the cognitive continuum in episodic memory from health to Alzheimer disease. Am J Geriatr Psychiatry. 2004;12(6):648–52.
98. Tromp D, et al. Episodic memory in normal aging and Alzheimer disease: insights from imaging and behavioral studies. Ageing Res Rev. 2015;24(Pt B):232–62.
99. Ostergaard AL. Episodic, semantic and procedural memory in a case of amnesia at an early age. Neuropsychologia. 1987;25(2):341–57.
100. Tulving E. In: Tulving E, editor. Elements of episodic memory. London: Oxford University Press; 1985.
101. Kihlstrom JF. Posthypnotic amnesia for recently learned material: interactions with "episodic" and "semantic" memory. Cogn Psychol. 1980;12(2):227–51.
102. Tune S, Asaridou SS. Stimulating the semantic network: what can TMS tell us about the roles of the posterior middle temporal gyrus and angular gyrus? J Neurosci. 2016;36(16):4405–7.
103. Fletcher PD, Warren JD. Semantic dementia: a specific network-opathy. J Mol Neurosci. 2011;45(3):629–36.
104. Irish M. Elucidating a core semantic network in the brain-implications for disorders of semantic cognition. J Neurosci. 2016;36(23):6144–6.
105. Fransen E, Lansner A. A model of cortical associative memory based on a horizontal network of connected columns. Network. 1998;9(2):235–64.
106. Haberly LB, Bower JM. Olfactory cortex: model circuit for study of associative memory? Trends Neurosci. 1989;12(7):258–64.
107. Fisk AD, et al. Age-related retention of skilled memory search: examination of associative learning, interference, and task-specific skills. J Gerontol B Psychol Sci Soc Sci. 1995;50(3):P150–61.

108. Page RA, von Merten S, Siemers BM. Associative memory or algorithmic search: a comparative study on learning strategies of bats and shrews. Anim Cogn. 2012;15(4):495–504.
109. Mayes AR, Montaldi D. Exploring the neural bases of episodic and semantic memory: the role of structural and functional neuroimaging. Neurosci Biobehav Rev. 2001;25(6):555–73.
110. Starr JM, et al. Episodic and semantic memory tasks activate different brain regions in Alzheimer disease. Neurology. 2005;65(2):266–9.
111. Martin-Ordas G, Atance CM, Caza JS. How do episodic and semantic memory contribute to episodic foresight in young children? Front Psychol. 2014;5:732.
112. Nyberg L, McIntosh AR, Tulving E. Functional brain imaging of episodic and semantic memory with positron emission tomography. J Mol Med (Berl). 1998;76(1):48–53.
113. Lambon Ralph MA, Patterson K. Generalization and differentiation in semantic memory: insights from semantic dementia. Ann N Y Acad Sci. 2008;1124:61–76.
114. Lambon Ralph MA, et al. Semantic memory is impaired in both dementia with Lewy bodies and dementia of Alzheimer's type: a comparative neuropsychological study and literature review. J Neurol Neurosurg Psychiatry. 2001;70(2):149–56.
115. Farah MJ, et al. Category-specificity and modality-specificity in semantic memory. Neuropsychologia. 1989;27(2):193–200.
116. Farah MJ, McClelland JL. A computational model of semantic memory impairment: modality specificity and emergent category specificity. J Exp Psychol Gen. 1991;120(4):339–57.
117. Tippett LJ, Grossman M, Farah MJ. The semantic memory impairment of Alzheimer's disease: category-specific? Cortex. 1996;32(1):143–53.
118. McCarthy RA, Warrington EK. Disorders of semantic memory. Philos Trans R Soc Lond Ser B Biol Sci. 1994;346(1315):89–96.
119. McCarthy RA, Warrington EK. Past, present, and prospects: reflections 40 years on from the selective impairment of semantic memory (Warrington, 1975). Q J Exp Psychol (Hove). 2016;69(10):1941–68.
120. Cohen G, Johnston R, Plunkett K. In: Cohen G, editor. Exploring cognition: damaged brains and neural networks. Exploring cognition: Damaged brains and neural networks. Erlbaum: Psychology Press; 2002.
121. Kable JW, et al. Conceptual representations of action in the lateral temporal cortex. J Cogn Neurosci. 2005;17(12):1855–70.
122. Thompson-Schill SL, et al. Role of left inferior prefrontal cortex in retrieval of semantic knowledge: a reevaluation. Proc Natl Acad Sci U S A. 1997;94(26):14792–7.
123. Baddeley A. Working memory: looking back and looking forward. Nat Rev Neurosci. 2003;4(10):829–39.
124. Constantinidis C, Klingberg T. The neuroscience of working memory capacity and training. Nat Rev Neurosci. 2016;17(7):438–49.
125. Zhou YD, Ardestani A, Fuster JM. Distributed and associative working memory. Cereb Cortex. 2007;17(Suppl 1):i77–87.
126. Granger R, Lynch G. Higher olfactory processes: perceptual learning and memory. Curr Opin Neurobiol. 1991;1(2):209–14.
127. Magnussen S, Greenlee MW. The psychophysics of perceptual memory. Psychol Res. 1999;62(2–3):81–92.
128. Hedge C, Oberauer K, Leonards U. Selection in spatial working memory is independent of perceptual selective attention, but they interact in a shared spatial priority map. Atten Percept Psychophys. 2015;77(8):2653–68.
129. Koshino H, Olid P. Interactions between modality of working memory load and perceptual load in distractor processing. J Gen Psychol. 2015;142(3):135–49.
130. Weil RS, et al. Opposite effects of perceptual and working memory load on perceptual filling-in of an artificial scotoma. Cogn Neurosci. 2012;3(1):36–44.
131. Cowan N. What are the differences between long-term, short-term, and working memory? Prog Brain Res. 2008;169:323–38.

132. Rhodes S, Cowan N. Attention in working memory: attention is needed but it yearns to be free. Ann N Y Acad Sci. 2018;1424(1):52–63.
133. Cowan N. Multiple concurrent thoughts: the meaning and developmental neuropsychology of working memory. Dev Neuropsychol. 2010;35(5):447–74.
134. Ericsson KA, Kintsch W. Long-term working memory. Psychol Rev. 1995;102(2):211–45.
135. Oberauer K, et al. What limits working memory capacity? Psychol Bull. 2016;142(7):758–99.
136. Buszard T, Masters RS, Farrow D. The generalizability of working-memory capacity in the sport domain. Curr Opin Psychol. 2017;16:54–7.
137. Glassman RB. A working memory "theory of relativity": elasticity in temporal, spatial, and modality dimensions conserves item capacity in radial maze, verbal tasks, and other cognition. Brain Res Bull. 1999;48(5):475–89.
138. Unsworth N, Robison MK. A locus coeruleus-norepinephrine account of individual differences in working memory capacity and attention control. Psychon Bull Rev. 2017;24(4):1282–311.
139. Moran TP. Anxiety and working memory capacity: a meta-analysis and narrative review. Psychol Bull. 2016;142(8):831–64.
140. Fukuda K, Awh E, Vogel EK. Discrete capacity limits in visual working memory. Curr Opin Neurobiol. 2010;20(2):177–82.
141. Fuster JM, Bressler SL. Cognit activation: a mechanism enabling temporal integration in working memory. Trends Cogn Sci. 2012;16(4):207–18.
142. Bodner M, et al. Patterned firing of parietal cells in a haptic working memory task. Eur J Neurosci. 2005;21(9):2538–46.
143. Collette F, Van der Linden M. Brain imaging of the central executive component of working memory. Neurosci Biobehav Rev. 2002;26(2):105–25.
144. Passingham D, Sakai K. The prefrontal cortex and working memory: physiology and brain imaging. Curr Opin Neurobiol. 2004;14(2):163–8.
145. Heck A, et al. Converging genetic and functional brain imaging evidence links neuronal excitability to working memory, psychiatric disease, and brain activity. Neuron. 2014;81(5):1203–13.
146. Hartley AA, Speer NK. Locating and fractionating working memory using functional neuroimaging: storage, maintenance, and executive functions. Microsc Res Tech. 2000;51(1):45–53.
147. Hsieh LT, Ranganath C. Frontal midline theta oscillations during working memory maintenance and episodic encoding and retrieval. Neuroimage. 2014;85(Pt 2):721–9.
148. Jensen O. Maintenance of multiple working memory items by temporal segmentation. Neuroscience. 2006;139(1):237–49.
149. Nyberg L, Eriksson J. Working memory: maintenance, updating, and the realization of intentions. Cold Spring Harb Perspect Biol. 2015;8(2):a021816.
150. Brewin CR. Episodic memory, perceptual memory, and their interaction: foundations for a theory of posttraumatic stress disorder. Psychol Bull. 2014;140(1):69–97.
151. Schacter DL. Perceptual representation systems and implicit memory. Toward a resolution of the multiple memory systems debate. Ann N Y Acad Sci. 1990;608:543–67.. discussion 567–71
152. Robinson L, Platt B, Riedel G. Involvement of the cholinergic system in conditioning and perceptual memory. Behav Brain Res. 2011;221(2):443–65.
153. Hampton RR. Monkey perirhinal cortex is critical for visual memory, but not for visual perception: reexamination of the behavioural evidence from monkeys. Q J Exp Psychol B. 2005;58(3–4):283–99.
154. Khan ZU, Martin-Montanez E, Baxter MG. Visual perception and memory systems: from cortex to medial temporal lobe. Cell Mol Life Sci. 2011;68(10):1737–54.
155. Miyashita Y. Inferior temporal cortex: where visual perception meets memory. Annu Rev Neurosci. 1993;16:245–63.
156. Super H. Cognitive processing in the primary visual cortex: from perception to memory. Rev Neurosci. 2002;13(4):287–98.

157. Miller E. The affective nature of illusion and hallucination. Part Ii: eidetic imagery. J Neurol Psychopathol. 1931;12(45):1–13.
158. Saltzman SS, Machover S. An inquiry into eidetic imagery with particular reference to visual hallucinations. Am J Psychiatry. 1952;108(10):740–8.
159. Haber RN, Standing LG. Direct measures of short-term visual storage. Q J Exp Psychol. 1969;21(1):43–54.
160. Giray EF, et al. The incidence of eidetic imagery as a function of age. Child Dev. 1976;47(4):1207–10.
161. Lanke J, et al. Spatial memory and stereotypic behaviour of animals in radial arm mazes. Brain Res. 1993;605(2):221–8.
162. O'Keefe J. Hippocampus, theta, and spatial memory. Curr Opin Neurobiol. 1993;3(6):917–24.
163. Kessels RP, et al. Varieties of human spatial memory: a meta-analysis on the effects of hippocampal lesions. Brain Res Brain Res Rev. 2001;35(3):295–303.
164. O'Keefe J, Dostrovsky J. The hippocampus as a spatial map. Preliminary evidence from unit activity in the freely-moving rat. Brain Res. 1971;34(1):171–5.
165. Moser EI, Paulsen O. New excitement in cognitive space: between place cells and spatial memory. Curr Opin Neurobiol. 2001;11(6):745–51.
166. Eichenbaum H, et al. The hippocampus, memory, and place cells: is it spatial memory or a memory space? Neuron. 1999;23(2):209–26.
167. Buchsbaum MS, et al. Temporal and hippocampal metabolic rate during an olfactory memory task assessed by positron emission tomography in patients with dementia of the Alzheimer type and controls. Preliminary studies. Arch Gen Psychiatry. 1991;48(9):840–7.
168. Coureaud G, et al. Pheromone-induced olfactory memory in newborn rabbits: involvement of consolidation and reconsolidation processes. Learn Mem. 2009;16(8):470–3.
169. Staubli U, et al. Antagonism of NMDA receptors impairs acquisition but not retention of olfactory memory. Behav Neurosci. 1989;103(1):54–60.
170. Rochefort C, et al. Enriched odor exposure increases the number of newborn neurons in the adult olfactory bulb and improves odor memory. J Neurosci. 2002;22(7):2679–89.
171. Wilson DA, Stevenson RJ. Olfactory perceptual learning: the critical role of memory in odor discrimination. Neurosci Biobehav Rev. 2003;27(4):307–28.
172. Wilson DA, Sullivan RM. Cortical processing of odor objects. Neuron. 2012;72(4):506–19.
173. Mizuno K, Ueda A. Antenatal olfactory learning influences infant feeding. Early Hum Dev. 2004;76(2):83–90.
174. Pautassi RM, et al. Acute ethanol counteracts the acquisition of aversive olfactory learning in infant rats. Alcohol. 2005;36(2):99–105.
175. Kaylor CW, Davidon RS. Accuracy of recall as a function of eidetic imagery. Percept Mot Skills. 1979;48(3 Pt 2):1143–8.
176. Brainerd CJ, Reyna VF, Ceci SJ. Developmental reversals in false memory: a review of data and theory. Psychol Bull. 2008;134(3):343–82.
177. Loftus EF. Memory distortion and false memory creation. Bull Am Acad Psychiatry Law. 1996;24(3):281–95.
178. Kurkela KA, Dennis NA. Event-related fMRI studies of false memory: an Activation Likelihood Estimation meta-analysis. Neuropsychologia. 2016;81:149–67.
179. Bookbinder SH, Brainerd CJ. Emotion and false memory: the context-content paradox. Psychol Bull. 2016;142(12):1315–51.
180. Reyna VF, Brainerd CJ. Fuzzy-trace theory and false memory: new frontiers. J Exp Child Psychol. 1998;71(2):194–209.
181. Bremner JD, et al. Neural mechanisms in dissociative amnesia for childhood abuse: relevance to the current controversy surrounding the "false memory syndrome". Am J Psychiatry. 1996;153(7 Suppl):71–82.
182. Foster TC. Dissecting the age-related decline on spatial learning and memory tasks in rodent models: N-methyl-D-aspartate receptors and voltage-dependent Ca2+ channels in senescent synaptic plasticity. Prog Neurobiol. 2012;96(3):283–303.

183. Wu WW, Oh MM, Disterhoft JF. Age-related biophysical alterations of hippocampal pyramidal neurons: implications for learning and memory. Ageing Res Rev. 2002;1(2):181–207.
184. Lipsitt LP. The study of sensory and learning processes of the newborn. Clin Perinatol. 1977;4(1):163–86.
185. Ghose GM. Learning in mammalian sensory cortex. Curr Opin Neurobiol. 2004;14(4):513–8.
186. Boldin AM, Geiger R, Emberson LL. The emergence of top-down, sensory prediction during learning in infancy: a comparison of full-term and preterm infants. Dev Psychobiol. 2018;60(5):544–56.
187. Li G, et al. Mapping longitudinal hemispheric structural asymmetries of the human cerebral cortex from birth to 2 years of age. Cereb Cortex. 2014;24(5):1289–300.

Chapter 3
Experimental Models and Strategies for Studying Associative Learning and Memory

Abstract As discussed in Chap. 2, there are various memory patterns that influence cognitions and emotions in human life. It is essential to reveal the cellular and molecular mechanisms underlying the formation and retrieval of these memory patterns, which is important to develop approaches to enhance memory capacity in normal population and improve memory capability in memory-deficit patients. To fulfill this goal, animal models that mimic memory formation and retrieval, especially associative learning and memory, are needed. The most commonly used animal models in relevance to associative learning and memory include classical conditioning, operant conditioning, spatial learning, social learning, and associative memory retrieval with a reciprocal form. It is great expectation to discuss these animal models in terms of their principle, procedures, impacts, and validation. With the animal models, another critical issue is to apply comprehensive strategies to search neuronal substrates for information storage and retrieval. In order to reveal the causal relationship of molecular and cellular units in memory traces to memories, three aspects should be met: parallel changes between memory formation and memory traces are observed in a quantitative manner after associative learning; the downregulation of memory trace emergence or function attenuates memory formation being newly learned or memories previously formed; and the upregulation of neuronal substrate recruitment or function facilitates memory formation or strengthens memory previously formed. This proportional alternation between the emergence and quantity of neural substrates in memory traces and the maintenance and strength of memories grants their causal relationship. In this chapter, the author intends to review animal models and strategies that help to reveal the cellular and molecular profiles for associative learning and memory.

Keywords Animal model · Classical conditioning · Operant conditioning · Spatial memory and associative learning

J.-H. Wang, *Associative Memory Cells: Basic Units of Memory Trace*,
https://doi.org/10.1007/978-981-13-9501-7_3

3.1 Introduction

Cellular and molecular mechanisms underlying the acquisition, memory, and retrieval of information have been investigated for decades. Numerous molecules and cell activities involved in memory formation have been presented in the conclusions of various studies. In these reports, some experiment results are not reproducible or the conclusions are controversial. Moreover, some cellular and molecular processes in these studies are generally observed in many brain regions and are shared by other neuron-related events, i.e., nonspecific to memory formation. To these inconsistencies, studies by using more animal models well validated for mimicking learning and memory as well as comprehensive strategies are expected to be conducted in searching neuronal substrates in memory traces [1]. The revelation of neural substrates for memory formation and retrieval is critically important to develop useful approaches that enhance memory capacity in normal population and improve memory capability in memory-deficit patients.

As we know, the acquisition of information, knowledge, and experiences is associative in nature, called as associative learning in that multiple signals are jointly acquired. The integrative storages of these associated signals confer their retrieval in reciprocal manner, i.e., a signal induces the recall of its associated signal, or the other way around as well as the retrieval of signals with automatic conversion among cross-modalities. Based on these features of associative learning and memory, various animal models have been developed to address cellular and molecular profiles for associative memory, e.g., classical conditioning, operant conditioning, spatial learning, social learning, and so on [2–6]. However, these animal models do not show the reciprocal feature of memory retrievals, i.e., a signal induces the retrieval of its associated signals, or the other way around. For instance, to eyeblink conditioning and fear memory in rats or mice, it is unknown about whether the air-puffing to the cornea and the electrical shock to the feet will induce their feeling of listening sound, though the sound induces their eyeblinking and their body freezing in relevance to the fear of electrical shock. Recently, there is a mouse model about associative learning and memory, in which memory retrieval has been shown in the reciprocal form of associative memory [7–9]. Here, the author will discuss these models in terms of principle, procedures, impacts, and validation.

Looking through references in relevance to the studies about cellular and molecular mechanisms underlying learning and memory, one will easily see that the causal relationships between neuronal substrates and memories are studied by one of three strategies, i.e., parallel changes between memory formation and neural substrates are observed in a quantitative manner after associative learning; the down-regulation of memory trace emergence or function will decrease memory formation being newly learned or memories previously formed; and the upregulation of memory trace recruitment or function facilitates memory formation or strengthens memory previously formed. In the studies with partial strategies, that proportional changes between the emergence of neuronal substrates in memory traces and the maintained strength of memories grant their causal relationship cannot be observed.

In this regard, the experimental results may be inconsistent, and the conclusion is partially correct and even controversial if these studies select one of three strategies differently one another. Therefore, the application of all three strategies to reveal neural substrates of learning and memory should be encouraged, instead of labeling overly ambitious [10]. In the meantime, the uses of more animal models and strategies to reveal basic elements in memory traces should also be encouraged, in order to validate the role of potential neural substrates in memory formation [1].

As two issues are important to clearly reveal the cellular and molecular mechanisms underlying associative memory formation and retrieval, i.e., animal models and strategies for searching cell assemblies in engrams, the author intends to discuss animal models and strategies used to study associative learning and memory in this chapter.

3.2 Experimental Models Used to Track Engrams

Associative learning and memory are major style of information acquisition and storage in comparison with non-associative learning. As discussed in Chap. 2, associative memory can be classified into memories with the declarative versus nondeclarative retrievals, episodic memory versus semantic memory, working memory (operant conditioning with reinforcement and punishment) versus perceptual memory through the different sensory systems, eidetic memory versus false memory, spatial memory versus temporal memory, and memories to exogenous signals and endogenous signals as well as memories to unitary signals versus complicated signals organized from unitary signals. The development of experiment models to cover all of these features in human's memory formation and retrievals is worthy to be focused. Unfortunately, we only have the limited number of animal models being used to study neural substrates relevant to human learning and memory, some of which need to be validated in interpreting the principle, procedures, and impacts.

3.2.1 Classical Conditioning

In animal models of conditioned reflexes, classical conditioning has been extensively used to figure out the features and mechanisms underlying associative learning and memory, which includes Pavlov's conditioning, eyeblink conditioning, and fear memory. Classical conditioning, or Pavlovian or respondent conditioning, refers to a learning process, in which a biologically potent stimulus (food) is paired with a previously neutral stimulus (a bell). After this pairing, the neutral stimulus becomes able to elicit a response (salivation) that is originally elicited by the biologically potent stimulus. In the experiments of performing classical conditioning, a conditioned stimulus (CS) is paired with an unconditioned stimulus (US). The CS is usually a neutral stimulus (the sound from a tuning fork). The US has biological potency (the taste of food), and the unconditioned response (UR) to the US is an unlearned reflex (salivation). After repeated pairing, organisms demonstrate a conditioned response (CR) to the CS presented alone, which is originally elicited by the US (salivation). Although Edwin Twitmyer published relevant findings 1 year earlier, this fact was studied in detail by Ivan Pavlov through experiments with dogs [4], initially published in 1897. The conditioned response is often similar to the unconditioned response; they may have differences sometimes [11]. For instance, unlike the UR, the CR is acquired by experience and less permanent than the UR. As a result, most theorists propose that the conditioned stimulus as a signal induces the prediction about the forthcoming of the unconditioned stimulus, i.e., certain changes in thinking or mind, instead of purely memory [12, 13]. It is noteworthy that classical conditioning differs from operant conditioning. In classical conditioning, behaviors are modified by the association of stimulus signals, while in operant conditioning behaviors are modified by the effect they produce reward or punishment. Both operant conditioning and classical conditioning have become the behaviorism foundation in school psychology and impacts on psychological therapy.

This Pavlov's conditioning was discovered in an occasion. In his study of digestion physiology in dogs, he created a model to study digestive processes by redirecting dog's digestive fluids out of the body with measurements for a long period. Pavlov found that these dogs started to salivate in the presence of his assistants who often fed dogs, in addition to salivating in the presence of food. Pavlov named this phenomenon in dogs as anticipated salivation or psychic secretion. By transferring this random observation to an experiment study, Pavlov presented the metronome sound as a stimulation, which was followed by immediately giving dogs food. By giving this association for a few times, these dogs began to salivate in response to this metronome sound. Pavlov proposed that if a stimulation in the environment around dogs was present while food was given, this stimulation became associated with food to cause its salivation [4]. Pavlov found many basic facts about the conditioning. For example, learning occurs most rapidly when the interval between the CS and the US appearance was relatively short. Saliva produced by the CS differs in composition from that produced by the US, i.e., the CR is not simply replica. Repeated pairings are necessary for conditioning to emerge; however, many conditioned responses can be learned with a single trial, especially stronger stimulus for inducing fear conditioning and taste aversion learning.

The efficiency of associative learning is fastest in forward conditioning, where the onset of the CS precedes that of the US to signify the US following. Its two forms include delay conditioning and trace conditioning. In terms of delay conditioning, the presentation of the CS and the US coincided. For example, individuals listen to the buzzer for 5 s. In the meantime, the air is puffed onto the cornea in their eyes to induce their eyelid blinking. After a few pairings of the air-puff and buzzer, individuals will blink their eyelids in the presence of a buzzer sound alone. This is a typical example of delay conditioning. If the CS and the US do not coincide, it is called as trace conditioning. The CS usually starts and stops before the US is given. The interstimulus interval, or stimulus-free period, is called as trace interval or conditioning interval. If the air-puff as the US comes 1 s after the buzzer sound as the

CS stops, the trace conditioning has a trace interval of 1 s. If the CS and the US are presented and terminated at the same time, it is called as simultaneous conditioning. If individuals hear a bell plus air-puff onto eyes at the same time, the repeated pairings lead to their eyelid blinking when they hear the bell without the air-puff.

This conditioning can be second-order or higher-order in nature, which follows two-step procedures. In step one, the neutral stimulus (CS1) is associated with the unconditioned stimulus (US), which leads to the forward conditioning. Subsequently, the second neutral stimulus (CS2) is associated with the CS1, which leads to responses induced by the US initially and the CS1 followingly. For instance, the bell sound is associated with food until this bell sound can elicit the salivations. Next, the light is associated with the bell sound. Eventually the light may elicit salivation as well. In this secondary order conditioning, the salivation is the UR, the food is the US, the bell sound is the CS1, and the light becomes the CS2.

There are other types of conditioning in timing, such as temporal conditioning, backward conditioning, and zero contingence procedure. (1) Backward conditioning refers to a situation that the CS immediately follows the US, in contrast to forward conditioning that the CS precedes the US. The conditioned reflex of backward conditioning in response to the CS often leads to inhibitory effect. The mechanism underlying forward conditioning, or conventional conditioning, is thought to be that the CS induces a prediction about the forthcoming of the US [6]. In this conditioning, the CS functions as a signal that the US has stopped, instead of as a signal that the US is about to forthcoming. For example, an air-puff toward individual's eyes may be followed by the buzzer sound. With this crossed function, the conditioning tends to be inhibitory. This interpretation appears to be farfetched. In fact, after associative memory is formed, mutual synapse innervations are generated among the coactivated brain areas. In addition to their functions are initiated each other, the interconnected sensory cortices facilitate their functions each other. The neurons in the pre-activated cortex will elevate the excitatory state of the neurons in its connected cortices, such that these pre-activated neurons facilitate the activation of their interconnected neurons [1]. In the forward conditioning, the situation that the CS precedes the US enables the CS to facilitate US-induced UR as well as to induce the UR. On the other hand, the backward conditioning that the US precedes the CS is impossible for the CS to facilitate US-induced UR. The UR disappears when the US ends. A fact that the UR is not efficiently converted into the CR induced by the CS may result from the functional adaptation of those neurons, which have been actively functioned in US-induced UR. (2) In temporal conditioning, the US is presented at regular intervals, for instance, every 10 min. Conditioning response tends to occur in response to the CS, that is, shortly before each US. Although the result implies a biological clock that can function as the CS in animal, the mechanism underlying this temporal conditioning is like that for the forward conditioning. This approach has been used to examine timing capability in animals. In an example of temporal conditioning, the US (food to a hungry mouse) is provided on the regular basis of time schedule, e.g., every 1 min. After enough exposure, the mouse begins to salivate before the food delivery. This enables this mouse temporal conditioning as the mouse is conditioned to the passage of time. (3) In zero contingency procedure, the

CS is paired with the US, but the US also occurs at other times. It is predicted that the US is likely to occur in the absence of the CS, that is, the CS does not induce the prediction of the US forthcoming. In this case, conditioning response fails and the CS does not come to elicit a CR. This finding indicates that the prediction rather than CS-US pairing is the key to conditioning responses, which greatly influences subsequent conditioning research and theory. Another possibility is that the functional association based on the mutual synapse innervations is maintained by the coactivity of interconnected cortices [1].

During the acquisition of conditioning, the CS and US are paired. The strength of such conditioning may be tracked by test, by which the CS is given alone and the CR is merited. Although the single CS-US pairing may lead to the CR sufficiently, multiple pairings are often needed. There is a gradual increase in the conditioned responses to the CS. This repeated number of trials may be proportional to the strength and frequency of the CR gradually. The speed of conditioning depends on the nature and strength of the CS and US, the previous experience, and the animal's motivation state. The process slows down as it nears the plateaus [6]. In terms of extinction, the CS is presented repeatedly in the absence of the US, also called as "extinguishing." When this procedure is given after the CS has been conditioned by one of the methods, the CR occurrence eventually returns to pretraining levels. However, this extinction does not entirely abolish the established conditioning, since it recovers spontaneously. The cessation or long temporal delay of CS-US pairings results in a learned decision to stop responding to this CS since there is a sufficient change in the rate at which the CS is being reinforced. This phenomenon can be explained by the accumulation of inhibition when a weak stimulus is given. It is noteworthy about the difference between extinction and external inhibition. External inhibition is observed if a strong or unfamiliar stimulus is presented just before or at the same time as the CS, which causes an attenuation in the conditioned response to the CS [6].

Several ways lead to the recovery of the CR that have been conditioned and then extinguished, which also indicates that the extinction procedure does not eliminate the effect of conditioning completely. (1) Reacquisition is a situation in that if the CS is repaired with the US and the CR is reemergent, but this second acquisition often occurs much faster than the first one. (2) Spontaneous recovery is defined to be the comeback of the previously extinguished conditioned response after a rest period. The CS elicits the CR again after its extinction without CS-US repairing. This renewed CR is usually much weaker than the CR observed prior to extinction. (3) Disinhibition is featured as a temporary recovery of the conditioned response to the CS if the CS is tested just after extinction and an intense but associatively neutral stimulus has occurred. (4) Reinstatement is a situation that the CS often elicits a response when it is tested later, if the US used in conditioning is presented to a subject in the same place where conditioning and extinction occurred without the CS-US pairing. (5) Renewal is defined as a reappearance of the conditioned response after extinction when an animal returns to the environment where the conditioned response has been acquired. These phenomena in the recovery of conditionings may be based on neural circuits that are formed during initial learning and memory formation, as well as are still present and functionally silent during the extinction [14, 15]. Reinstatement and renewal may be caused by the association of CS-US pairing with the third signal "place," i.e., associative learning with multiple signals [10].

There are certain phenomena relevant to stimulations to be the CS. Stimulus generalization is a term in that after a CS can elicit a CR, another similar stimulus can elicit the same CR. More similarities between the CS and the test stimulus will lead to greater CR in response to the test stimulus. On the other hand, more difference between the CS and the test stimulus will lead to weaker CR in response to the test stimulus, i.e., the CR differs from previous response. Stimulus discrimination is featured when one stimulus (CS1) elicits one CR and another stimulus (CS2) elicits either another CR or no CR at all. This can be brought about by pairing CS1 with an effective US and giving the CS2 without the US. Latent inhibition refers to a situation that it takes longer for a familiar stimulus to become a CS than it does for a novel stimulus to become a CS, when the stimulus is subsequently paired with an effective US. Conditioned suppression is one of the most common approaches to measure the strength of learning in the classical conditioning. For instance, rats learn to push a pedal by operant conditioning. After the rats perform a series of trials, these rats are exposed to light or noise as the CS, which is followed by mild electric shock as the US. An association between the CS and the US develops, and the fear conditioned response (body freezing) forms. These rats slow or stop their pedal pushing when the CS comes on. The rate of their pedal pushing during the CS is used to measure the strength of this classical conditioning. The slower the rat pushes, the stronger the association of the CS and the US forms. This slow pedal pushing indicates fear conditioning, i.e., an example of conditioned emotional responses. This conditioned suppression can be explained by associative memory cells including glutamatergic and GABAergic cells, where glutamatergic associative cells activate GABAergic associative memory cells to inhibit neighboring associative memory cells through lateral inhibition [1, 9, 16]. In another possibility, the number and encoding capability of associative memory cells for fear memory are much greater than associative memory cells for operant conditioning, such as memory output cells are occupied by associative memory cells for fear memory [1]. However, the intensive muscle contraction by body freezing response may attenuate the precise operations, such as pushing pedal.

Based on the classical conditioning, the neural substrates of learning and memory have been paid attention to be studied. In Pavlov's proposal, the conditioning involves the connection between brain centers for conditioned and unconditioned stimulations [6], which appears to be proved by recent studies [7, 9, 17, 18]. However, this classical conditioning plus others, such as fear conditioning, eyeblink conditioning, and operant conditioning, is also widely applied for this purpose. Fear conditioning and eyeblink conditioning are featured by pairing a tone as neutral stimulus with an unconditioned stimulus. In the eyeblink conditioning, the air-puff to the cornea is the unconditioned stimulus. In fear conditioning, the threatening or aversive, such as an electrical foot shock and aversive taste, is the unconditioned stimulus. Fear and eyeblink conditioning do not involve the overlapping of neural circuitry but share molecular mechanisms. Fear conditioning occurs in the basolateral amygdala that receives glutaminergic input directly from thalamic afferents, as well as indirectly from prefrontal cortical projections. The anterior cingulate and hippocampus may also involve fear conditioning [19–21]. The presynaptic activation of protein kinase A and postsynaptic activation of NMDA receptors and its signal transduction pathway are necessary for conditioning-relevant plasticity. CREB may also be necessary for conditioning-relevant plasticity [22, 23].

In addition to the application of classical conditioning to address neuronal substrates for associative memory, certain therapies are associated with using classical conditioning, such as aversion therapy, systematic desensitization, and flooding. Aversion therapy has been designed for making patients to stop an undesirable practice by pairing this habitation with a strong unhappy unconditioned stimulus. The unpleasant stimulus used for the pairing in medication can be aversive tastes, such as an association of alcohol taste with stomach upset. The systematic desensitization has been widely used for the treatment for phobias. In this desensitization, the patients are trained to be relaxed while they are exposed to anxiety-provoking stimulations with progressively more strength, such as threat objects and annoying words. This desensitization belongs to the counterconditioning. The desensitization of phobias or anxiety intends to associate fear stimulations with relaxation response that is not compatible with anxiety or counterconditioning. The flooding is a strengthening form of the desensitization. The flooding attempts to remove specific phobia and anxiety by repetitive exposures to the extremely distressed stimulations until a lack of reinforcement of the anxiety response can be used, which causes its extinction. The flooding often involves actual exposure to stimulations. The implosion refers to imagined exposure. Both terms have been used synonymously sometimes [6]. The mechanism underlying behavioral desensitization may be based on the functional adaptation of associative memory cells, since both functional downregulation and upregulation of associative memory cells have been detected in vivo [1, 24].

In conditioned drug response, the stimulus associated with a drug administration or consumption may lead eventually to a conditioned physiological response that mimics drug effect. In a typical example, coffee drinkers with habituation may feel alertness and be more conscious when they smell coffee. In fact, the associative memory cells for drinking coffee, smelling coffee, and their effects, which are recruited during their previous associations, are partially activated by smelling coffee [1]. This conditioned response appears to be a compensatory effect that tends to counterbalance drug effectiveness. If the drug induces the body to become less sensitive to pain stimuli, the compensatory conditioned effect may induce the drug users more sensitive to pain. Therefore, this compensatory conditioned effect may be related to drug tolerance. The drug users may increase drug dosage to feel better or to feel drug effects. In case of the absence of the CS, the overdose reaction may occur. If the drug administration is always associated with the same place or room, signals from that place or room as the CS may induce conditioned compensatory effect. If the drug administration is given in different locations, i.e., the conditioned stimulations are absent, the overdose reaction may happen [6]. This coordinate effect of the US and the CS indicates their interaction based on mutual synapse innervations among sensory cortices in relevance to these US and CS [1].

In conditioned emotional response, the effect of classical conditioning can be seen in emotional responses such as phobia, disgust, nausea, anger, and sexual arousal. A common example is a conditioned nausea, in which the CS is the sight or the smell of specific food that has previously caused an unconditioned stomach upset. The sight or smell of the food with the shape or odor similarly to that inducing stomach discomfort will lead to stomach upset. Similarly, the associations of the sight of dogs as the CS and the pain bitten by the dogs as the US will result in the conditioned fear to dogs. Conditioned emotional responses can be an example of conditioned suppression. As an adaptation, the emotional conditioning assists to protect people from harmfulness or prepare them for important biological events. For instance, the stimulation that has occurred comes to induce sexual arousal before the sexual interaction, and this arousal prepares individuals for sexual contact [6].

Pavlovian-instrumental transfer is a phenomenon that occurs when a conditioned stimulus (CS, known as a cue) that has been associated with rewarding or aversive stimulations via classical conditioning alters motivational salience and operant behavior. In a typical experiment, a rat is given with sound-food pairings (classical conditioning), and this rat learns to press a lever to get food (operant conditioning). Test sessions show that the rat presses the lever faster in the presence of the sound than in silence, although the sound has never been associated with lever pressing. Pavlovian-instrumental transfer is suggested to play a role in the differential outcomes, the procedure that enhances operant discrimination by pairing stimulations with specific outcomes [25]. The mechanism for Pavlovian-instrumental transfer may be based on the formation of associative memory cells that encode these three signals, since associative memory cells that encode multiple signals have been identified in sensory cortices and the prefrontal cortex [1, 10, 24, 26].

In fact, the association of multiple signals may lead to each of these signals to be mutually conditioned. For instance, the repeated pairings of whisker, odor, and tail stimulations cause the following situations that odor stimuli lead to odorant-induced whisker motion and tail retraction, tail stimuli lead to tail-induced whisker motion and tail-induced olfactory response, as well as whisker stimuli lead to whisker-induced olfactory response and whisker-induced tail retraction [10]. Mutual synapse innervations have been formed among these sensory cortices [27, 28].

3.2.2 *Operant Conditioning*

Operant conditioning, or instrumental conditioning, is one form of associative learning, in which behavioral strength is modified by reinforcement (behavioral increase) or punishment (behavioral decrease). The reinforcement includes positive and negative support. In positive reinforcement, appetitive stimulation is given after behaviors are correct. Negative reinforcement is featured to be escape by removing noxious stimulation after correct behaviors or active avoidance that behaviors avoid noxious stimulation. On the other hand, positive punishment is the addition of noxious stimulation after behaviors, and negative punishment is the removal of appetitive stimulation after behaviors. Although both classical conditioning and operant conditioning encompass the behaviors influenced by environment stimulations, they show obvious difference. In the operant conditioning, the stimulation is given while the operation is rewarded or punished to influence that operation. For instance, children learn to avoid touching the hot stove or to open the box for getting candies inside, in which the stove and the box are discriminative stimulations. The operant behaviors are thought to be voluntary. The children may have to make choices between opening the box and petting a puppy. In contrast, classical conditioning is thought to be involuntary behaviors, whose formation is based on the association of stimulations with biologically significant activities. For instance, the sound of door bang may signify an angry parent, which causes children to tremble. The sight of candies may induce children to salivate. Trembling and salivation are not operations, not reinforced by consequences and not voluntarily chosen. The investigations of associative learning with animal models are mainly conducted by the analyses of classical and operant conditioning, which remain the core of behavioral analysis [6, 29, 30].

The history about operant conditioning has been reviewed [25]. Operant conditioning, or instrumental learning, was first extensively studied by Thorndike, in which cats tried to escape from homemade puzzle boxes [5]. Each cat could escape from the box by pulling a cord or pushing a pole. With an initial constraining, these cats spend long time to get out. After the repeated trials make ineffective effort occurred infrequently and successful effort occurred frequently, the cats escaped more quickly. Edward Thorndike generalized his finding as a rule that behaviors with satisfied consequences tend to be repeated and behaviors with unpleasant consequences are less likely repeated. By plotting escape time against trial number, he produced the first animal learning curves in this procedure. Humans appear to learn numerous simple behaviors by Thorndike's ways, called as operant conditioning [25]. The responses are retained when they lead to a successful outcome and discarded when they do not or they produce aversive effects. Further studies were given by Skinner who worked for operant conditioning and its application to human and animal behaviors [31, 32]. Skinner believed that classical conditioning was too simplistic to be used to describe things as complicated as human behaviors. Operant conditioning better described human behavior as it examined reasons and effects of intentional behaviors [33, 34]. Without the appreciation of Thorndike's view about unobservable mental states such as satisfaction, Skinner designed an operant conditioning chamber or Skinner box. Subjects, e.g., rats and pigeons, were isolated and placed in this box, so that the precisely controlled stimulations can be given to these subjects. Different from Thorndike's puzzle box, the placement of these subjects into this box allowed them to perform one or two simple repeatable responses. The rate of their responses became Skinner's primary measurement of animal behaviors. Another innovation in Skinner's box was to implant an accumulative recorder that graphically recorded animal behaviors to evaluate their response rates. The reinforcement schedule delivers reinforcement to a subject based on well-defined rules. The effects of schedules became the basic findings from which Skinner developed his account of operant conditioning, including the fiction of peaceful, happy, and productive community as well as verbal behaviors [31–34]. Skinner devoted his main effort to the application of operant conditioning to human behavior [33, 34].

Operant behaviors are not initially elicited by any specific stimulation. Its emergence is original from new bodily structure in variation and selection. Individual behaviors vary from moment to moment as the specific motion involved, the amount of force used, or the timing of responses. Variations that cause reinforcement are strengthened. If the reinforcement is consistent, the behavior tends to remain stable. However, behavioral variability can itself be altered by the manipulation of certain variables. The reinforcement and punishment are core tools by which operant behavior is modified to be either positive or negative. Both positive reinforcement and negative reinforcement raise the probability of behaviors that they follow. Positive punishment and negative punishment reduce the probability of behaviors that they follow. The extinction refers to a situation in that a behavior previously reinforced is no longer strengthened by either positive or negative reinforcement. Once the behavior becomes less probable, the extinction occurs. Occasional reinforcement can result in the long delay before behavior extinction, once the learning factors of repeated occurrence become necessary to get reinforcement, compared with the reinforcement being given at each opportunity before extinction [33–35].

When a behavior is rewarding or is followed by another rewarding stimulus, positive reinforcement occurs with increased frequency of that behavior. For example, if a rat in a Skinner box gets food when it presses a lever, its rate of pressing goes up, often called simply reinforcement. When a behavior is followed by the removal of aversive stimulus, negative reinforcement (escape) occurs with increased frequency of original behavior. In the Skinner box, the aversive stimulus may be continuous loud noise; negative reinforcement would happen when the rat presses a lever to turn off the noise. On the other hand, positive punishment (punishment by contingent stimulation) occurs when a behavior is followed by aversive stimulus. Pain from a spanking often results in a decrease in that behavior. Negative punishment (penalty, or punishment by contingent withdrawal) occurs when a behavior is followed by the removal of a stimulus. Taking away a child's toy following an undesired behavior by him/her leads to a decrease in the undesirable behavior. Extinction refers to a situation that a behavior that has been reinforced previously becomes no longer effective. For example, a rat is initially given food as a reward many times for pushing a pedal. When the experimenter no longer gives food, this rat would push this pedal less often and then stop. This pedal pushing would be called as extinguishment. It is noteworthy that animals demonstrate to be reinforced, punished, or extinguished by behaviors that are thought of. Natural events can also reinforce, punish, or extinguish behaviors and are not always planned or delivered on purpose [33–35].

Schedules for reinforcement refer to rules that control reinforcement delivery.

1. Schedule in fixed interval: Reinforcement arises after response one when a fixed time has elapsed from previous reinforcement. This schedule makes subject's responses featured as break-run way. For example, subjects often interrupt their responses after the reinforcement, and followingly their responses become quickly for next reinforcement methods.
2. Schedule in fixed ratio: Reinforcement arises after a fixed number of responses has been released from a previous reinforcement. This schedule may lead to a situation that subjects interrupt their responses temporally after a reinforcement; followingly their responses become in the high rate. Whether subjects interrupt their responses depends upon the rate of response requirement.
3. Schedule in variable interval: Reinforcement arises after the first response when a variable time has elapsed from previous reinforcement. This schedule often leads to a relatively stable rate of responses that vary with average time between reinforcements.
4. Schedule in variable ratio: Reinforcement arises after a variable number of responses have been released from the previous reinforcement. This schedule often leads to a very high and persistent rate of response.
5. Continuous reinforcement: Reinforcement arises after each response. Subjects often respond as rapidly as they can in order to give the time for obtaining and enjoying reinforcement until they are satiated [25, 29, 30].

Factors influencing the efficiency of reinforcement and punishment are listed below.

1. Satisfaction versus deprivation: If individuals have received sufficient positive or appetitive stimulation to satisfy their appetite, their responses to that stimulation will be reduced. For instance, subjects with full stomach wouldn't feel as motivated as hungry ones. On the other hand, the opposition may arise if individuals are deprived from that stimulation, i.e., the effectiveness of that stimulation is increased.
2. Immediacy: An immediate outcome is more effective than a delayed outcome. If people treat a dog for sitting within 5 s, this dog will learn faster, in comparison to the outcome by treating for 30 s.
3. Contingency: Reinforcement should happen consistently after responses, in order to be most effective. If reinforcement occurs irregularly, associative learning may be slower. However, responses reinforced intermittently are often slower to be extinguished than responses reinforced consistently.
4. Volume: The number of a stimulation influences its potency as reinforcer. In the analysis of cost-benefit, this rule is suitable for animals and human beings. For instance, a smaller volume of food to rats is not likely a worthwhile reward for their great effort to push pedal. A pile of coins from the slot machine to gamblers will make them pull this lever longer than a coin. Most of these factors influence biochemical processes and physiological functions in the body [29, 30]. The satisfaction will enable subjects to maintain the stable internal environment to be

homeostasis. When subjects are deprived of sugar, the taste of sugar is effective reinforcer. When the subject's blood sugar exceeds a physiological level, the taste of sugar becomes less effective or even aversive [25].

In terms of the shaping of operant conditioning, methods used by trainers in animal training and in human nonverbal teaching depend upon operant variability and reinforcement. Trainers initially identify the desired target behaviors. Subsequently, trainers select a behavior that humans and animals have expressed with some probability. With successive trials, the form of this behavior is progressively changed to reinforce behaviors that approach the target behavior closely. Once this target behavior is emitted, it can be recovered, strengthened, and preserved by the schedule of reinforcement. To the stimulation types and strength in operant conditioning, operants are controlled by stimulations that are present when behavior is reinforced, i.e., discriminative stimulations, which set for responses that generate reward or punishment. For instance, rats are trained to push a pedal when the light is on. Dogs rush to the kitchen when the rattle of food bag is heard. Children reach for candies when they see some candies on the table. A few terms need to be clarified. (1) Discrimination refers to a situation that the response is reinforced only in the presence of specific stimulations. For instance, if pigeons are fed for pecking under the red light and not at the green light, the pigeons peck in red and ends pecking in green. Moreover, many complicated organizations of stimulations have been investigated. For instance, subject may be reinforced on an interval schedule in the presence of a stimulus and be reinforced on a ratio schedule in the presence of another stimulus. (2) Generalization refers to the tendency of response to stimulations that are like discriminative stimulations previously trained. Although pigeons are trained by pecking in red, these pigeons may also peck in pink with less strongly. (3) Context stands for the stimulations that are persistently existing, such as wall, tables, and chairs in the space or interior environment in an operant conditioning box. Context stimulations may influence behaviors as discriminative stimulations, despite usually weaker. Behaviors learned in one context may be absent or altered in another [25, 30].

In terms of behavioral outcomes, there are a few situations. In conditioned reinforcement, most behaviors cannot easily be described in individual responses reinforced one by one. The scope of operant analysis is expanded by the idea of behavioral chains, i.e., a series of responses bound together by three-term contingencies. The chaining is based on experiment fact that a discriminative stimulus not only sets for subsequent behavior but also reinforces a behavior that precedes it. In other words, a discriminative stimulus is also a conditioned reinforcer. For example, the light that sets for lever pressing may be used to reinforce "turning around" in the presence of a noise, which leads to a sequence "noise, turnaround, light, press lever, and food." Much longer chains can be built by adding more stimuli and responses. In escape learning, a behavior terminates an aversive stimulus. For example, shielding the eyes from sunlight ends aversive bright light in the eyes or negative reinforcement. Behaviors that are executed by preventing a stimulus is called as avoidance. A discriminated avoidance experiment involves a series of trials in which a neutral stimulus (light) is followed by an aversive stimulus (shock). After the neutral stimulus occurs, an operant response (such as a lever press) prevents or terminates this aversive stimulus. In initial trials, the subject does not make the response until the aversive stimulus comes on or escapes trials. As learning progresses, this subject starts to respond to neutral stimulus and prevents the forthcoming of aversive stimulus. In the avoidance trials, classical conditioning is involved, since a neutral CS (conditioned stimulus) is paired with the aversive US (unconditioned stimulus). In free-operant avoidance, a subject periodically receives an aversive stimulus (often electric shock) unless an operant response is made, and the response delays the onset of the shock. In this situation, unlike discriminated avoidance, no prior stimulus signifies the shock. Two crucial time intervals determine the rate of avoidance learning. First is the S-S (shock-shock) interval. This is the time between successive shocks in the absence of a response. The second interval is the R-S (response-shock) interval. This specifies the time by which an operant response delays the onset of the next shock [25, 30].

In comparison with animal model of operant conditioning, reinforcement and punishment are ubiquitous in human's social interactions. Many applications of operant principles have been implemented, such as behavioral analysis and training as well as addiction and dependence relief. In behavior analysis, Skinner applies conditioning principle to modify social behaviors in human beings. A conditioned stimulation influences behaviors developed by classical conditioning, such as emotion reaction. In Skinner's three-term contingency, the discriminative stimulation sets for the responses that lead to reinforcement. This protocol is effective when the tools of operant conditioning are used to modify human behaviors [35]. The record of behaviors may function as a reinforcement. What an individual sees a pattern of weight loss will reinforce a maintenance in the behavioral weight loss program. Individuals may perceive reinforcement which tends to be positive as negative or vice versa. If the record of weight loss reminds how individuals are heavy, this record may act as negative reinforcement. In many cases, practitioners use some operant skills to develop the constructively and socially acceptable behaviors to replace aberrant behaviors. The techniques of behavior analysis have been effectively applied to these things, such as early intensive behavioral interventions for kids with autism spectrum disorder, criminal behavior, conservation of natural resources, education, health exercise, gerontology, industrial safety, language acquisition, littering, medical procedures, parenting, seatbelt uses, psychotherapy, severe mental disorders, sports, substance abuse, phobias, and zoo management [25, 29, 35].

Positive and negative reinforcement plays the central role in the development and maintenance of addiction and drug dependence. An addictive drug is an intrinsical reward that functions as a primary positive reinforcer of drug use. The brain's reward system assigns incentive salience as wanted or desired, such that an addiction develops and the drug deprivation leads to craving. In addition, stimulations associated with drug use, e.g., the sight of a syringe and the location of use, become associated with the intense reinforcement induced by the drug. These previously neutral stimulations acquire several new properties. For instance, their appearance can induce craving. These stimuli become conditioned positive reinforcers of continued use. If addicted individuals encounter one of these drug cues, a craving for the associated drug may reappear. Anti-drug agencies previously used some posters with the pictures of drug paraphernalia as an attempt to show the danger of drug abuse. These posters are no longer used because of the effects of incentive salience in causing relapse upon the sight of the stimulations illustrated in the posters [36]. In drug-dependent individuals, negative reinforcement arises when the drug is self-administered to alleviate or escape the symptoms about physical dependence (tremor and sweating) and psychological dependence (anhedonia, restlessness, irritability, and anxiety) that express during the state of drug withdrawal [36].

In addition, positive reinforcement has been used in parent management training by providing the positive reinforcement for appropriate child behaviors. Normally, parents become known to reward the appropriate behaviors by certain social rewards, such as smiles, praise, kiss, and hugs, as well as by concrete rewards, such as stickers or points toward larger rewards. These rewards become parts of the incentive system created to be collaborative with children [37]. Moreover, parents become able to select simple behaviors as initial attentions and to reward each of the small progresses, by which their children achieve toward reaching larger goal. This technique has been a concept as successive approximations [37].

Other psychological manipulations are listed below. The positive reinforcement includes praise, superficial charm, superficial sympathy (crocodile tears), excessive apology, money, gifts, approval, attention, facial expression (a forced laugh or smile), and public recognition. The negative reinforcement includes the removal of one from these negative situations. Intermittent or partial negative reinforcement generates the effective climate of fear and doubt. Intermittent or partial positive reinforcement can urge victims to keep up. For instance, in most types of gambling, gamblers are happy to win now and again though they lose money overall. On the other hand, the punishment can be done by silent treatment, yelling, nagging, intimidation, threats, emotional blackmail, swearing, guilt trip, sulking, crying, and playing victims. In addition, traumatic one-trial learning can be done by explosive anger, verbal abuse, or other threatening behaviors to produce superiority or dominance. Even one occasion of these behaviors can train or remind victims to avoid upsetting, confronting, or contradicting from manipulators [25, 35].

As the common seeing of operant conditioning in life and its extensive application of modifying behaviors, operant conditioning as an animal model of associative learning has been used to address the mechanism underlying memory formation, though many efforts have been done for understanding this specific type of learning. The first experiment showing neurons in relevance to encoding conditioned stimulations was conducted by DeLong [38–41]. The neurons in the nucleus basalis whose released acetylcholine influences cerebral cortices are activated in relevance to arm movement shortly after a conditioned stimulus or after primary reward with no conditioned stimulus. These neurons are equally active for positive and negative reinforcers and are related to visuomotor tasks and movement orientation [42, 43]. There is considerable evidence that dopamine participates in both reinforcement and aversive learning, in which dopamine pathways project densely onto frontal cortical areas. On the other hand, cholinergic pathway is dense in posterior cortical regions including the primary visual cortex. The study of patients with Parkinson's disease, a condition attributed to the insufficient action of dopamine, further illustrates the role of dopamine in positive reinforcement [44]. Moreover, when subjects receive a reinforce stimulation, neural pathways including dopamine in the central nervous system are activated, such that neurons in such pathways release short pulses of dopamine to a lot of regions, spreading reinforcement signal to postsynaptic neurons globally [45]. These dopamine molecules facilitate those recently activated synapses to be strengthened in their sensitivity to signals. Therefore, there is an increased probability of occurrence for those recent responses that precede the reinforcement. When the use of reinforcement is either less immediate or less contingent, the ability of dopamine to act on the appropriate synapses is reduced.

In addition, operant conditioning includes reinforcement and punishment for subject's cognitive events and emotional reactions, such as thinking, preference, dislike, judgment, and decision-making. This type of associative learning and memory may demand the participations of the primary associative memory cells in sensory cortices and secondary associative memory cells in cognition- and emotion-relevant brain areas, as well as the interaction among them [1, 27, 28]. In operant conditioning, repeated trials make less occurrence of ineffective effort and frequent occurrence of successful effort [25, 35]. That is, subjects remember more preferred things to do and more aversive things to avoid. Activity-dependent positive cycle in the recruitment and refinement of associative memory cells [1, 28] may work for this repetition-facilitated memory and behaviors. Behaviors in operant conditioning are dependent upon cognition and emotion, so that operant conditioning is not simply a process of learning and memory. In other words, behavioral outputs may not be as good as for tracking memory traces; instead, the detection of memory cells in response to learning cues is needed (please see the next section).

3.2.3 Spatial Learning and Memory

Spatial learning stands for information acquisition from environments through navigation. Spatial memory refers to the information storage regarding subjects' spatial location and orientation of their environment as well as spatial relations among objects [46–49], which is one type of associative memories critically for the survival [1]. For example, spatial memory in individuals is needed to navigate and be familiar with their environments, as well as in animals is required for learning the location of food at the arm end of a maze or of platform under the water of a maze. In both humans and animals, spatial memories are summarized as a cognitive map. There are specific regions of the brain associated with spatial memory. Many methods are used to measure spatial memory in children, adults, and animals [48, 49]. Spatial memory can be classified into short-term spatial memory, working spatial memory, and long-term spatial memory.

Working memory can be described as a limit capacity system that allows subjects to temporarily store and process information. This temporary store enables to work on complex tasks while being able to keep information in mind. For instance, the ability to work on a complicated mathematical problem utilizes subject's working memory. The newest version in the Baddeley's and Hitch's model of working memory includes four subcomponents, called as phonological loop, visuospatial sketchpad, central executive, and episodic buffer. The visuospatial sketchpad is presumably responsible for the temporary storage, maintenance, and manipulation of visual and spatial information. In contrast to this multicomponent model, short-term memory can be viewed as a unitary construct. Visual, spatial, and verbal information are thought to be organized by levels of representation rather than their storages. However, studies in visuospatial memory construct have been conducted in accordance to Baddeley's and Hitch's paradigm [50–52].

In long-term spatial memory, people remember the general layout of a particular space and then cue target locations within that spatial set. This paradigm includes an ordinal scale of features that an individual must attend in order to inform his or her cognitive map. The recollection of spatial details is a top-down procedure that requires an individual to recall the superordinate, ordinate, and subordinate features of a cognitive map in sequence. Two spatial features are prominent in navigating a path: general layout and landmark orienting [53]. People are not only capable of learning the spatial layout of their environment but also piece together novel routes and new spatial relations through inference, which may be based on previous exposure to similar environments. A cognitive map is a mental model of objects' spatial configuration that permits navigation along optimal path between arbitrary pairs of points, which is built upon two fundamental bedrocks, layout (route knowledge) and landmark orientation. It is fairly clear that people use both the layout of a particular space and the presence of orienting landmarks in order to navigate. People perceive objects in their environment relative to other objects in the same environment, i.e., landmarks and layout are complimentary systems for spatial recall. However, it is unknown how these two systems interact when both types of information are available [51, 52, 54].

Through virtual reality, researchers have designed three different virtual towns, each of which had its own unique road layout and a unique set of five stores. The overall footprint of the different maps was exactly same size. In this experiment, participants have to partake in two different sets of trials. Participants were initially assigned two of the three virtual landscapes and tasked with the role of a taxi driver. The participant's avatar picked up a total of 25 passengers and dropped them off at random locations throughout the map. By measuring participants' delivery path length in both sets of the first trial, researchers found that there was a clear decrease in path length and an increase in the number of deliveries in a town. The participants' improved route length correlated with the increasing amount of experience that they got driving in virtual towns. However, the knowledge that participants gleaned from the first town did not aid them in their travels around the second town. The second set of trials involved the same task for the participants but with some changes to the context of the environment. Researchers added two more maps and made them smaller. It is noteworthy that the second map set in this trial was standardized with the same layout and landmarks, i.e., all participants would experience this map on their second run. Moreover, three of the five maps replaced some set of the landmarks with novel landmarks, while the layout of target locations remains unchanged. Other two maps were either identical to the second traversed landscape of the second trial or both the landmarks and their relative locations were changed. Participants in the group that confronted the changed building representations in the same locations showed the highest level of outline knowledge transfer from the first town. The results from trial one show that participants can learn the spatial outline of interactive environments. The results from trial two show that participants can still find the novel shortest routes in spite of a change in landmark presence. In these trials, the participants do not travel across previously memorized routes and can perform well based on excess path length standard, indicating some higher-order survey representation of the environment [55].

In terms of studies in animal models, recent studies in rodents have shown that they are able to use spatial memory to locate previously hidden food. Experiments using the radial maze have allowed researchers to control the number of variables, such as the type of food hidden, the location of food hidden, the retention interval, and odor cue that could skew results of memory research [56]. Studies have indicated that rats memorize where they have hidden food and what type of food they have hidden. These rats select in going more often to the arms of the maze where they have previously hidden preferred food than to arms with less preferred food or no food hidden. This study provides evidence for the spatial memory of some species of animals, such as the rats that do use spatial memory to locate and retrieve hidden food stores. With GPS to track where domestic cats go and when their owners let them outside, the cats demonstrate exceptional long-term spatial memory [57].

The visual system plays an important role in identifying spatial signals, though the auditory and olfactory systems do so. It has been proposed that visuospatial sketchpad includes visual and spatial subcomponents, i.e., the visual cache and the inner scribe. The visual cache is a temporary visual store for the dimensions of color and shape. The inner scribe is a rehearsal mechanism for visual information and works for information of movement sequences [58]. A growing amount of evidences shows that two components are separate and serve distinct functions. Visual memory works for retaining visual shapes and colors, while spatial memory processes information about locations and movement. This division of visual and spatial memory components is experimentally found by dual-task paradigm. For instance, the retention of visual shape or colors is disrupted by the presentation of irrelevant pictures or dynamic visual noise. The retention of location and orientation is disrupted only by spatial tracking tasks, spatial tapping tasks, and eye movements [59–62].

There are a variety of tasks that are used to measure spatial memory on adults, children, and animal models. These tasks allow professionals to identify cognitive irregularities as well as administer varying types of drugs and or lesions in participants to measure consequential effects on spatial memory. (1) Corsi Span Test is commonly used to study the visual-spatial memory span and implicit visual-spatial learning ability [51, 63, 64]. Participants sit with nine 3 × 3 cm blocks fastened before them on a 25 × 30 cm baseboard in standard random order. Experimenter taps onto blocks a sequence pattern which participants must replicate. The blocks are numbered on the experimenter side to allow for efficient pattern demonstration. The sequence length increases each trial until participants are no longer able to replicate the pattern correctly. The test can be used to measure both short-term and long-term spatial memory, depending on the length of time between test and recall. (2) Visual pattern span is similar to the Corsi block-tapping test but regarded as a pure test of visual short-term recall [65]. Participants are presented with a series of matrix patterns that have half their cells colored and the other half blank. The matrix patterns are arranged in a way that is difficult to code verbally, forcing participants to rely on visual-spatial memory. Starting with a small 2 × 2 matrix, participants copy the matrix pattern from memory into an empty matrix. Matrix patterns are increased in size and complexity at a rate of two cells until participants' ability to replicate them breaks down. On average, participants' performance tends to break down at 16 cells. (3) Pathway span task is designed to measure spatial memory ability in children [66]. Experimenter asks participants to visualize a blank matrix with a little man. Through a series of directional instructions such as forward, backward, left, or right, experimenter guides participant's little man on a pathway throughout the matrix. At the end, participants are asked to indicate on a real matrix where the little man that he or she visualized finished. The length of the pathway varies depending on the level of difficulty (1–10), and the matrices themselves may vary in length from 2 × 2 cells to 6 × 6. (4) Dynamic mazes are intended for measuring spatial ability in children [67]. With this test, an experimenter presents the participants with a drawing of a maze with a picture of a man in the center. While participants watch, experimenter uses his or her finger to trace a pathway from the opening of the maze to the drawing of the man. Participants then are expected to replicate the demonstrated pathway through the maze to the drawing of the man. Mazes vary in the complexity as difficulty increases [68, 69].

In terms of the measurement of spatial learning and memory in animals, a few mazes have been developed. (1) Radial arm maze is designed by Olton et al. [3, 57, 70] to test the spatial memory capabilities of rats. Mazes typically include a center platform and a varying number of arms branching off with food placed at the ends. These arms are usually shielded from each other in some ways but not to the extent that external cues cannot be used as references. A rat is placed in the maze's center to explore each arm individually and retrieve food while simultaneously remembering which arms it has already pursued. The maze is set to force the rat returning to the center of the maze before pursuing another arm. Measures are usually taken to prevent the rat from using its olfactory sense to navigate such as placing extra food throughout the bottom of the maze. (2) Morris water navigation task is a classic test for studying spatial learning and memory in rats and first developed by Richard Morris [2]. In this designed water maze, a rat or mouse is placed in a round tank of translucent water with walls that are too high for it to climb out and water that is too deep for it to stand. The walls of the tank are decorated with visual cues to serve as references. A hidden platform just below the surface of water allows rat or mouse to climb up, with which the rat or mouse swim around the pool and by chance discovers this platform. In the experiments, rats or mice initially swim around the edge of the pool before venturing into maze's center in a meandering pattern before stumbling upon the hidden platform. Because the time spent in the pool increases their experiences, the time needed to locate the platform shortens, such that veteran rats or mice swim directly to the platform almost immediately after being placed in the water maze.

To neurobiological basis for spatial learning and memory, the hippocampus is thought of a critical area, i.e., the hippocampus may confer animals the spatial map of their environment [71, 72], in which O'Keefe recorded the activities of unitary neurons in the hippocampus from free-moving rats and found a correlation between the hippocampus and space movement. It stores information regarding non-egocentric space (egocentric refers to body position in space) and grants the view-point independence in spatial memory. The hippocampus is important for long-term spatial memory of allocentric space (external cues in space) and makes the use of reference and working memory that plays the important role of processing information about spatial locations [73, 74]. For instance, blocking plasticity in this area leads to problems in goal-directed navigation and impairs the ability to remember precise location. Amnesic patients in hippocampal damage cannot learn or remember spatial layouts, and patients having undergone hippocampal removal are severely impaired in spatial navigation. Monkeys with hippocampal lesions cannot learn object-place associations, and rats also display spatial deficits by not reacting to spatial change. Rats with hippocampal lesions shown time-independent retrograde amnesia that is resistant to the recognition of a learned platform task only when the entire hippocampus is lesioned, but not when it is partially lesioned. Deficits in spatial memory are also found in spatial discrimination tasks [75–80]. In terms of hippocampal sublocation, lesions to the ventral hippocampus have no effect on spatial memory, while the dorsal hippocampus is required for retrieval, processing short-term memory and transferring memory from short-term to long-term memory [77]. The infusion of amphetamine into the dorsal hippocampus has also been shown to enhance memory for spatial locations learned previously. These findings indicate that there is a functional dissociation between the dorsal and ventral hippocampus [81]. In addition, the London taxi drivers were asked to recall complex routes around the city as well as famous landmarks for which the drivers had no knowledge of their spatial location. This leads to an activation of the right hippocampus solely during recall of the complex routes, indicating the right hippocampus for the navigation in large-scale spatial environments [82]. The hippocampus may include two separate memory circuits. One circuit, the entorhinal-CA1 system, is used for the recollection-based place recognition memory. Another system including hippocampal entohinal-dentate-CA3-CA1 loop is used for place recall memory, and the facilitation of plasticity at the entorhinal-dentate synapse in mice is sufficient to enhance place recall [83–85]. It is noteworthy that other cortical areas, in addition to the hippocampus, may be involved in spatial learning and memory, such as posterior parietal cortex, entorhinal cortex, perirhinal cortex, and prefrontal cortex [86–93]. The entorhinal cortex contributes to the processing and the integration of geometric properties in the environments. For instance, lesions to this region impair the use of distal landmarks during the navigation and produces delay-dependent deficit in spatial memory that is proportional to the delay length, as well as create retention deficits for tasks learned up to 4 weeks prior to the lesions. The medial prefrontal cortex processes spatial information, short-term spatial memory used to guide planned searching behavior, and the temporal organization of information [86, 88, 92, 94].

In terms of cellular mechanisms underlying spatial learning and memory, long-term plasticity of synaptic transmission in the hippocampus, especially CA1 and CA3 areas, has been thought to be involved. The activation of NMDA and AMPA receptors in postsynaptic densities plays an initiative role in long-term synaptic plasticity and spatial memory [95–97]. It is noteworthy that spatial learning and memory need the integrations of exogenous signals in intramodal and cross-modal manners; associative memory cells work presumably for integrative storage and reciprocal retrieval of associated signals [27, 28].

3.2.4 Associative Learning and Memory in Reciprocal Manner

Experiment models of associative learning and memory listed above demonstrate the conditioning induced by cues or conditioned stimulus. In fact, the retrieval of associated signals in humans and animals is characterized in a reciprocal manner [27]; one signal induces the representation of its associated signals or other way around. In order to reveal mechanisms underlying reciprocal retrieval of associated signals by cues or automatic conversions among different modalities, animal model with reciprocal retrievals of associated signals is needed [7–9]. Recently, a mouse model by pairing whisker, olfactory, and tail stimulations has been established, in which the reciprocal retrieval of associated signals is a major feature [10, 16, 98].

The following thoughts have been taken into account in designing this mouse model [8, 24]. (1) Signals to be associated in experiments should not occur in natural association, or pairing, in order to prevent the association of signals in control group during postnatal development. In sensory systems, the most common associations include visual with auditory, olfactory, and gustatory sensations. Somatosensory signals are less associated with olfactory signals naturally. (2) Sensory cortices to be studied in experiments should have no naturally mutual connections, in order to prevent natural connections in control group during postnatal development, except for that the paired training induces mutual innervations among sensory cortices in the experimental group. It has been found that there is no connection between the barrel cortex and the piriform cortex up to 3 months after postnatal development in mice [99]. (3) The afferent pathways for different sensory signals are not convergent to pass through the thalamus, so that there is less chance to be the subcortical integration of afferent pathways. There is no any subcortical connection of afferent pathways for whisker tactile signal and olfactory signal [100]. The barrel cortex encodes whisker tactile sensation [101], whereas the piriform cortex receives odor signal [102–104]. (4) Sensory cortical areas are easily accessed for the experimental convenience, such as dorsal surface areas of the cerebral cortex for their accessibility without injury in subcortical areas during experimental manipulations. The barrel cortex meets this requirement. Taken all points together, sensory cortices for encoding somatosensory tactile and olfactory signals meet less natural association, no naturally mutual connection, no subcortical convergence, and their feasible access. So, the artificial association of multiple signals may induce memory formation with reciprocal retrieval as well as mutual synapse innervations among coactivated cortices [8, 24]. In addition, signals in associative learning are unitary signals in these sensory systems, since most of complicated signals to be learned in life are organized from these unitary signals previously learned [24].

Recently, a series of studies have been conducted by pairing somatosensory signals with olfactory signal in mice. After the simultaneous association of whisker and olfactory stimulations is given, mice demonstrate odorant-induced whisker motion as well as whisking-induced olfactory responses [7, 8]. The training paradigms are described below. The mice in postnatal day 20 are placed into three groups that receive the following treatments in whisker stimulus (5 Hz mechanical stimulation) and odor stimulus (butyl acetate). One group of mice is given a simultaneous pairing of odor stimulus and whisker stimulus, unpairing of odor and whisker stimulations (unpaired group), and no odor and whisker stimulations (naïve control group). Whisker and odor stimuli were given by a digital multiple sensory modal stimulator (MSMS). Odor stimulus is given by switching on the butyl acetate-contained tube and generating a small liquid drop in front of the mouse noses. The intensity of butyl acetate odor is sufficient to induce the response at olfactory bulb neurons detected by two-photon cellular imaging. Whisker stimulus to long whiskers is given to the contralateral side of the barrel cortices where associative memory cells are tracked. The intensity of whisker stimulus is sufficient to trigger whisker fluctuation (i.e., whisker-induced whisker motion). The parameters to train each of the mice by giving whisker stimulus and odor stimulus are 20 s each training and five times per day with 2-h intervals for 10 days. This training period was based on the fact that the onset of odorant-induced whisker motion and whisking-induced olfactory response reached to the plateau level approximately 10 training days. Stimulation intensity, duration, and frequency are precisely controlled by MSMS, which are fixed for each trial and each mouse. In the training, each of the mice is placed in homemade cage, where their running and motion are restricted but their body and arms can be freely extended. There are no circadian disturbance and stress conditions, such as noise, light, unusual odor, and experimenter motions. Mice are placed in the cage for 10 min every day about 1 week to have these mice habituated to experiment conditions before trainings and placed into the cage about 5 min prior to each training for their quiet adaptation during the training. Cares are also used in the odor test procedure. It is noteworthy that the mice in naïve group are placed in these homemade cages but do not receive whisker stimulus and odor stimulus [7–9].

Mouse whisker motion tracks are monitored by a digital video camera with the high speed. All images are digitized and converted into whisker motion traces. Whisker motions are quantified by public software (MB-Ruler, v5.0 by Markus Bader, MB-Softwaresolution, Germany), including whisker retraction time, whisking frequency, and fluctuation magnitude. Whisker retraction is defined as backward motion greater than five degrees away from original position at 0.5 s. Whisker fluctuation magnitudes are defined as absolute changes of whisking angles [105]. Mouse whisker motions in response to the odor test (butyl acetate toward the noses for 20 s) are recorded before the training and 1 h after the end of each training day up to day 10 to quantify the onset time and level of odorant-induced whisker motion and whisking-induced olfactory responses. The odorant-induced whisker motion is accepted if whisker motions meet the criteria listed below. Patterns of odorant-induced whisker motion are similar to typical whisker motions induced by whisker stimulus, but not spontaneous whisking at low magnitudes. Whisking frequency, whisking angle, and whisker retraction time increase significantly, compared with control and before the training. This odorant-induced whisker motion is originally induced by whisker stimulus, in which the odor signal induces the recall of whisker signal and leads to whisker motion, i.e., conditioning responses. It is noteworthy that odorant-induced whisker motion is not related to mouse sniffing, since the sniffing changes the baseline of whisker motion trace [7–9].

Onsets of whisker-induced olfactory responses are examined after the trainings for 10 days. During the test, mice are placed in the central arm of a T-maze [106], their assigned whiskers are stimulated similar to the training paradigm, and their motions toward other two arms are monitored, where the two arms include an object coated with butyl acetate versus an object coated with glycerol, respectively. This test is based on these principles. (1) Distances between these objects and central zone in T-maze are set by moving two objects symmetrically away from the central zone to the positions where mice move into either of two arms just to be an equal chance (no statistical difference in moving into either of two arms) before training, in control and without whisker stimulus. That is, mice that reached the central zone under these conditions are not able to smell butyl acetate from butyl acetate-coated objects [106]. This belief is based on a fact that the odorant diffusion in air and its dilution by air causes odor concentration to be gradually reduced with the distance. This principle allows to measure olfactory sensitivity based on distances from animals to odorants and to read out the minimal concentration of odorants just before the animals respond to the source of the odorants correctly, i.e., the threshold of olfactory sensitivity [106]. After pairing whisker and odor stimulations, olfactory responses (i.e., mice moved toward or away from the odorants) by stimulating their whiskers are due presumably to that these mice smell the odorants. This method to test the presence of whisker-induced olfaction response is based on a belief that mice became able to smell the training odorant (butyl acetate) by whisker stimuli after associative learning. In other words, whisker stimulations to mice with associative learning induce their recall of this trained odorant while seeing the increase of their olfactory sensitivity triggered by stimulating whiskers. (2) Two objects are randomly placed in either arm to prevent their memory to previous reaction toward either left or right and to make sure their motions based on the odorant smell. (3) The rates that mice correctly select the arms are calculated from ten times of the test for each mouse and averaged from the groups of the mice. (4) As butyl acetate is an aversive stimulus to mice, they intend to move into an arm without butyl acetate-coated object after associative learning. The onset of whisker-induced olfactory responses is warranted if percentages of selecting the arm including the object without coating butyl acetate increase significantly after associated training [7–9].

In the care of experiments, T-maze is cleaned by 70% ethanol and then water-wet papers before each trail to remove any odors adhering on T-maze walls. Fresh blocks are used for each trail to maintain the consistency of odor concentration. Experiments are done in a 60 m^2 room with good and constant ventilations. The movement of experimenters in the room is restricted to prevent making odor plume and noise. Mice moving into either arm above 30 cm and staying above 5 s are counted as their entrance [106]. Long whiskers (such as arcs 1–2) on the same side and same rows are assigned for the training and test by mechanical whisker stimulus. It is noteworthy that short whiskers are not trimmed since the whisker trimming increases barrel cortical excitability [107], which might affect the conditioning responses.

In addition to the reciprocal retrieval of associated signals, whisker stimulus and odor stimulus, after their paring (associative learning), associative memory with multiple signals can be formed after the association of these signals, for instance, the associative pairing of whisker stimulus (5 Hz mechanical stimulus), odor stimulus (butyl acetate), and tail stimulus (TS, the heat plate to touch the tail around 45 °C). The associative pairing of whisker, odor, and tail stimulations are given by the multiple-sensory modal stimulator, where their intensity, duration, and interval are precisely set. Odor stimulus is given by switching on a butyl acetate-contained tube and generating a small liquid drop on its tip in front of the mouse noses without air pressure. The intensity of butyl acetate stimulus is sufficient to induce the responses of olfactory bulb neurons detected by two-photon imaging. Whisker stimulus to long whiskers is given to the contralateral side of barrel cortices where memory cells are tracked. The intensity of whisker stimulus is enough to trigger whisker fluctuations after the end of stimulation, i.e., whisker-induced whisker motion. The tail stimulus to mice is given by using a heat plate that touches to distal ends of their tails. The intensity of tail stimulus is about 45 ± 2 °C that is surely to evoke mouse tail swing away from this heat plate within 10 s. This temperature does not make the injury of thick skin on mouse tail. These parameters to train each mouse are 20 s each training and five times every day in 2-h intervals for 10 days. The stimulus intensities, duration, and frequency were precisely controlled by this multiple-sensory modal stimulator, which were fixed in each trial for mice. During the training, each of the mice is placed in a homemade cage, in which their running and movement are restricted but their bodies and arms extended freely. Cares in the experiment of three-stimulus association are given similar to two-stimulus association discussed above. After this associative learning, the mice demonstrate odorant-induced whisker motion and tail swing, whisker-induced olfactory response and tail swing, as well as tail-induced whisker motion and olfactory response, i.e., the reciprocal retrieval of associated signals [10].

Associative learning by pairing these sensory signals has the following features. (1) The onset of associative memory to plateau levels takes about 10 days on average. However, its onset to the plateau level appears variable from training day 6 to 2 weeks among mice. Learning and memory efficiency in some mice is high and in others is low, so that learning and memory efficacies are classified into high learning efficiency and low learning efficiency [108]. (2) The efficacy of associative learning and memory appears to be bell shape during postnatal development. The efficacy of associative learning and memory for a start training day is the highest during postnatal day 22, in comparison with developmental stages before and after this period. There is a gradual increase in this efficacy with the start training day before postnatal day 22, as well as a gradual decay of this efficacy after this period [8]. (3) This associative memory appears an automatic decay without training for 1 week. There is a quick re-onset of this associative memory by an additional day of paired training, indicating no complete extinction of this associative memory [14]. (4) This associative memory demonstrates the unilateral learning in one side of the hemisphere and bilateral memory in both sides of hemispheres, indicating the transferring of memorized associative signals between both cortical hemispheres [17]. (5) Abilities to this associative learning and memory in mice can be transgenerational inheritance, i.e., can be transmitted to their offspring, which appears to be female dominant [109]. (6) An inhibition of microRNAs that regulate expressions of genes and proteins in relevance to axonal prolongation and synapse formation significantly blocks the formation of this associative memory in mice [10, 110]. These features correlated the recruitment and refinement of associative memory cells in these sensory cortices (please refer to Chap. 5).

3.3 Strategies Used to Track Memory Cells in Engrams

There are two strategies used to study mechanisms underlying learning and memory, theoretical modeling to anticipate memory traces and experimental manipulation in vivo to find out neural substrates for memory [1, 111–115]. Historically, before experimental manipulation can be conducted in animals to track memory traces, studies through psychological analysis are used to figure out the theoretical modeling of memory trace or engram. In general, theologists in the field of learning and memory focus on drawing potential units for information storage in the brain, such as memory traces, engrams, and cell assemblies. Experimenters make efforts to figure out neural substrates including molecular components and cellular architectures for memory formation. These two approaches are parallelly conducted by psychologists and neurobiologists. Regardless theoretical modeling and experimentation, the causal relationship between neural substrates and memory formation should be met to conclude the discovery of memory traces, engrams, or cell assemblies in relevance to memory. Although it is difficult for theologists to make a diagram of memory traces with this causal relationship, experimental studies for this conclusion can be obtained by demonstrating the causal relationship between neural substrates and memory formation. Three principles should be taken into account in order to conclude the causal relationship between neural substrates and memory formation by experiments. The emergence of new neural substrates and cellular architectures is parallel to memory formation. The downregulation of these newly formed neural substrates and cellular architectures through various approaches can weaken behavior relevant to previously formed memory or prevent the formation of new memory-relevant behavior. The upregulation of these newly emerged neural substrates and cellular architecture can significantly strengthen previously formed memory or facilitate new memory formation [10, 27, 28, 110]. The proportional change between neural substrates (manipulated factor) and memory formation (consequence) constitutes the correlations similar to a mathematical relation between changeable factors and function [1].

3.3.1 Theoretical Modeling from Memory Traces to Cell Assemblies

Memories to various events, objects, and facts appear to maintain some things likely traces sculptured in the brain that can be recalled, represented, or recollected in the future [1]. The term memory traces has been proposed for information storage coined by ancient Greeks, though its natures and functional role in learning and memory are largely unknown. In ancient China, the theory of Chinese medicine book in "Inner Canon of Yellow Emperor" assumed that memories are based on three factors, i.e., "髓 (shui)" that can be translated as core material (the brain, spinal, and medulla), "精 (jing)" that can be translated as the spirit, as well as "血 (xue)" that can be translated as the blood, since these three factors largely influence memory capability and capacity. Although Chinese medicine knows about molecular and cellular substrates as well as structural identity in relevance to memory formation many centuries ago, the brain and mind seem to be essential for memory, and the blood supply influences the brain functions including memory. This assumption is purely based on or limited by the cognition that such three elements are critical for consciousness activities during that period.

Richard Semon, a zoologist and evolution biologist as well as renowned theologist in learning and memory, has coined two terms, engram and ecphory [116]. The engram corresponds to memory trace, while the ecphory refers to memory retrievals [114, 117]. Based on his suggestion, memory retrievals involve the interaction between engrams and retrieval cues. Provided engram-awakening cues are similar to original stimulus, these incomplete retrieval cues are efficiently enough to activate engrams. The further expansion of his view will be that the interaction between retrieval cues and engrams may induce the production of new engrams relevant to the strengthening of existing memory and the storage of endogenous signals generated from cognitive events, such as logical reasoning and associative thinking [1]. Awakening the original engram may generate a new engram related to memory to this event. The association of these old and new engrams by contiguity and rehearsal strengthens original memory. Moreover, the simultaneous activation of multiple engrams that memorize similar contents and their associations by coactivation, i.e., resonance among engrams, will constitute foundations for the storages of endogenous signals produced from cognitive events and emotional reactions [1] as well as for the complicated cognitive processes, such as abstraction, generalization, knowledge, and formulation [118]. This is the first hypothesis that awakening engrams is dynamic and use-dependent, which is currently updated to be the positive cycle between the recruitment and refinement of associative memory cells in engram circuits for memory consolidation, strengthening, and new memories [1]. Although the engram termed by Semon lacks experimental evidences and neural correlates during that period, by proposing engrams to control memory behaviors and describing memory regulations with engrams, his framework about the engram is consistent with the natures of memory activities. That is, frequent representations lead to deeper memory, as well as the repeatedly simultaneous recalls of similar memory contents induce them to be summarized, conceptualized, and generalized.

Donald Hebb, another renowned theologist in physiology, calls memory traces as cell assemblies to be the foundation of memory-relevant behaviors. In his stage, the surgical ablation of brain tissues has been used in human beings for medical treatments as well as the experimental removal of brain tissues in animals has been introduced to the field of learning and memory. For instance, Hebb and Penfield have observed that the surgical ablation of large amounts of cerebral cortices in the patients produces little effect on memory [119, 120]. Lashley's experiments that the ablation of widespread cortices in animals does not induce parallel lesions in memory-relevant behaviors [121, 122]. Hebb has proposed cell assemblies to be components of memory traces, or engrams, that are the widely distributed neural substrates for memory. Each cell ensemble is a group of cells that are interconnected and strengthened during their simultaneous activities [123, 124]. The interconnection of these cell assemblies makes their activities in this circuit to be maintained briefly after learning events, i.e., short-term memory. Recurrent activities for sufficient periods in this cell ensemble will induce growth or metabolic change that strengthens those interconnections among ensemble cells, leading to the conversion of short-term memory into longer-term memory [124]. The strengthening of connections between presynaptic and postsynaptic nerve cells in their simultaneous activity confers these neurons a property of firing together and strengthening together. The strengthening of neuron connection has been shown in long-term potentiation of synaptic transmission [125]. More interconnections among cells allow entire ensembles to be activated if a subset of cells is activated by the process of pattern completion that induces memory retrieval. As Hebb's cell assemblies are widely distributed across brain areas, lesions to a small proportion of cells may not lead to the complete destroy of memory traces, i.e., graceful network degradation, which may account for Lashley's experimental results. Compared with Semon's engrams, Hebb's cell assemblies in wide distributions have overlapped multiple spatial scales from the strengthening of interconnected synapses (a microscale level) to the enhancement of cell assemblies in networks (a mesoscale level). Current experiments have proved that associative memory cells can be located primarily in sensory cortices through their mutual innervations and secondarily in the prefrontal cortex and hippocampus through receiving convergent and divergent synapse innervations from sensory cortices [1]. It is pointed out that Hebb's hypothesis is likely influenced by the knowledge about neurons, synapses, and their interactions, especially action potentials on various neurons [126] and synaptic facilitation at neuromuscular junctions induced by their intensive activation during that period [127, 128]. In other words, the knowledge from experiment studies has been guiding the formation of new theoretical modeling.

In the third stage, the theoretical modeling about memory formation is made based on experimental data as well as helps from computational simulations for the activity integration of numerous cell assemblies in relevance to memory behaviors. Experimental data are mainly from neuronal activities recorded by electrophysiology in the brain. The detection of various patterns in neuronal activities during learning and memory retrieval has been thought to be the replays of memory-relevant cells [129]. For instance, the ripple wave type of neuronal spiking is thought to be linked to memory replay [130]. Coordinate interactions from hippocampal neurons to prefrontal cortical neurons and associative cortical neurons are detected to be involved in spatial memory tasks [131, 132]. The cortical-hippocampal-cortical neuronal circuit is critical for memory consolidation [133]. Hippocampal cell assemblies can trigger neuronal activities in the ventral striatum for the replay of place-reward message [134]. The acquisition of associative memory in the hippocampus initiates a gradual-to-stable encoding process in neurons of the medial prefrontal cortex without continued trainings [135]. The representation of emotional memory is linked with the reactivation of the hippocampus-amygdala system during the sleep state [136]. Based on these data and others, the computational simulation of neural substrates for learning and memory has been used to deliver the theoretical model of memory traces, engrams, or cell assemblies. In the study of neuronal and synaptic architectures for memory traces and memory-relevant behaviors, there are clear indications about the involvement of neuronal ensemble and synaptic plasticity in processes of learning and memory in spite of a lack of evidences about synapses, neurons, and their plasticity specifically correlated to memory formation [118, 137–140].

The theoretical modeling of memory formation has generated the great frameworks that can provide useful guideline for revealing cellular mechanisms underlying learning and memory. However, such models about memory traces and cell assemblies have not indicated insights about the integrative storage of associated signals and need to be proved by experiments. In the experimental studies to reveal cell assemblies in relevance to memory formation and retrieval, three strategies can be used to ensure the causal relationship between cell assemblies and memory-relevant events. Cell assemblies in memory trace should be detected in memory formation and cue-induced memory retrieval, i.e., parallel change between them. The downregulation of memory cell assemblies should deteriorate memory in a prevention manner or to interrupt memory previously formed. The upregulation of memory cell assemblies is able to facilitate memory formation or strengthen memory previously formed [10, 27, 28, 110]. There are two usual methods to track memory traces or cell assemblies, i.e., the detection of memory cells during learning and memory and the activation of memory cells to retrieve the memory-relevant behaviors. The detection of memory cells is to see their responses to memory cues by electrophysiological recording and two-photon calcium imaging as well as to localize their distribution through AAV-carried fluorescent neural tracing after memory formation. The activation of memory cells can be done by electrical, pharmacological, optogenetic, or chemogenetic stimulations to induce the emergence of memory-related events [28]. It is noteworthy that memory traces are widely distributed in the brain and are interconnected across brain areas. The stimulation may lead to the antegrade and retrograde activation of neural pathways, as well as the indirect activation of memory traces is unable to localize primary versus secondary allocations for memory formation.

As the associative learning of multiple signals is the most common approach of information acquisition in life, mechanisms underlying the integrative storage of these associated signals should be addressed experimentally by using appropriate animal models featured by association. A few of animal models have been used to the study of associative learning and memory, such as classical conditioning that includes Pavlov's conditioned reflex, eyeblink conditioning, and fear conditioning in rodents and withdraw reflex in *Aplysia* as well as operant conditioning with a reinforcement (e.g., operation plus reward and place plus reward) or punishment in the mammalians [25, 29, 134, 141–153]. In these models, a stimulus is unconditioned, and another stimulus is conditioned (please see the experimental models in the first part of this chapter for details). In human beings, the memory of associated signals occurs by the signal inducing the recall of its associated signals or the other way around. This reciprocal retrieval of associated signals is essential for cognitive processes, such as associative thinking, logical reasoning, computation, and imagination in forward and backward manners. Thus, these animal conditioning models do not signify whether air-puffing to the cornea or electric shock to the feet is able to induce the recall of sound signal after the onset of eyeblink conditioning or fear conditioning. In other words, these conditioning models may not be ideally used to study associative memory. Moreover, electrical shock may activate the whole brain by spreading electrical current in the body, such that the association is not region-specific in the brain [27, 28]. Compared to electrical stimulations used in the study of fear memory, physical and psychological stresses in social interaction are closer to various conditions in real life [154–156]. An animal model has been recently introduced to the society for memoriology in order to investigate associative memory in that the association of whisker and olfactory stimulations in mice can lead to odorant-induced whisker motion and whisker-induced olfactory responses, a typical example of the reciprocal retrieval of associated signals [7, 8, 10, 16, 98, 157].

3.3.2 Parallel Changes Between Engrams and Memory-Relevant Behaviors

The emergence of new neural substrates and architectures is parallel to memory formation. Parallel changes between cell assemblies including functional and structural alternations and memory formation or memory retrieval have been searched by different approaches, such as electrophysiology, two-photon cell imaging, spine quantity, and tracing axons/synapses. These studies have brought important messages for the characters of engrams in relevance to learning and memory [1].

In terms of the identification of neurons in a particular engram, their activities in response to specific cues relevant to this memory are expectedly detected. Neuronal activities are featured by electrical signals generated on cell membrane as well as calcium signals raised in cells, such that the recording of electrical signals and the imaging of intracellular calcium dynamics can be used to monitor the functional activity of cell assemblies related to memory formation and retrieval, i.e., the detection of the replays of memory-relevant cells [8, 129]. Electrophysiological recordings by electrodes or electrode array have been used to monitor the replays of neurons in the hippocampus, sensory cortices, amygdala, and ventral tegmental areas under different conditions, such as retrieval cue, wakefulness, and sleep state [158–172]. For example, the coordinate interactions from hippocampal neurons to prefrontal cortical neurons and associative cortical neurons are involved in spatial memory tasks [131, 132]. The cortical-hippocampal-cortical neuronal circuit is critical for memory consolidation [133]. Hippocampal cell assemblies can trigger neuronal activities in the ventral striatum during the replay of place-reward message [134]. The acquisition of associative memory in the hippocampus initiates a gradual-to-stable encoding process in neurons of the medial prefrontal cortex without continued trainings [135]. The representation of emotional memory is associated with the reactivation of the hippocampus-amygdala system during the sleep state [136]. These data from functional study are granted by anatomical evidences among the prefrontal cortex, hippocampus, and thalamic nucleus [173].

Memory-relevant neurons in a particular engram have been recently identified in cerebral cortices by using two-photon cell calcium imaging in vivo [174–176]. For instance, the gradual emergence of neuronal activity relevant to spatial memory in the retrosplenial cortex, which is the major recipient of the hippocampus, depends upon the intact hippocampus. The indirect connections between the retrosplenial cortex and the hippocampus indicate the mutual influence between the hippocampus and neocortex via polysynaptic neural pathways, i.e., the hippocampus and cerebral cortices include memory-relevant neurons in the widely distributed engrams [177]. Repetitive motor learning induces the formation of new dendritic spines in vivo [178]. Associative memory cells in response to retrieval cues are identified in primary sensory cortices, the prefrontal cortex, the hippocampus, and the motor cortex [8, 10, 24, 26]. Therefore, memory-relevant neurons or cell assemblies can be tracked by electrophysiological and imaging recordings based on their activities in response to the retrieval cues and during memory-relevant events.

The functional identification of memory-relevant neurons or cell assemblies should be proved and validated by the morphological approach, i.e., their morphology is quantified and their distribution is localized. Two ways can be used for this purpose, the trace of their synapse innervations from axon inputs that are carrying the learned signals as well as the labeling of memory cell assemblies by molecules specifically relevant to memory. In the identification of associative memory cells about their mutual synapse innervations, the expression of fluorescent proteins mediated by adeno-associated viruses (AAV) in memory relevant neurons and their axons has been conducted by injecting the AAVs tagged with genes of these fluorescent proteins into the source side of predicted memory traces as well as by detecting their axon terminals and target on associative memory cells or other way around [7, 10, 17]. Such associative memory cells are characterized by receiving both innate and new synapse innervations. It is noteworthy that the combination of tracing new synapse contact and labeling memory cell assemblies with memory-relevant molecules would be an ideal way to identify memory cell assemblies.

The activity of neurons and synapses can induce the alternation of their included molecules [179–181]. These molecules may be relevant to recruit neurons as memory cells during learning. Thus, these molecules may be used to label memory cells to indicate the allocation of memory cell assemblies. The stimulations to neurons couple with the expression of immediate early genes [182]. Their expression in dendrites is regulated by synapse activities [183]. The immediate early gene *Arc* is specifically linked to neuronal encoding [184]. Some immediate early genes are widely expressed in the brain after fear memory, in which the number of labeled cells is positively correlated to fear memory [150, 185]. There may be the association between the expression of immediate early genes and the active strength of memory cells. Their parallel change leads to a thought that cells labeled by immediate early gene expression are likely engram cells, in which their morphology and function can be studied [176, 186–189]. However, an upregulated expression of immediate early genes is also observed in those neurons with hyperactivity, such as seizure discharge in epilepsy [190–193] and neuron toxicity in brain ischemia [194–196]. In these regard, immediate early genes may be suitable for identifying all of the neurons that have highly active strength. Genes and proteins specifically linked to memory cell assemblies and their memory contents remain to be explored [111, 197]. It is noteworthy that there are around 10,000 types of proteins in living cells [198], which is much less than unit signals remembered in life, such as words, unitary images, odorants, and so on. As more than ten billion of neurons reside in the central nervous system, those neurons with synapse interconnections, or associative memory cells, should be basic units for memory traces, instead of the possibility in a specific protein for a given content of memory.

3.3.3 Downregulation of Neural Substrates

The downregulation of newly emerged neural substrates and cellular architectures substantially attenuates memory formation through the approaches of surgical ablations to brain tissues, pharmacological block to neuronal activities, and genetic knockout/mutagenesis to molecules in nerve cells or synapses [10, 27, 28, 110]. In this regard, the detection of structural and functional changes in newly emerged cell assemblies in relevance to memory formation can be combined with the downregulation of these neural substrates in experiments in order to examine whether there are proportional changes between neural substrates and memory formation toward negative side.

If particular memory contents depend upon memory cells recruited during learning, the downregulation of certain molecules critical for memory cells by pharmacological blockers, gene knockout, or optogenetic approaches should prevent or attenuate the emergence of memory-relevant behaviors. This is a common strategy to address the causal relationship among molecular substrates, cellular architectures, and memory events. The first use of surgical ablation to search the distribution of engrams was done by Lashley. Although he failed to localize memory traces, his studies imply the wide distribution of engram cells in the cerebral brain [121, 122, 199]. In subsequent studies, the removal of the temporary lobe in human beings leads to the loss of recent memory due to the impairment of the hippocampus [76, 120, 200, 201]. In the study of memory cells by pharmacological reagents, recent memory can be blocked by the intracerebral injection of puromycin [202–204], in which Flexner is thought to be a pioneer to study molecular mechanisms, such as protein synthesis, underlying learning and memory. These studies reveal the causal relationship between memory cells in wide brain areas and memory formation though these memory cells specific for content-relevant events remain to be tracked. With the advanced molecular biology, the downregulation of gene expressions by gene knockout [205] and optogenetics [206, 207] has been successfully applied to find out inverse correlations among molecules, memory cells, and behaviors. These studies provide strong evidences for the causal relationships among molecular substrates, cellular architectures, and memory formation.

3.3.4 Upregulation of Neural Substrates

The upregulation of the newly emerged neural substrates and cellular architectures significantly facilitates memory formation through the approaches of pharmacological, electrical, or optogenetic stimulations to nerve cells and gene overexpression in neurons and synapses [10, 27, 28, 110]. In this regard, the detection of structural and functional changes in newly emerged cell assemblies in relevance to memory formation can be combined with the upregulation of these neural substrates in experiments in order to examine whether there are proportional changes between neural substrates and memory formation toward positive side.

Because particular memory retrieval may be based on engram cells recruited during learning, the activation of memory cells to induce the emergence of memory-relevant behaviors should be used to study memory formation. This strategy is based on a positive correlation between memory cells and memory behaviors. If some neurons store particular memory contents, the activation of these cells by electrical, pharmacological, and optogenetic approaches should induce memory formation to these contents or strengthen these particular memories. Electrical stimulation to engram cells in certain areas of the brain has been used by Penfield who originally expected to localize the source of epilepsy [208]. The stimulation to engram cells in the temporal lobe induces memory recalls in wakeful epileptic patients [209, 210]. The pharmacological stimulations to activate serotonin or norepinephrine systems by using these transmitters facilitate memory formation successfully [211, 212]. Optogenetic stimulations to activate memory engrams induce fear memory or false memory [213–217]. There is a positive correlation between engram cells and memory-relevant behavior. It is noteworthy that direct optogenetic activation to neurons without increases of synaptic strength and dendritic spine density leads to memory retrievals [218], implying nonspecific neuron activation. Because these engram cells are widely distributed in the brain and the brain regions are interconnected, stimulations may lead to the antegrade and retrograde activation of neural pathways as well as the indirect activation of memory cells. Methods in stimulations to those engram cells are unable to localize primary versus secondary allocations for memory formation.

These strategies and approaches to identify memory cells need to be evaluated and validated in terms of their advantage and disadvantage. In logical analyses, parallel change, negative correlation, as well as positive correlation between functions and changeable factors should be met to ensure their causal relationship. What the manipulations of molecules and/or cells lead to the changes of memory-relevant events in these three criteria should be combinedly used to figure out memory cells recruited during learning, through which consistent results are expected to obtained for the conclusion [1]. However, inconsistent results may occur in these studies. For instance, silencing and stimulating patriate cortical neurons lead to inconsistent results during memory retrievals. The parietal lesions do not normally yield severe episodic memory deficits, whereas parietal activations are seen frequently in functional neuroimaging studies of episodic memory [219]. These separate categories of evidence suggest that an answer to this puzzle requires to distinguish the contributions of dorsal versus ventral parietal regions and the influence of top-down versus bottom-up attention on memory. The natures of memory trace cells identified in these studies include that these engram cells can encode the trained signals, receive synapse inputs, and undergo synaptic plasticity [112–114, 220]. The activation of memory cells induces strong memory retrievals. Memory events are upregulated by norepinephrine and serotonin. How the memory traces encode multiple signals associatively learned remains to be revealed (please refer to Chap. 5).

References

1. Wang J-H. Searching basic units of memory traces: associative memory cells. F1000Res. 2019;8(457):1–28.
2. Morris R. Developments of a water-maze procedure for studying spatial learning in the rat. J Neurosci Methods. 1984;11(1):47–60.
3. Olton DS, Isaacson RL. Importance of spatial location in active avoidance tasks. J Comp Physiol Psychol. 1968;65(3):535–9.
4. Pavlov I. Conditioned reflexes: an investigation of the physiological activity of the cerebral cortex. Translated by Anrep GV. Nature. 1927;121(3052):662–4.
5. Thorndike EL. Animal intelligence: an experimental study of the associative processes in animals. Psychol Rev Monogr Suppl. 1901;2:1–109.

6. Wasserman EA, Miller RR. What's elementary about associative learning? Annu Rev Psychol. 1997;48:573–607.
7. Liu Y, et al. Piriform cortical glutamatergic and GABAergic neurons express coordinated plasticity for whisker-induced odor recall. Oncotarget. 2017;8(56):95719–40.
8. Wang D, et al. Neurons in the barrel cortex turn into processing whisker and odor signals: a cellular mechanism for the storage and retrieval of associative signals. Front Cell Neurosci. 2015;9:320.
9. Yan F, et al. Coordinated plasticity between barrel cortical glutamatergic and GABAergic neurons during associative memory. Neural Plast. 2016;2016(ID5648390):1–20.
10. Feng J, et al. Barrel cortical neuron integrates triple associated signals for their memory through receiving epigenetic-mediated new synapse innervations. Cereb Cortex. 2017;27(12):5858–71.
11. Rescorla RA, Wagner AR. A theory of Pavlovian conditioning: variations in the effectiveness of reinforcement and nonreinforcement. In: Prokasy BA, editor. Classical conditioning II: current theory and research. New York: Appleton-Century; 1972.
12. Rescorla RA. Behavioral studies of Pavlovian conditioning. Annu Rev Neurosci. 1988;11:329–52.
13. Rescorla RA. Pavlovian conditioning. It's not what you think it is. Am Psychol. 1988;43(3):151–60.
14. Guo R, et al. Associative memory extinction is accompanied by decayed plasticity at motor cortical neurons and persistent plasticity at sensory cortical neurons. Front Cell Neurosci. 2017;11(168):1–12.
15. Wang J-H, Guo R, Wei Z. Associative memory extinction is accompanied by decays of associative memory cells and their plasticity at motor cortex but not sensory cortex. Soc Neurosci. 2017;81(09):10385.
16. Wang J-H, et al. Both glutamatergic and gabaergic neurons are recruited to be associative memory cells. Biophys J. 2016;110(3):supplement 481a.
17. Gao Z, et al. Associations of unilateral whisker and olfactory signals induce synapse formation and memory cell recruitment in bilateral barrel cortices: cellular mechanism for unilateral training toward bilateral memory. Front Cell Neurosci. 2016;10(285):1–16.
18. Wang D, et al. Neurons in the barrel cortex turn into processing whisker and odor signals: a cellular mechanism for the storage and retrieval of associative signals. Front Cell Neurosci. 2015;9(320):1–12.
19. Ehrlich I, et al. Amygdala inhibitory circuits and the control of fear memory. Neuron. 2009;62:757–71.
20. Marchand A, et al. A role for anterior thalamic nuclei in contextual fear memory. Brain Struct Funct. 2014;219(5):1575–86.
21. Maren S. The amygdala, synaptic plasticity, and fear memory. Ann N Y Acad Sci. 2003;985:106–13.
22. Maren S, Quirk GJ. Neuronal signalling of fear memory. Nat Rev Neurosci. 2004;5(11):844–52.
23. Pape HC, Stork O. Genes and mechanisms in the amygdala involved in the formation of fear memory. Ann N Y Acad Sci. 2003;985:92–105.
24. Wang JH, et al. Secondary associative memory cells and their plasticity in the prefrontal cortex. Biophys J. 2019;116(3):427a.
25. Staddon JE, Cerutti DT. Operant conditioning. Annu Rev Psychol. 2003;54:115–44.
26. Wang J-H, et al. Prefrontal cortical neurons are recruited as secondary associative memory cells for associative memory and cognition. Biophys J. 2018;114(3):155a.
27. Wang JH, Cui S. Associative memory cells: formation, function and perspective. F1000Res. 2017;6:283.
28. Wang JH, Cui S. Associative memory cells and their working principle in the brain. F1000Res. 2018;7:108.

29. Blough DS, Millward RB. Learning: operant conditioning and verbal learning. Annu Rev Psychol. 1965;16:63–94.
30. Dalla C, Shors TJ. Sex differences in learning processes of classical and operant conditioning. Physiol Behav. 2009;97(2):229–38.
31. Skinner BF. A discrimination without previous conditioning. Proc Natl Acad Sci U S A. 1934;20(9):532–6.
32. Skinner BF. The behavior of organisms: an experimental analysis. In: Skinner BF, editor. The behavior of organisms: an experimental analysis. New York: Appleton-Century-Crofts; 1938.
33. Skinner BF. Are theories of learning necessary? Psychol Rev. 1950;57(4):193–216.
34. Skinner BF. Teaching machines; from the experimental study of learning come devices which arrange optimal conditions for self instruction. Science. 1958;128(3330):969–77.
35. Skinner BF. The operant side of behavior therapy. J Behav Ther Exp Psychiatry. 1988;19(3):171–9.
36. Edwards S. Reinforcement principles for addiction medicine; from recreational drug use to psychiatric disorder. Prog Brain Res. 2016;223:63–76.
37. Kazdin AE. In: Kazdin AE, editor. Problem-solving skills training and parent management training for oppositional defiant disorder and conduct disorder. Evidence-based psychotherapies for children and adolescents. 2nd ed. New York: Guilford Press; 2010.
38. DeLong MR. Activity of pallidal neurons during movement. J Neurophysiol. 1971;34(3):414–27.
39. DeLong MR. Activity of basal ganglia neurons during movement. Brain Res. 1972;40(1):127–35.
40. DeLong MR. Putamen: activity of single units during slow and rapid arm movements. Science. 1973;179(4079):1240–2.
41. Richardson RT, DeLong MR. Electrophysiological studies of the function of the nucleus basalis in primates. In: Napier TC, Kalivas P, Hamin I, editors. The basal forebrain: anatomy to function (Advances in experimental medicine and biology, vol. 295). New York: Plenum; 1991.
42. Mitchell SJ, et al. The primate globus pallidus: neuronal activity related to direction of movement. Exp Brain Res. 1987;68(3):491–505.
43. Mitchell SJ, et al. The primate nucleus basalis of Meynert: neuronal activity related to a visuomotor tracking task. Exp Brain Res. 1987;68(3):506–15.
44. Vitek JL, et al. Neuronal activity in the basal ganglia in patients with generalized dystonia and hemiballismus. Ann Neurol. 1999;46(1):22–35.
45. Schultz W. Predictive reward signal of dopamine neurons. J Neurophysiol. 1998;80(1):1–27.
46. Barnes CA. Spatial learning and memory processes: the search for their neurobiological mechanisms in the rat. Trends Neurosci. 1988;11(4):163–9.
47. Conrad CD. A critical review of chronic stress effects on spatial learning and memory. Prog Neuro-Psychopharmacol Biol Psychiatry. 2010;34(5):742–55.
48. Foster TC. Dissecting the age-related decline on spatial learning and memory tasks in rodent models: N-methyl-D-aspartate receptors and voltage-dependent Ca2+ channels in senescent synaptic plasticity. Prog Neurobiol. 2012;96(3):283–303.
49. Grossberg S. From brain synapses to systems for learning and memory: object recognition, spatial navigation, timed conditioning, and movement control. Brain Res. 2015;1621:270–93.
50. Baddeley A. Working memory: looking back and looking forward. Nat Rev Neurosci. 2003;4:829–39.
51. Della Sala S, et al. Pattern span: a tool for unwelding visuo-spatial memory. Neuropsychologia. 1999;37(10):1189–99.
52. Hitch GJ, et al. Competition for the focus of attention in visual working memory: perceptual recency versus executive control. Ann N Y Acad Sci. 2018;1424(1):64–75.
53. Kahana MJ. The cognitive correlates of human brain oscillations. J Neurosci. 2006;26(6):1669–72.

54. Becker JT, Walker JA, Olton DS. Neuroanatomical bases of spatial memory. Brain Res. 1980;200(2):307–20.
55. Baddeley A, et al. The brain decade in debate: I. Neurobiology of learning and memory. Braz J Med Biol Res. 2000;33(9):993–1002.
56. Bird LR, et al. Spatial memory for food hidden by rats (*Rattus norvegicus*) on the radial maze: studies of memory for where, what, and when. J Comp Psychol. 2003;117(2):176–87.
57. Olton DS, Papas BC. Spatial memory and hippocampal function. Neuropsychologia. 1979;17(6):669–82.
58. Klauer KC, Zhao Z. Double dissociations in visual and spatial short-term memory. J Exp Psychol Gen. 2004;133(3):355–81.
59. Mammarella N, Cornoldi C, Donadello E. Visual but not spatial working memory deficit in children with spina bifida. Brain Cogn. 2003;53(2):311–4.
60. Mammarella N, et al. Aging and intrusion errors in an active visuo-spatial working memory task. Aging Clin Exp Res. 2009;21(4–5):282–91.
61. Passolunghi MC, Mammarella IC. Selective spatial working memory impairment in a group of children with mathematics learning disabilities and poor problem-solving skills. J Learn Disabil. 2012;45(4):341–50.
62. Passolunghi MC, Mammarella IC, Altoe G. Cognitive abilities as precursors of the early acquisition of mathematical skills during first through second grades. Dev Neuropsychol. 2008;33(3):229–50.
63. Garcia RB, et al. Visuospatial working memory for locations, colours, and binding in typically developing children and in children with dyslexia and non-verbal learning disability. Br J Dev Psychol. 2014;32(1):17–33.
64. Mammarella IC, Cornoldi C. Sequence and space: the critical role of a backward spatial span in the working memory deficit of visuospatial learning disabled children. Cogn Neuropsychol. 2005;22(8):1055–68.
65. Lancia S, et al. Are ventrolateral and dorsolateral prefrontal cortices involved in the computerized Corsi block-tapping test execution? An fNIRS study. Neurophotonics. 2018;5(1):011019.
66. Thomas E, et al. Spatial sequence memory and spatial error monitoring in the Groton Maze Learning Task (GMLT): a validation study of GMLT sub-measures in healthy children. Child Neuropsychol. 2016;22(7):837–52.
67. Pickering SJ, et al. Development of memory for pattern and path: further evidence for the fractionation of visuo-spatial memory. Q J Exp Psychol A. 2001;54(2):397–420.
68. Guariglia CC. Spatial working memory in Alzheimer's disease: a study using the Corsi block-tapping test. Dement Neuropsychol. 2007;1(4):392–5.
69. Moscovitch M, et al. The cognitive neuroscience of remote episodic, semantic and spatial memory. Curr Opin Neurobiol. 2006;16(2):179–90.
70. Olton DS. Spatial memory. Sci Am. 1977;236(6):82–4. 89–94, 96, 98.
71. O'Keefe J, Dostrovsky J. The hippocampus as a spatial map. Preliminary evidence from unit activity in the freely-moving rat. Brain Res. 1971;34(1):171–5.
72. O'Keefe J. Hippocampus, theta, and spatial memory. Curr Opin Neurobiol. 1993;3(6):917–24.
73. Squire LR, et al. Activation of the hippocampus in normal humans: a functional anatomical study of memory. Proc Natl Acad Sci U S A. 1992;89(5):1837–41.
74. Squire LR. Memory and the hippocampus: a synthesis from findings with rats, monkeys, and humans. Psychol Rev. 1992;99(2):195–231.
75. Martin SJ, de Hoz L, Morris RG. Retrograde amnesia: neither partial nor complete hippocampal lesions in rats result in preferential sparing of remote spatial memory, even after reminding. Neuropsychologia. 2005;43(4):609–24.
76. Milner B, Penfield W. The effect of hippocampal lesions on recent memory. Trans Am Neurol Assoc. 1955;(80th Meeting): 42–8.
77. Moser MB, Moser EI. Distributed encoding and retrieval of spatial memory in the hippocampus. J Neurosci. 1998;18(18):7535–42.

78. Penfield W. Observations on the anatomy of memory. Folia Psychiatr Neurol Neurochir Neerl. 1950;53(2):349–51.
79. Squire LR, Knowlton B, Musen G. The structure and organization of memory. Annu Rev Psychol. 1993;44:453–95.
80. Squire LR, McKee R. Influence of prior events on cognitive judgments in amnesia. J Exp Psychol Learn Mem Cogn. 1992;18(1):106–15.
81. McGaugh JL. Memory – a century of consolidation. Science. 2000;287(5451):248–51.
82. Maguire EA, Frackowiak RS, Frith CD. Recalling routes around London: activation of the right hippocampus in taxi drivers. J Neurosci. 1997;17(18):7103–10.
83. Brun VH, et al. Place cells and place recognition maintained by direct entorhinal-hippocampal circuitry. Science. 2002;296(5576):2243–6.
84. Goodrich-Hunsaker NJ, Hunsaker MR, Kesner RP. The interactions and dissociations of the dorsal hippocampus subregions: how the dentate gyrus, CA3, and CA1 process spatial information. Behav Neurosci. 2008;122(1):16–26.
85. Saab BJ, et al. NCS-1 in the dentate gyrus promotes exploration, synaptic plasticity, and rapid acquisition of spatial memory. Neuron. 2009;63(5):643–56.
86. Cho YH, Kesner RP. Involvement of entorhinal cortex or parietal cortex in long-term spatial discrimination memory in rats: retrograde amnesia. Behav Neurosci. 1996;110(3):436–42.
87. Colby CL, Goldberg ME. Space and attention in parietal cortex. Annu Rev Neurosci. 1999;22:319–49.
88. Kesner RP, Holbrook T. Dissociation of item and order spatial memory in rats following medial prefrontal cortex lesions. Neuropsychologia. 1987;25(4):653–64.
89. Liu P, Bilkey DK. The effect of excitotoxic lesions centered on the hippocampus or perirhinal cortex in object recognition and spatial memory tasks. Behav Neurosci. 2001;115(1):94–111.
90. Sasaki T, Leutgeb S, Leutgeb JK. Spatial and memory circuits in the medial entorhinal cortex. Curr Opin Neurobiol. 2015;32:16–23.
91. Save E, Moghaddam M. Effects of lesions of the associative parietal cortex on the acquisition and use of spatial memory in egocentric and allocentric navigation tasks in the rat. Behav Neurosci. 1996;110(1):74–85.
92. Witter MP, Moser EI. Spatial representation and the architecture of the entorhinal cortex. Trends Neurosci. 2006;29(12):671–8.
93. Wirt RA, Hyman JM. Integrating spatial working memory and remote memory: interactions between the medial prefrontal cortex and hippocampus. Brain Sci. 2017;7(4).
94. Kesner RP. The posterior parietal cortex and long-term memory representation of spatial information. Neurobiol Learn Mem. 2009;91(2):197–206.
95. Lee I, Kesner RP. Differential contribution of NMDA receptors in hippocampal subregions to spatial working memory. Nat Neurosci. 2002;5(2):162–8.
96. Morris RG, et al. Selective impairment of learning and blockade of long-term potentiation by an N-methyl-D-aspartate receptor antagonist, AP5. Nature. 1986;319:774–6.
97. Morris RG. Synaptic plasticity and learning: selective impairment of learning rats and blockade of long-term potentiation in vivo by the N-methyl-D-aspartate receptor antagonist AP5. J Neurosci. 1989;9(9):3040–57.
98. Wang JH, et al. Upregulation of glutamatergic receptor-channels is associated with cross-modal reflexes encoded in barrel cortex and piriform cortex. Biophys J. 2014;106(2):supplement 191a.
99. Aronoff R, et al. Long-range connectivity of mouse primary somatosensory barrel cortex. Eur J Neurosci. 2010;31(12):2221–33.
100. Haberly LB, et al. Distribution and ultrastructure of neurons in opossum piriform cortex displaying immunoreactivity to GABA and GAD and high-affinity tritiated GABA uptake. J Comp Neurol. 1987;266(2):269–90.
101. Petersen CCH. Functional organization of the barrel cortex. Neuron. 2007;56:339–55.
102. Barkai E, Saar D. Cellular correlates of olfactory learning in the rat piriform cortex. Rev Neurosci. 2001;12(2):111–20.

103. Wilson DA. Receptive fields in the rat piriform cortex. Chem Senses. 2001;26(5):577–84.
104. Wilson DA, Sullivan RM. Cortical processing of odor objects. Neuron. 2012;72(4):506–19.
105. Ni H, et al. Upregulation of barrel GABAergic neurons is associated with cross-modal plasticity in olfactory deficit. PLoS One. 2010;5(10):e13736.
106. Ye B, et al. The functional upregulation of piriform cortex is associated with cross-modal plasticity in loss of whisker tactile inputs. PLoS One. 2012;7(8):e41986.
107. Zhang G, et al. Upregulation of excitatory neurons and downregulation of inhibitory neurons in barrel cortex are associated with loss of whisker inputs. Mol Brain. 2013;6(1):2.
108. Zhao X, et al. Coordinated plasticity among glutamatergic and GABAergic neurons and synapses in the barrel cortex is correlated to learning efficiency. Front Cell Neurosci. 2017;11(221):1–12.
109. Liu Y, et al. Activity strengths of cortical glutamatergic and GABAergic neurons are correlated with transgenerational inheritance of learning ability. Oncotarget. 2017;8(68):112401–16.
110. Lei Z, et al. Synapse innervation and associative memory cell are recruited for integrative storage of whisker and odor signals in the barrel cortex through miRNA-mediated processes. Front Cell Neurosci. 2017;11(316):1–11.
111. Asok A, et al. Molecular mechanisms of the memory trace. Trends Neurosci. 2019;42(1):14–22.
112. Dubnau J, Chiang AS, Tully T. Neural substrates of memory: from synapse to system. J Neurobiol. 2003;54(1):238–53.
113. Josselyn SA, Kohler S, Frankland PW. Finding the engram. Nat Rev Neurosci. 2015;16(9):521–34.
114. Josselyn SA, Kohler S, Frankland PW. Heroes of the engram. J Neurosci. 2017;37(18):4647–57.
115. Lisman J, et al. Memory formation depends on both synapse-specific modifications of synaptic strength and cell-specific increases in excitability. Nat Neurosci. 2018;21(3):309–14.
116. Semon RW. In: Semon RW, editor. Mnemic psychology. London: Allen, Unwin; 1923.
117. McGaugh JL. The search for the memory trace. Ann N Y Acad Sci. 1972;193:112–23.
118. McClelland JL, McNaughton BL, O'Reilly RC. Why there are complementary learning systems in the hippocampus and neocortex: insights from the successes and failures of connectionist models of learning and memory. Psychol Rev. 1995;102(3):419–57.
119. Hebb DO. Spontaneous neurosis in chimpanzees; theoretical relations with clinical and experimental phenomena. Psychosom Med. 1947;9(1):3–19.
120. Penfield W. Bilateral frontal gyrectomy and postoperative intelligence. Res Publ Assoc Res Nerv Ment Dis. 1948;27(1 vol.):519–34.
121. Lashley KS. Mass action in cerebral function. Science. 1931;73(1888):245–54.
122. Lashley KS. Structural variation in the nervous system in relation to behavior. Psychol Rev. 1947;54(6):325–34.
123. Hebb DO. The organization of behavior, a neuropsychological theory. New York: Wiley; 1949.
124. Hebb DO. Animal and physiological psychology. Annu Rev Psychol. 1950;1:173–88.
125. Bliss TVP, Lynch MA. Long-term potentiation of synaptic transmission in the hippocampus: properties and mechanisms. In: Landfield PW, Deadwyler SA, editors. Long-term potentiation: from biophysics to behavior. New York: Alan R. Liss; 1988. p. 3–72.
126. Hodgkin AL, Huxley AF. Action potentials recorded from inside a nerve fiber. Nature. 1939;144:710–1.
127. Feng TP. Studies on the neuromuscular junction. XXVI. The changes of the end-plate potentail during and after prolonged stimulation. Chin J Physiol. 1941;16:341–72.
128. Feng TP. Looking back, looking forward. Annu Rev Neurosci. 1988;11:1–12.
129. Zhao J, Wang D, Wang JH. Barrel cortical neurons and astrocytes coordinately respond to an increased whisker stimulus frequency. Mol Brain. 2012;5:12.

130. Buzsaki G. Network properties of memory trace formation in the hippocampus. Boll Soc Ital Biol Sper. 1991;67(9):817–35.
131. Jones MW, Wilson MA. Theta rhythms coordinate hippocampal-prefrontal interactions in a spatial memory task. PLoS Biol. 2005;3(12):e402.
132. Khodagholy D, Gelinas JN, Buzsaki G. Learning-enhanced coupling between ripple oscillations in association cortices and hippocampus. Science. 2017;358(6361):369–72.
133. Rothschild G, Eban E, Frank LM. A cortical-hippocampal-cortical loop of information processing during memory consolidation. Nat Neurosci. 2017;20(2):251–9.
134. Lansink CS, et al. Hippocampus leads ventral striatum in replay of place-reward information. PLoS Biol. 2009;7(8):e1000173.
135. Takehara-Nishiuchi K, McNaughton BL. Spontaneous changes of neocortical code for associative memory during consolidation. Science. 2008;322(5903):960–3.
136. Girardeau G, Inema I, Buzsaki G. Reactivations of emotional memory in the hippocampus-amygdala system during sleep. Nat Neurosci. 2017;20(11):1634–42.
137. Katkov M, Romani S, Tsodyks M. Memory retrieval from first principles. Neuron. 2017;94(5):1027–32.
138. Jahans-Price T, et al. Computational modeling and analysis of hippocampal-prefrontal information coding during a spatial decision-making task. Front Behav Neurosci. 2014;8:62.
139. Lansner A. Associative memory models: from the cell-assembly theory to biophysically detailed cortex simulations. Trends Neurosci. 2009;32(3):178–86.
140. Schwindel CD, McNaughton BL. Hippocampal-cortical interactions and the dynamics of memory trace reactivation. Prog Brain Res. 2011;193:163–77.
141. Bracha V, et al. The cerebellum and eye-blink conditioning: learning versus network performance hypotheses. Neuroscience. 2009;162(3):787–96.
142. Burhans LB, Smith-Bell C, Schreurs BG. Conditioning-specific reflex modification of the rabbit's nictitating membrane response and heart rate: behavioral rules, neural substrates, and potential applications to posttraumatic stress disorder. Behav Neurosci. 2008;122(6):1191–206.
143. Davis M, et al. Fear-potentiated startle: a neural and pharmacological analysis. Behav Brain Res. 1993;58(1–2):175–98.
144. Glanzman DL. The cellular basis of classical conditioning in Aplysia californica – it's less simple than you think. Trends Neurosci. 1995;18(1):30–6.
145. Hawkins RD. A cellular mechanism of classical conditioning in Aplysia. J Exp Biol. 1984;112:113–28.
146. Lechner HA, Baxter DA, Byrne JH. Classical conditioning of feeding in Aplysia: I. Behavioral analysis. J Neurosci. 2000;20(9):3369–76.
147. Maren S. Pavlovian fear conditioning as a behavioral assay for hippocampus and amygdala function: cautions and caveats. Eur J Neurosci. 2008;28(8):1661–6.
148. Pennypacker HS, et al. An apparatus and procedure for conditioning the eye-blink reflex in the squirrel monkey. J Exp Anal Behav. 1966;9(5):601–4.
149. Perkowski JJ, Murphy GG. Deletion of the mouse homolog of KCNAB2, a gene linked to monosomy 1p36, results in associative memory impairments and amygdala hyperexcitability. J Neurosci. 2011;31(1):46–54.
150. Reijmers LG, et al. Localization of a stable neural correlate of associative memory. Science. 2007;317(5842):1230–3.
151. Sears RJ, Baker JS, Frey PW. The eye blink as a time-locked response: implications for serial and second-order conditioning. J Exp Psychol Anim Behav Process. 1979;5(1):43–64.
152. Theios J, Brelsford JW Jr. A Markov model for classical conditioning: application to eye-blink conditioning in rabbits. Psychol Rev. 1966;73(5):393–408.
153. Woodruff-Pak DS, Disterhoft JF. Where is the trace in trace conditioning? Trends Neurosci. 2008;31(2):105–12.
154. Martinez M, Calvo-Torrent A, Pico-Alfonso MA. Social defeat and subordination as models of social stress in laboratory rodents: a review. Aggress Behav. 1998;24:241–56.

155. Tsankova NM, et al. Sustained hippocampal chromatin regulation in a mouse model of depression and antidepressant action. Nat Neurosci. 2006;9(4):519–25.
156. Vasconcelos M, Stein DJ, de Almeida RM. Social defeat protocol and relevant biomarkers, implications for stress response physiology, drug abuse, mood disorders and individual stress vulnerability: a systematic review of the last decade. Trends Psychiatry Psychother. 2015;37(2):51–66.
157. Wang J-H, et al. Neurons in barrel cortex turn into processing whisker and odor signals: a novel form of associative learning. Soc Neurosci. 2013;653(14):WW11.
158. Carr MF, Jadhav SP, Frank LM. Hippocampal replay in the awake state: a potential substrate for memory consolidation and retrieval. Nat Neurosci. 2011;14(2):147–53.
159. Harris KD, et al. Organization of cell assemblies in the hippocampus. Nature. 2003;424(6948):552–6.
160. Ji D, Wilson MA. Coordinated memory replay in the visual cortex and hippocampus during sleep. Nat Neurosci. 2007;10(1):100–7.
161. Jadhav SP, et al. Awake hippocampal sharp-wave ripples support spatial memory. Science. 2012;336(6087):1454–8.
162. Kay K, et al. A hippocampal network for spatial coding during immobility and sleep. Nature. 2016;531(7593):185–90.
163. Kudrimoti HS, Barnes CA, McNaughton BL. Reactivation of hippocampal cell assemblies: effects of behavioral state, experience, and EEG dynamics. J Neurosci. 1999;19(10):4090–101.
164. McNaughton BL, Barnes CA, O'Keefe J. The contributions of position, direction, and velocity to single unit activity in the hippocampus of freely-moving rats. Exp Brain Res. 1983;52(1):41–9.
165. Penfield W, Perot P. The brain's record of auditory and visual experience. A final summary and discussion. Brain. 1963;86:595–696.
166. Sirota A, et al. Communication between neocortex and hippocampus during sleep in rodents. Proc Natl Acad Sci U S A. 2003;100(4):2065–9.
167. Skaggs WE, McNaughton BL. Replay of neuronal firing sequences in rat hippocampus during sleep following spatial experience. Science. 1996;271(5257):1870–3.
168. Wilson MA, McNaughton BL. Dynamics of the hippocampal ensemble code for space. Science. 1993;261(5124):1055–8.
169. Wilson MA, McNaughton BL. Reactivation of hippocampal ensemble memories during sleep. Science. 1994;265(5172):676–9.
170. Wirth S, et al. Single neurons in the monkey hippocampus and learning of new associations. Science. 2003;300(5625):1578–81.
171. Yu JY, et al. Specific hippocampal representations are linked to generalized cortical representations in memory. Nat Commun. 2018;9(1):2209.
172. Yokose J, et al. Overlapping memory trace indispensable for linking, but not recalling, individual memories. Science. 2017;355(6323):398–403.
173. Varela C, et al. Anatomical substrates for direct interactions between hippocampus, medial prefrontal cortex, and the thalamic nucleus reuniens. Brain Struct Funct. 2014;219(3):911–29.
174. Nikolenko V, Poskanzer KE, Yuste R. Two-photon photostimulation and imaging of neural circuits. Nat Methods. 2007;4(11):943–50.
175. Stosiek C, et al. In vivo two-photon calcium imaging of neuronal networks. Proc Natl Acad Sci USA. 2003;100(12):7319–24.
176. Wang KH, et al. In vivo two-photon imaging reveals a role of arc in enhancing orientation specificity in visual cortex. Cell. 2006;126(2):389–402.
177. Mao D, et al. Hippocampus-dependent emergence of spatial sequence coding in retrosplenial cortex. Proc Natl Acad Sci U S A. 2018;115(31):8015–8.
178. Fu M, et al. Repetitive motor learning induces coordinated formation of clustered dendritic spines in vivo. Nature. 2012;483(7387):92–5.
179. Sugiura H, et al. Transducing neuronal activity into dendritic spine morphology: new roles for p38 MAP kinase and N-cadherin. Neuroscientist. 2009;15(1):90–104.

180. Thiagarajan TC, Piedras-Renteria ES, Tsien RW. Alpha- and beta-CaMKII. Inverse regulation by neuronal activity and opposing effects on synaptic strength. Neuron. 2002;36:1103–14.
181. Wong-Riley MT. Cytochrome oxidase: an endogenous metabolic maker for neuronal activity. Trends Neurosci. 1989;12(3):94–101.
182. Morgan JI, Curran T. Stimulus-transcription coupling in neurons: role of cellular immediate-early genes. Trends Neurosci. 1989;12(11):459–62.
183. Link W, et al. Somatodendritic expression of an immediate early gene is regulated by synaptic activity. Proc Natl Acad Sci U S A. 1995;92(12):5734–8.
184. Guzowski JF, et al. Environment-specific expression of the immediate-early gene Arc in hippocampal neuronal ensembles. Nat Neurosci. 1999;2(12):1120–4.
185. Tayler KK, et al. Reactivation of neural ensembles during the retrieval of recent and remote memory. Curr Biol. 2013;23(2):99–106.
186. Bissiere S, et al. Electrical synapses control hippocampal contributions to fear learning and memory. Science. 2011;331(6013):87–91.
187. Grundemann J, Luthi A. Ensemble coding in amygdala circuits for associative learning. Curr Opin Neurobiol. 2015;35:200–6.
188. Tanaka KZ, McHugh TJ. The hippocampal engram as a memory index. J Exp Neurosci. 2018;12:1179069518815942.
189. Tonegawa S, et al. Memory engram cells have come of age. Neuron. 2015;87(5):918–31.
190. Kiessling M, Gass P. Immediate early gene expression in experimental epilepsy. Brain Pathol. 1993;3(4):381–93.
191. Meldrum BS. Concept of activity-induced cell death in epilepsy: historical and contemporary perspectives. Prog Brain Res. 2002;135:3–11.
192. Wang X, et al. Persistent hyperactivity of hippocampal dentate interneurons after a silent period in the rat pilocarpine model of epilepsy. Front Cell Neurosci. 2016;10:94.
193. Simonato M, et al. Differential expression of immediate early genes in the hippocampus in the kindling model of epilepsy. Brain Res Mol Brain Res. 1991;11(2):115–24.
194. Abe H, Nowak TS Jr. Induced hippocampal neuron protection in an optimized gerbil ischemia model: insult thresholds for tolerance induction and altered gene expression defined by ischemic depolarization. J Cereb Blood Flow Metab. 2004;24(1):84–97.
195. Bokesch PM, et al. Dextromethorphan inhibits ischemia-induced c-fos expression and delayed neuronal death in hippocampal neurons. Anesthesiology. 1994;81(2):470–7.
196. Kiessling M, et al. Differential transcription and translation of immediate early genes in the gerbil hippocampus after transient global ischemia. J Cereb Blood Flow Metab. 1993;13(6):914–24.
197. Sossin WS. Molecular memory traces. Prog Brain Res. 2008;169:3–25.
198. Milo R. What is the total number of protein molecules per cell volume? A call to rethink some published values. BioEssays. 2013;35(12):1050–5.
199. Lashley KS. Cerebral organization and behavior. Res Publ Assoc Res Nerv Ment Dis. 1958;36:1–4; discussion 14–18
200. Penfield W, Milner B. Memory deficit produced by bilateral lesions in the hippocampal zone. AMA Arch Neurol Psychiatry. 1958;79(5):475–97.
201. Penfield W. Engrams in the human brain. Mechanisms of memory. Proc R Soc Med. 1968;61(8):831–40.
202. Flexner JB, Flexner LB, Stellar E. Memory in mice as affected by intracerebral puromycin. Science. 1963;141(3575):57–9.
203. Flexner LB, Flexner JB, Roberts RB. Memory in mice analyzed with antibiotics. Antibiotics are useful to study stages of memory and to indicate molecular events which sustain memory. Science. 1967;155(3768):1377–83.
204. Flexner JB, Glexner LB. Studies on memory: evidence for a widespread memory trace in the neocortex after the suppression of recent memory by puromycin. Proc Natl Acad Sci U S A. 1969;62(3):729–32.

205. Silva AJ, et al. Impaired spatial learning in alpha-calcium-calmodulin kinase II mutant mice. Science. 1992;257(10 July):206–11.
206. Liu X, et al. Identification and manipulation of memory engram cells. Cold Spring Harb Symp Quant Biol. 2014;79:59–65.
207. Ramirez S, Tonegawa S, Liu X. Identification and optogenetic manipulation of memory engrams in the hippocampus. Front Behav Neurosci. 2013;7:226.
208. Penfield W. Some mechanisms of consciousness discovered during electrical stimulation of the brain. Proc Natl Acad Sci U S A. 1958;44(2):51–66.
209. Rasmussen T, Penfield W. The human sensorimotor cortex as studied by electrical stimulation. Fed Proc. 1947;6(1 Pt 2):184.
210. Penfield W, Welch K. Instability of response to stimulation of the sensorimotor cortex of man. J Physiol. 1949;109(3–4):358–65, illust
211. Ferry B, Roozendaal B, McGaugh JL. Role of norepinephrine in mediating stress hormone regulation of long-term memory storage: a critical involvement of the amygdala. Biol Psychiatry. 1999;46(9):1140–52.
212. Meneses A, Liy-Salmeron G. Serotonin and emotion, learning and memory. Rev Neurosci. 2012;23(5–6):543–53.
213. Liu X, et al. Optogenetic stimulation of a hippocampal engram activates fear memory recall. Nature. 2012;484(7394):381–5.
214. Liu X, Ramirez S, Tonegawa S. Inception of a false memory by optogenetic manipulation of a hippocampal memory engram. Philos Trans R Soc Lond Ser B Biol Sci. 2014;369(1633):20130142.
215. Roy DS, et al. Memory retrieval by activating engram cells in mouse models of early Alzheimer's disease. Nature. 2016;531(7595):508–12.
216. Ramirez S, et al. Creating a false memory in the hippocampus. Science. 2013;341(6144):387–91.
217. Ramirez S, et al. Activating positive memory engrams suppresses depression-like behaviour. Nature. 2015;522(7556):335–9.
218. Ryan TJ, et al. Memory. Engram cells retain memory under retrograde amnesia. Science. 2015;348(6238):1007–13.
219. Cabeza R, et al. The parietal cortex and episodic memory: an attentional account. Nat Rev Neurosci. 2008;9(8):613–25.
220. Poo MM, et al. What is memory? The present state of the engram. BMC Biol. 2016;14:40.

Chapter 4
Cellular and Molecular Changes in Associative Memory

Abstract Basic units in the brain include neurons, glia cells, and their synaptic connections. Neuronal codes for the execution of various brain functions, such as memory, cognition, and emotion, are programmed by these neurons and synapses. While acquiring exogenous signals and processing endogenous signals, these signals are memorized in the brain essentially for guiding cognitions and behaviors. Memory formation and memory-relevant behavior emergence may be based on the recruitment and/or refinement of these neurons and synapses. The recruitment of neurons to be the basic units in memory traces will be discussed in Chap. 5. Here, author intends to summarize the plasticity of neurons and synapses, which is presumably relevant to memory formation. The featured function of neurons is to produce action potentials, or spikes, once excitatory synaptic signals drive their membrane potentials to a threshold potential for firing spikes or their membrane potentials fluctuate to this threshold potential. The patterns of neuronal spikes constitute the digital signals to program various signals and manager neuronal functions. In this regard, the plasticity at neurons is mainly characterized by changes in their spike patterns and/or threshold potentials that may move closely or away from the resting membrane potential. In terms of chemical synapses, their signal transmission includes transmitter release from presynaptic boutons as well as interactions between transmitters and their receptors in postsynaptic density. The plasticity at chemical synapses includes the changes in the capacity and release efficacy of presynaptic transmitters as well as in the number and responsiveness of postsynaptic receptors, which may control the conversions between inactive synapses and active synapses or between silent synapses and functional synapses. Long-term plasticity at synapses and neurons is believed to be cellular mechanisms underlying memory formation, which leads to long-term changes in memory-relevant behaviors.

Keywords Synaptic plasticity · Neuronal plasticity · Neuronal homeostasis and neuronal compatibility

J.-H. Wang, *Associative Memory Cells: Basic Units of Memory Trace*,
https://doi.org/10.1007/978-981-13-9501-7_4

4.1 Introduction

In addition to the recruitment of memory cells in engrams for information storage, the refinement of neurons and synapses is presumably involved in memory formation and memory-related behaviors, especially long-term potentiation and depression of synaptic transmission and/or neuronal excitability [1–4]. Numerous studies have provided evidences for the relationship between long-term plasticity and memory formation [5]. For instance, long-term potentiation of synaptic transmission is associated to memory formation. The downregulation of intracellular calcium-/calmodulin-dependent protein kinases by pharmacological reagents and gene knockout attenuates both memory formation and long-term synaptic potentiation. The upregulation of intracellular protein kinase C by pharmacological regents and gene overexpression facilitates both long-term synaptic potentiation and memory formation. These experimental data suggest that synapse plasticity is featured as engrams for memory formation [6, 7], though certain studies argue against this conclusion.

Synaptic plasticity is classified into short-term and long-term maintenance. Short-term synaptic plasticity includes synaptic augmentation, facilitation, and post-tetanic potentiation [8], which may result from the increased release of presynaptic transmitters and the decreased desensitization of postsynaptic receptors [3, 9]. Long-term potentiation of synaptic transmission may be influenced by the following factors, such as the increase of presynaptic neurotransmitter release through the facilitations of vesicle docking, vesicle discharge, and vesicle recycling, as well as the increase of postsynaptic receptors in their number and responsiveness [3], so that silent synapses are converted into functional synapses [10] and inactive synapses are converted into active synapses [11]. The increase of functional synapses in a given neuron will strongly drive its membrane potential toward threshold to produce digital spikes that are programmed neuronal codes and manager neuronal function execution. In addition, the transmission patterns of unitary synapses are variable among facilitation, parallel, and depression. The intensive activity of synapses converts this fluctuation into stabilization via intracellular calcium-calmodulin signaling pathway. This conversion enables neuronal encoding efficacy and timing precision to be strengthened [12]. On the other hand, long-term depression of synaptic transmission by the depletion of transmitter release and the desensitization of receptors attenuates neuronal spikes.

Neuronal plasticity results from either the enhancement of spiking capability or the decrement of threshold potential [4, 13], which may upregulate spike frequency and timing to facilitate neuronal encoding efficiency and timing precision. Therefore, both synaptic potentiation and neuronal upregulation strengthen the efficiency and timing precision of neuron encoding. If these alternations occur in memory cells, the strengthened efficiency and timing precision of neuronal encoding may facilitate the accurate retrieval of memorized information from memory cells in the brain [14]. It is noteworthy that both synaptic potentiation and neuronal upregulation are regulated by calcium-calmodulin signaling pathway [4, 15, 16].

In addition, a few types of the coordinated plasticity at synapses, neurons, and axons have been discovered. There is a coordinated plasticity among subcellular compartments in cortical excitatory neurons. This homeostatic plasticity among excitatory glutamatergic synapses, inhibitory GABAergic synapses, cell body, and axon in a given neuron upregulates the timing precision of neuronal encoding [15]. The axons and their branches of neurons are a critical coordinator for neuronal spike output. Spikelet generated in the soma [17] can be amplified into complete spikes by the high dense of voltage-gated sodium channels in the initial segment of axons [18]. On the other hand, if spike frequency is overly high, the axon propagation of sequential action potentials will decay in an activity-dependent manner [19–22]. Moreover, there is a mutual influence between spike propagation in presynaptic axon branches and spiking capability in postsynaptic neurons as well as between presynaptic transmitter release and postsynaptic receptor responsiveness [23]. This functional compatibility between presynaptic and postsynaptic partners is critically important for coordinated activities among network neurons [24].

4.2 Synapses, Neurons, and Their Functional Changes

Synapses work for the communications between neurons that constitute neural network to execute the brain functions [25–31]. The strength of the driving force from numerous synapses to each of the neurons influences neuronal spike encoding in their capacity and timing precision. A chemical synapse consists of the presynaptic bouton that releases neurotransmitters and the postsynaptic density that embeds receptor channels. Interactions between transmitters and receptors at chemical synapses fulfill the signal transmissions among neurons [32, 33]. Thus, the efficacy of synaptic transmission (functional state) and the number of chemical synapses on each of the neurons influence the strength of synaptic driving force to neurons, the coordination of these neurons, and the working efficiency of their constituted network to execute the brain functions, such as memory, cognition, emotion, and behaviors.

The signal transmission of chemical synapses includes the activation of voltage-gated calcium channels, the increase of intracellular calcium, the docking of synaptic vesicle, and the release of neurotransmitters in the presynaptic bouton as well as the interaction of transmitters with receptors in the postsynaptic density [34–37]. Numerous intracellular signaling pathways regulate these processes to influence the efficacy of synaptic transmission, such as calcium-/calmodulin-dependent protein kinase, cAMP-dependent protein kinase, calcium-/phospholipid-dependent protein kinase, CREBs, and other molecules. The activation of these signal pathways may strengthen synaptic transmission and short-term and long-term potentiation of synaptic transmission [16, 38–43].

With monitoring the activities of unitary synapses between a presynaptic neuron and a postsynaptic neuron by the electrophysiological pair recording between synapse-connected neurons, we are able to observe postsynaptic responses, i.e., postsynaptic currents or potentials, which are elicited by presynaptic action potentials [12]. A few characteristics in these unitary synapses are listed below. (1) The amplitudes of synaptic currents are variable, and synaptic responses may not be elicited by presynaptic spikes sometime. The fluctuation and failure of synapse transmission indicate that mechanisms underlying synaptic functions are not reliable. In other words, there are some potential chances for synapses to be converted into the reliably functional state, such as the conversion of inactive synapses into active synapses or the conversion of silent synapses into functional synapses [10, 11]. (2) The patterns of synaptic transmission in response to sequential spikes (spike patterns at the most neurons) appear to be fluctuated among facilitation, parallel, and depression. That is, the transmission to sequential neuronal signals is unreliable. The conversion of unreliable into reliable patterns of synaptic transmission in response to sequential neuronal spikes can strengthen the efficiency of neuronal communications [12, 44]. (3) Presynaptic neurotransmitter release and postsynaptic receptor response appear to be functionally compatible, i.e., the high probability of transmitter release is associated to the high responsiveness of receptors or vice versa. This functional compatibility between presynaptic and postsynaptic partners supports the efficiency of neuronal communication in network with less energy consumption [23, 24] (Fig. 4.1).

In summary, the increases in the reliability of synaptic transmission, the release probability of transmitters, and the quantity of transmitter in each vesicle in presynaptic boutons will raise the total amount of neurotransmitters to activate postsynaptic receptors maximally. The increases in the number and responsiveness of receptor channels in postsynaptic density will raise the amplitudes of postsynaptic currents. The increases of the number of excitatory functional synapses or active synapses will strengthen synaptic driving forces to neurons for their efficient encoding of digital spikes. All of these factors lead to synaptic potentiation in long-term or short-term maintenance [3]. If these synapses are located on memory cells with functional and molecular identification, long-term synaptic potentiation should be involved in memory-related cognition, emotion, and behavior [45].

Neurons are characterized to produce sequential action potentials for their digital code, another encoding style in the brain, in addition to analogue codes at synapses [12]. The programming of digital spike codes in neurons is driven by synaptic currents and influenced by threshold potentials by themselves, as well as is controlled by their capability to encode sequential spikes intrinsically. In addition to factors above, the amplitude of synaptic currents is set by the cable property of cellular membrane that is affected by the membrane capacitance and resistance. The changeable resistance of cell membrane is influenced by various ion channels opened under different membrane potentials [46]. Voltage-gated sodium channels are thought to be a key controller of threshold potential to initiate spikes [47–50]. The coordination among those voltage-gated sodium channels, voltage-gated potassium channels, and voltage-gated transient calcium channels may influence the capability to generate sequential spikes [51]. Therefore, the regulation of these ion channels by intracellular signaling pathways will influence neuronal spiking in capacity and timing precision [12, 15]. If this situation occurs in memory cells with functional and

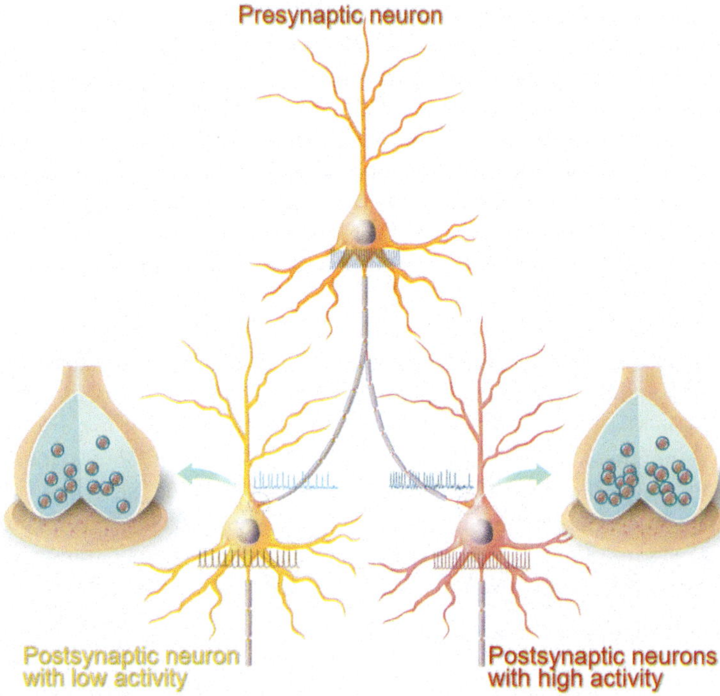

Fig. 4.1 Functional compatibility between presynaptic axonal branches and postsynaptic neuron partners. The axons of presynaptic neurons have two branches that synaptically innervate two postsynaptic neurons, respectively. The sequential spikes are induced by the depolarization pulse. These spikes are propagated to axonal branches and further toward terminals, as well as trigger presynaptic vesicles (blue circles) to release neurotransmitters (red dots). These neurotransmitters act on postsynaptic receptors (red dots). When electrical signals at presynaptic terminals and their postsynaptic neuron partners are recorded simultaneously, two presynaptic axon branches demonstrate spike patterns with high and low frequencies, respectively. Two postsynaptic neurons in their partners show spike patterns high and low frequencies, respectively, as well as excitatory postsynaptic potentials with high and low amplitudes and frequencies, respectively. The presynaptic axon branches with high spike frequency and transmitter release probability connect postsynaptic neurons with high spike frequency and receptor responsiveness or vice versa. There are linear correlations between presynaptic axon branches and postsynaptic neuron partners in their spike frequency, as well as between presynaptic transmitter release probability and postsynaptic partner responses. Therefore, there is the compatibility between presynaptic axon branches and their postsynaptic partner neurons in spike-encoding capability and synapse transmission. Spike frequency denotes spike-encoding capability. The number of presynaptic vesicles denotes transmitter release probability. The density of postsynaptic receptors denotes postsynaptic responses

molecular identification, long-term neuronal plasticity should be involved in the memory-relevant cognition, emotion, and behaviors.

It is noteworthy that the excitability and threshold potential change inversely at the cell body and the axon of individual neurons after their intensive activity [15]. Some cortical neurons express a weakened excitability at their soma and a strengthened excitability at their axon. Other cortical neurons express a strengthened excitability at their soma and a weakened excitability at their axon. Although it is unclear about the inverse changes among network neurons, the homeostasis of neural networks, or network balance, is maintained by this coordination among them. At individual neurons, the inverse regulation of cell body and axon will maintain cellular homeostasis without over-excitation. This inverse change is regulated by intracellular calcium, calmodulin-dependent kinase, and phosphatase [15].

4.3 Synaptic Plasticity

In terms of the chemical synapses, their plasticity can be classified into potentiation and depression based on spatial changes at synapses. Synaptic plasticity can also be classified into short-term and long-term in temporal changes at synapses [3]. As synaptic depression may be caused by transmitter depletion and/or receptor desensitization [52, 53], author intends to review synaptic upregulation. Short-term synaptic potentiation includes augmentation, facilitation, and post-tetanic potentiation that occur in a range of minutes [54]. Long-term potentiation of synaptic transmission, or LTP, is usually defined as synaptic potentiation maintenance above 30 min [5].

4.3.1 Synaptic Facilitation

This term refers to the gradual increase in the amplitude of synaptic transmission in response to repeated presynaptic stimulations or to presynaptic sequential spikes [8, 55]. Although its mechanisms may be complicated, its functional role is to boost membrane potential toward the threshold potential for producing spikes [12]. In the central nervous system, synapses in most neurons in response to presynaptic sequential spikes appear as initial synaptic facilitation and subsequent decay. This up-to-down changes may be caused by the upregulation and then downregulation of presynaptic transmitter release as well as the sensitization and then desensitization of postsynaptic receptors [9, 56, 57]}. In addition to synaptic facilitation, some unitary synapses constituted from a pair of presynaptic and postsynaptic neurons show depression pattern in response to presynaptic sequential spikes [3, 12, 58, 59]. Moreover, the transmission of presynaptic sequential spikes at distinct unitary synapses appear to be one of three patterns, facilitation, parallel, and depression, on the average of response amplitudes over time [12]. The signal transmission at a given unitary synapse in a pair of neurons can express all of these three patterns. The fluctuation among facilitation, parallel, and depression appears to be chaos, i.e., irregular with time, despite one pattern in dominance [12]. As signal transmissions between the neurons also carry the temporal messages for their programming of neuronal codes to manage brain functions at that moment via the efficient communication of pair neurons, the synaptic transmission patterns should not be evaluated based on the average of synaptic responses over time [12].

In terms of molecular mechanisms underlying synaptic facilitation, numerous studies have been conducted in a few synapse models, such as neuromuscular junction and synapses in the hippocampus, cerebellum, and cortices [55, 60]. The most influencing experiment was initially done at neuromuscular junctions by Katz and Miledi, in which the double concentration of extracellular calcium was washed to preparations. They observed that the initial responses to presynaptic stimulations were increased, but not for subsequent responses [61]. They proposed the accumulation of presynaptic calcium in response to sequential presynaptic stimuli, also called as residual calcium hypothesis, for synaptic facilitation. They further explained that the plateau level of synaptic transmission by washing extracellular calcium is due to the reach of residual calcium accumulation to be saturated in triggering transmitter release. An obvious possibility has been ignored by them and other followers, that is, extracellularly used calcium may also affect molecules in postsynaptic neurons, such as membrane receptors and intracellular signaling pathways [56]. This possibility has been examined by infusing calcium chelator or calmodulin-dependent protein kinase inhibitor into postsynaptic neurons, by which synaptic facilitation is elevated. Moreover, the inhibition of glutamate receptor desensitization can enhance synaptic facilitation [56]. A diagram has been drawn for synaptic facilitation. The presynaptic mechanism generates synaptic facilitation, and the postsynaptic mechanism modulates the strength of synaptic facilitation. A chain reaction from postsynaptic calcium, calmodulin-dependent protein kinase, and glutamate receptor desensitization to the synaptic responses modulates synaptic facilitation [3, 56]. Therefore, both presynaptic and postsynaptic molecules work together to set up synaptic patterns in order to boost membrane potentials toward the threshold potential and trigger the neuronal encoding of digital spikes.

The postsynaptic modulation of synaptic facilitation is also proved by studying long-term potentiation of synaptic transmission (please see next section). With the enhancement of synaptic transmission, synaptic facilitation will be attenuated [57, 62]. The inhibition of postsynaptic calmodulin-dependent protein kinases can block long-term potentiation and the attenuation of synaptic facilitation [57]. Therefore, postsynaptic calcium and calmodulin signaling pathway are required for synaptic facilitation and long-term synaptic potentiation [3].

In addition to synaptic facilitation, the fluctuation of signal transmission pattern among synaptic facilitation, parallel, and depression at unitary synapses is influenced by the postsynaptic mechanism. There is a negative inverse relationship between response one and the difference of responses two minus one. A large response one is associated to either low synaptic facilitation or synaptic depression or vice versa. This feature of synaptic transmission patterns indicates the presence of postsynaptic limitation, or receptor desensitization, for synaptic facilitation [12]. The high initial synaptic responses with synaptic depression facilitate neuronal spike encoding in the early phase, while lower initial synaptic responses with synaptic facilitation allow synaptic responses to be accumulated toward the threshold potential for the delayed spike encoding. Some neurons encode spikes in early phase and others encode spikes in late phase. This temporal spike encoding among different neurons in a neural network may set up the coordinated neuronal encoding in the networks [12]. Moreover, the activation of postsynaptic calmodulin signaling pathway and/or the inhibition of postsynaptic calcineurin can drive the fluctuated transmission pattern toward a uniform pattern, e.g., synaptic depression [12]. In terms of physiological impact of the reliable uniform pattern versus the unreliable fluctuated transmission pattern, the study shows that there are the high capacity and timing precision of neuronal spikes driven by the currents integrated from synapses with uniform pattern, in comparison with the currents integrated from synapses with fluctuated transmission pattern [12]. If this uniform pattern occurs in associative memory cells after their plasticity [63], the information storage and memory retrieval will be highly efficient and precise [45].

4.3.2 Long-Term Potentiation of Synaptic Transmission

Long-term potentiation (LTP) refers to the increased efficacy of synaptic transmission after synapses experience an intensive activity [7, 64, 65]. In experiments, high-frequency stimulations or tetanic stimulations to presynaptic axons can induce the potentiated signal transmission at synapses made by the terminals of these axons [1, 66]. LTP can be electrophysiologically recorded in postsynaptic neurons, while the stimulations are given to a bundle of presynaptic axons [67, 68]. As LTP features are similar to the learning and memory in specificity, cooperativity, and association, as well as there are correlations between LTP and memory-relevant behaviors, LTP has been believed to be a cellular mechanism of learning and memory [5]. Donald Hebb has hypothesized that the interconnections between neurons may be strengthened, while both presynaptic and postsynaptic neurons are coincidently activated. These interconnected cell assemblies are critical for memory traces or engram [69, 70]. LTP well matches this hypothesis about the strengthening of interconnected cell assemblies. Such information leads to a belief from many neuroscientists that LTP is cellular substrate for learning and memory. However, the argument comes from different thoughts. LTP is not studied in the memory cells. LTP emergence in a specific pathway cannot explain the joint acquisition of the associated signals as well as their integrative storage and reciprocal retrieval, i.e., associative learning and memory that occurs in almost all of the situations.

In terms of characteristics of LTP induction, the specificity of LTP expression states that LTP is induced at specific axon-innervated synapses that have received high-frequency stimulation or tetanus [5]. The cooperativity among synapses in a neural pathway is believed to be important for LTP induction. For instance, high-frequency stimulations in weak intensity to a bundle of presynaptic axons, which presumably activates only an axon-made or a few of axon-made synapses, are usually unable to induce LTP expression at these synapses. On the other hand, high-frequency stimulations with strong intensity to this bundle of presynaptic axons are sufficient to induce LTP expression [5]. It is possible that tetanus-induced LTP may not be expressed at single synapse. This possibility is also supported by the study that high-frequency stimulations in maximal intensity cannot induce LTP [71], in which all axons in this bundle are presumably activated likely as a unitary synapse. In this regard, tetanus-induced LTP at the bundle of axons may not be due to LTP expression at these synapses, but due to the decrease of the axonal threshold to produce action potentials [4, 71]. The associativity of LTP induction refers to a situation that LTP can be induced at synapses in a pathway receiving weak stimulation when a strong stimulation is given to its neighboring pathway that may activate postsynaptic neurons [5]. This associativity of LTP induction can support the cooperativity among synapses, as well as granted by an observation that the association of postsynaptic neuron depolarization for NMDA receptor activation with stimuli to presynaptic axons [11, 72]. However, this associativity of LTP induction is different from the associative learning and memory of multiple signals based on the establishment of the newly mutual synapse innervations [14]. As specificity, cooperativity, and associativity are seen in memory formation, LTP induction and maintenance have been thought to be the neural substrates related to memory formation and consolidation. It is noteworthy that the associativity of LTP induction differs from a joint acquisition of associated signals, in which associative memory is characterized as independent pathways with coactivation in the intensity that is sufficient to activate reflexes, respectively [73, 74].

The correlation of LTP expression at synapses to memory formation is granted by the following experiment results. The blockade of LTP induction by pharmacological and genetic approaches can inhibit behavioral tasks that examine memory retrieval. The facilitation of LTP induction by pharmacological molecule activation and gene overexpression activation can strengthen behavioral tasks that show memory retrieval. This causal relationship between memory-related behavior and LTP induction grants a possibility that LTP is involved in learning and memory [5, 75, 76]. The following points should be taken into cautions. The expression of tetanus-induced LTP in a bundle of axons and their innervated synapses may be due to a decrease of their spiking threshold by this tetanus, instead of LTP expression at individual synapses. Once spike threshold decreases in some of bundle axons by their intensive activities, the intensity of original test stimulations may activate more axons and their innervated synapses to show the increase of synaptic responses [4]. If the strengthening of synaptic transmission expresses in a given neural pathway, this potentiation cannot explain the integrative storage of associated signals carried from their correspondent inputs [77]. Moreover, tetanus-induced LTP usually cannot express at synapses in unmatured animals [5], so that LTP is not relevant to learning and memory in early life when learning is efficient and memory is impressive [45].

There are various replicate models of synaptic potentiation that are eventually used to reveal mechanisms underlying long-term potentiation of synaptic transmission. The increase of extracellular calcium concentration for a few minutes is able to enhance the strength of synaptic responses for a long period, called as calcium-induced LTP [78]. This study supports the observation that calcium chelator inhibits LTP induction [68]. These two independent studies emphasize a causal relationship between LTP and intracellular calcium. Another study shows that the injection of calcium-calmodulin complex into postsynaptic neurons induces long-term enhancement of synaptic transmission, for which calmodulin-dependent protein kinases are required [79]. This result well matches studies that indicate the requirement of calmodulin- and calmodulin-dependent protein kinases for LTP induction and maintenance [16, 76, 80–83]. These independent studies show a causal relationship between LTP and intracellular calmodulin signaling in the side of calmodulin-dependent protein kinases. This point has been proved by the study that the inhibition of postsynaptic calmodulin-dependent phosphatase, or calcineurin, induces long-term potentiation of synaptic transmission [84]. In addition, the postsynaptic injection of protein kinase C [85] or its activator [86] induces synaptic potentiation. Together with the requirement of protein kinase C for LTP [43, 87, 88], there is a causal relationship between protein kinase C and LTP. A chain reaction from postsynaptic calcium, protein kinases (cAMP-dependent protein kinase, calcium-/calmodulin-dependent protein kinases, and calcium-/phospholipid-dependent protein kinase), to receptor phosphorylation is likely responsible for the long-term potentiation of synaptic transmission. A systemic review has presented the comprehensive diagram for the mechanisms underlying synaptic potentiation [3].

There are some debates about LTP induction versus maintenance and LTP expression in presynaptic versus postsynaptic locations [65, 89, 90]. Since the responses of the synapses to presynaptic action potentials rely on the interaction between transmitters and receptor, both presynaptic and postsynaptic components must be importantly required. For instance, the high volume of neurotransmitters from presynaptic terminals triggered by the tetanus activates postsynaptic receptors, especially NMDA type of glutamate receptor, intensively. The entry of calcium through NMDA receptor channels will elevate intracellular calcium and subsequently will activate calmodulin-dependent protein kinases and protein kinase C that phosphorylate glutamate receptors and strengthen receptor responsiveness to glutamates. In the meantime, calmodulin-activated retrograde messengers, such as arachidonic acid and nitro oxide [91–93], will trigger presynaptic terminals to release more neurotransmitters through a positive feedback process. In this regard, both presynaptic and postsynaptic mechanisms are involved in LTP induction and maintenance.

If tetanus-induced LTP is not reliable for individual synapses since numerous studies come by stimulating a bundle of axons and their innervated synapses, how are individual synapses functionally potentiated and recruited? Theoretically, the potentiation of signal transmission at unitary synapses may result from the following factors: the conversion of developmentally unmatured synapses into natured synapses, the conversion of silent synapses into functional synapses, the conversion of inactive synapses into active synapses, and the upregulated compatibility between presynaptic and postsynaptic partners. In postnatal development, the formation of glutamatergic synapses is featured by the emergence of NMDA receptors in the central area of postsynaptic density as well as the subsequent insertion of AMPA receptors in the peripheral of NMDA receptors [94–96]. The full formation of glutamatergic synapse includes the embedding of NMDA and AMPA receptors in postsynaptic density. These matured glutamatergic synapses are able to fulfill the functional plasticity at synapses. Even though both NMDA receptors and AMPA receptors are embedded in the postsynaptic density, their functions appear temporally to be different expression, in which AMPA receptors are activated by glutamates for fast synaptic transmission and the membrane depolarization assists the activation of NMDA receptors to mediate slow synaptic transmission [10]. This complete form of glutamatergic synapses demonstrates the feature about the conversion of silent synapses into functional synapses during synaptic plasticity, in which membrane depolarization activates NMDA receptors and subsequently AMPA receptors to have the glutamatergic synapses be converted from silence into function. In fact, even though synapses include both kinds of receptors, they remain to be inactive. Once calmodulin-dependent protein kinase is activated in postsynaptic neurons, these inactive synapses can be converted into active synapses (Fig. 4.2).

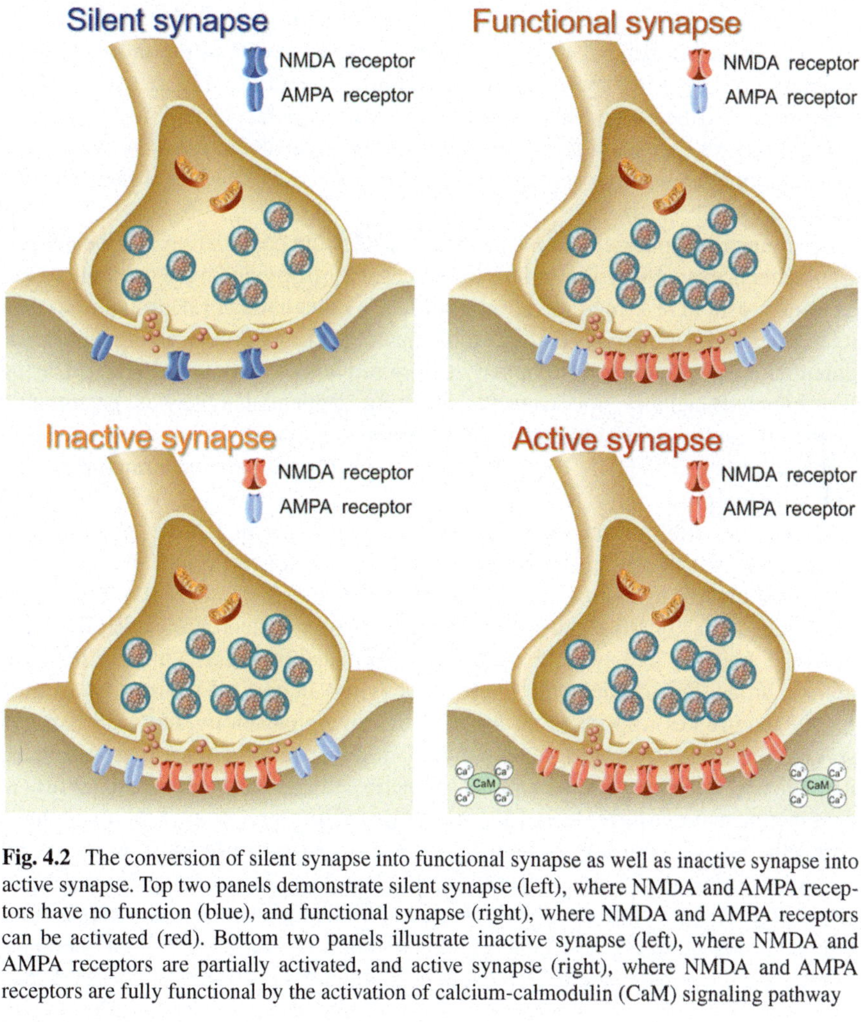

Fig. 4.2 The conversion of silent synapse into functional synapse as well as inactive synapse into active synapse. Top two panels demonstrate silent synapse (left), where NMDA and AMPA receptors have no function (blue), and functional synapse (right), where NMDA and AMPA receptors can be activated (red). Bottom two panels illustrate inactive synapse (left), where NMDA and AMPA receptors are partially activated, and active synapse (right), where NMDA and AMPA receptors are fully functional by the activation of calcium-calmodulin (CaM) signaling pathway

This process is matured during postnatal development [11]. Since this kinase plays a critical role in LTP induction and maintenance [16, 88], the conversion of inactive synapses into active synapses is a mechanism underlying synaptic potentiation. Finally, there is a functional compatibility between presynaptic and postsynaptic partners [24]. The upregulated compatibility between presynaptic and postsynaptic partners may be critical for the refinement of synapses.

4.4 Neuronal Plasticity

In addition to synaptic plasticity, the neuronal plasticity has been thought to be associated with learning and memory [97–101]. Neuronal plasticity includes functional and morphological refinements. The functional plasticity refers to a change of neuronal excitability. The structural plasticity mainly points to the morphological alternation of dendritic spines since the cell body, dendrites, and axonal branches have been well developed in the age of maturation [102]. In addition, there is a coordinated plasticity among the different subcellular compartments, which can maintain functional homeostasis in spike-encoding capacity and the increased spike precision [15].

The functional plasticity at the neurons includes the potentiation and the depression of their excitability that refers to the capacity of producing sequential spikes and the threshold of firing individual spikes. The potentiation of neuronal excitability is featured by the decrease of threshold potential closely to the resting membrane potential and the increase of capacity to fire sequential spikes [102–104]. During learning and memory, the capacity of firing sequential spikes is presumably increased, and the threshold potential of firing spikes is probably decreased at associative memory cells [105, 106]. The author aims to summarize neuronal potentiation in excitability.

The potentiation of neuronal excitability includes the increase of capability to fire sequential spikes and the decrease of threshold potential that initiate action potentials closely to the resting membrane potential [12, 15, 102–104]. There are two forms of factors that can induce the potentiation of neuronal excitability, i.e., neuronal activities and synaptic activities coordinated by presynaptic and postsynaptic neurons. For instance, the intensive activities in cortical neurons are able to induce themselves to possess an increase of spiking capability and a decrease of spiking threshold in a multigrade manner [4]. Synapse activations by coordinated activities between presynaptic and postsynaptic neurons enhance the excitability at presynaptic neurons [107, 108]. Moreover, the stronger activation of synapses in the cerebellar nuclear neurons can induce themselves to express the increases of intrinsic excitability and spiking patterns [13, 15, 109–111]. It is noteworthy that these forms of neuronal functional plasticity are not studied in memory-relevant neurons, which should be done based on the identification of memory cells. If memory-relevant cells express a potentiated excitability, memory retrievals will be facilitated by cues and/or in spontaneous manner. The upregulated excitability may also facilitate the coactivation of ready-recruited neurons for the conversion of the neurons into memory-relevant cells [77].

In addition, neuronal excitability can be potentiated in compensatory manner after the inhibition of voltage-gated sodium channels for 48 h is removed [112] or the inhibition of synaptic receptors is washed out [113–115]. These results indicate that the activity of neurons undergoes the potentiation that is induced by their intensive activity or driven by the strong activity of synapses. As these studies are conducted in cultural neurons, it is never known whether these neurons are memory-relevant cells. Currently, the cortical and hippocampal memory cells in vivo are functionally upregulated by input cues learned during the acquisition of associated signals [63]. Moreover, the potentiation of synapse function and neuron excitability has been detected in associative memory cells in vitro [105, 106, 116–119].

There is an important structure on the dendrites of neurons, called as spines that are presumable locations for the termination of presynaptic axon boutons. In other words, synapses are formed on dendritic spines. Numerous studies have indicated that the density and size of spines are upregulated in correspondence to synaptic plasticity by processes relevant to learning and memory [120–135]. In addition, the function of spines is upregulated in an experience-dependent manner [136–138]. On the other hand, the extinction procedure induces the pruning of dendritic spines in CA1 hippocampal field depending on strength of training in rats [139]. These studies led to a conclusion that new spines are recruited and upregulated for new memory [140]. It is pointed that the spine density in associative memory cells appears not increased though their size is enlarged [106, 118] and that the elimination of dendritic spines with long-term memory is specific to active circuits [141]. These data indicate the enlargement of spine size from filopodia to mushroom shapes plays an important role in memory formation. As the number and size of dendritic spines in ipsilateral barrel field cortex following unilateral whisker trimming in postnatal rat increased [142], dendritic spines are also regulated by the strength of sensory inputs in an inverse manner. Therefore, the number of spines may be responsible for raising the sensitivity to sensory inputs, and the size of spines may be responsible for raising the formation of functional synapses to ensure the consolidation storage of learned signals [45].

Neuronal plasticity may occur among different subcellular compartments in a coordinated manner [15]. For instance, the increase of intracellular calcium level by releasing calcium from intracellular stores can upregulate the signal transmission strength of both glutamatergic and GABAergic synapses, in spite of dominant potentiation in glutamatergic synapses. Moreover, this increased intracellular calcium attenuates the capability of firing sequential spikes. It is likely that the dominant increase of glutamatergic synapse activity, the partial increase of GABAergic synapses and the decrease of neuronal excitability make these compartments to be coordinated, such that neuronal spike encoding can be stabilized, or homeostasis among subcellular compartments. In fact, this homeostatic plasticity among subcellular compartments grants the precise encoding of neuronal sequential spikes [15]. If this homeostasis is not sufficient for neuronal encoding and signal outputs, an inverse feature of somatic plasticity and axonal plasticity will further ensure homeostasis in this neuron [15]. In terms of molecular mechanisms underlying this homeostatic plasticity among different subcellular compartments, intracellular calcium, calmodulin-dependent protein kinase, and protein kinase C are involved [15].

The plasticity of neuronal excitability and encoding capability can occur in presynaptic and postsynaptic partners in a coordinated manner. For instance, neurons may regulate themselves to be functionally compatible for presynaptic and postsynaptic partners if the interconnected neurons are initially not coordinated in their functional states. In addition to the functional compatibility between transmitter release from presynaptic axonal boutons and receptor responsiveness in postsynaptic densities, the functional compatibility has been observed in the excitability and spike encodings in presynaptic axonal branches and their postsynaptic partners [24] (Fig. 4.1). This functional compatibility between presynaptic axonal branches and their postsynaptic partners plays a critical role in the homeostasis of neural networks, instead of over-excitation or silence [24].

It is noteworthy that the learning and memory may be accompanied by synaptic and neuronal plasticity, but synaptic and neuronal may not indicate the occurrence of memory formation. As we have known, numerous physiological activities, such as compensational processes, memory, and various stressful conditions, are associated with neuronal and synaptic plasticity [106, 118, 143–146]. Moreover, certain pathological conditions are associated with synaptic and neuronal plasticity, such as major depression, anxiety, and epilepsy [147–151]. Scientists should not consider neural plasticity to be the formation of memories to specific signals, especially associated signals. The biggest effort should be made to quantify and identify the neural plasticity specifically responsible for learning and memory. If this duty is thought to be difficult, the experiments for studying synaptic and neuronal plasticity should be conducted at memory cells identified.

References

1. Bliss T, Lomo T. Long-lasting potentiation of synaptic transmission in the dentate area of the anaesthetized rabbit following stimulation of the perforant path. J Physiol Lond. 1973;232:331–56.
2. Stanton PK, Sejnowski TJ. Associative long-term depression in the hippocampus induced by hebbian covariance. Nature. 1989;339(6221):215–8.
3. Wang JH, Ko G, Kelly PT. Cellular and molecular bases of memory: synaptic and neuronal plasticity. J Clin Neurophysiol. 1997;14:264–93.
4. Zhang M, et al. Calcium signal-dependent plasticity of neuronal excitability developed postnatally. J Neurobiol. 2004;61:277–87.
5. Bliss TVP, Lynch MA. Long-term potentiation of synaptic transmission in the hippocampus: properties and mechanisms. In: Landfield PW, Deadwyler SA, editors. Long-term potentiation: from biophysics to behavior. New York: Alan R. Liss; 1988. p. 3–72.
6. Nicoll RA, Malenka RC. Contrasting properties of two forms of long-term potentiation in the hippocampus. Nature. 1995;377:115–8.
7. Poo MM, et al. What is memory? The present state of the engram. BMC Biol. 2016;14:40.

8. Zucker RS. Short-term synaptic plasticity. Ann Rev Neurosci. 1989;12:13–31.
9. Zucker RS, Regehr WG. Short-term synaptic plasticity. Annu Rev Physiol. 2002;25:355–405.
10. Liao D-Z, Hessler NA, Malinow R. Activation of postsynaptic silent synapses during pairing-induced LTP in CA1 region of hippocampal slice. Nature. 1995;375:400–4.
11. Wang J-H, Kelly PT. Ca2+/CaM signalling pathway up-regulates glutamatergic synaptic function in non-pyramidal fast-spiking neurons of hippocampal CA1. J Physiol Lond. 2001;533(2):407–22.
12. Wang JH, et al. The gain and fidelity of transmission patterns at cortical excitatory unitary synapses improve spike encoding. J Cell Sci. 2008;121(17):2951–60.
13. Daoudal D, Debanne D. Long-term plasticity of intrinsic excitability: learning rules and mechanisms. Learn Mem. 2003;10:456–65.
14. Wang JH, Cui S. Associative memory cells: formation, function and perspective. F1000Res. 2017;6:283.
15. Chen N, Chen X, Wang J-H. Homeostasis established by coordination of subcellular compartment plasticity improves spike encoding. J Cell Sci. 2008;121(17):2961–71.
16. Wang J-H, Kelly PT. Balance between postsynaptic Ca2+–dependent protein kinase and phosphatase activities controlling synaptic strength. Learn Mem. 1996;3:170–81.
17. Epsztein J, et al. Impact of spikelets on hippocampal CA1 pyramidal cell activity during spatial exploration. Science. 2010;327:474–7.
18. Chen N, et al. Axons amplify somatic incomplete spikes into uniform amplitudes in mouse cortical pyramidal neurons. PLoS One. 2010;5(7):e11868.
19. Bucher D, Goaillard JM. Beyond faithful conduction: short-term dynamics, neuromodulation, and long-term regulation of spike propagation in the axon. Prog Neurobiol. 2011;94(4):307–46.
20. Debanne D, et al. Axon physiology. Physiol Rev. 2011;91(2):555–602.
21. Yang Z, et al. Essential role of axonal VGSC inactivation in time-dependent deceleration and unreliability of spike propagation at cerebellar Purkinje cells. Mol Brain. 2014;7:1.
22. Yang Z, Wang JH. Frequency-dependent reliability of spike propagation is function of axonal voltage-gated sodium channels in cerebellar Purkinje cells. Cerebellum. 2013;12(6):862–9.
23. Wang JH, et al. Functional compatibility between Purkinje cell axon branches and their target neurons in the cerebellum. Biophys J. 2013;104(2):330a.
24. Yang Z, et al. Functional compatibility between Purkinje cell axon branches and their target neurons in the cerebellum. Oncotarget. 2017;8(42):72424–37.
25. Dubnau J, Chiang AS, Tully T. Neural substrates of memory: from synapse to system. J Neurobiol. 2003;54(1):238–53.
26. Kelso SR, Ganong AH, Brown TH. Hebbian synapses in hippocampus. Proc Natl Acad Sci U S A. 1986;83:5326–30.
27. Lee SH, Sheng M. Development of neuron-neuron synapses. Curr Opin Neurobiol. 2000;10(1):125–31.
28. Lichtman JW, Colman H. Synapse elimination and indelible memory. Neuron. 2000;25(2):269–78.
29. London M, et al. The information efficacy of a synapse. Nat Neurosci. 2002;5(4):332–40.
30. Mayford M, Siegelbaum SA, Kandel ER. Synapses and memory storage. Cold Spring Harb Perspect Biol. 2012;4(6):a005710.
31. Stevens CF, Wang Y-Y. Facilitation and depression at single central synapses. Neuron. 1995;14:795–802.
32. Borst A, Theunissen FE. Information theory and neural coding. Nat Neurosci. 1999;2(11):947–57.
33. Shepherd GM. Synaptic potentials and synaptic integration. In: Shepherd GM, editor. Neurobiology. New York: Oxford University Press; 1998. p. 97–117.
34. Borst JG, Sakmann B. Calcium influx and transmitter release in a fast CNS synapse. Nature. 1996;383(6599):431–4.
35. Borst JG. The low synaptic release probability in vivo. Trends Neurosci. 2010;33(6):259–66.

36. Sudhof T. The synaptic vesicle cycle: a cascade of protein-protein interactions. Nature. 1995;375:645–53.
37. Sudhof TC. The synaptic vesicle cycle revisited. Neuron. 2000;28:317–20.
38. Bito H, Deisseroth K, Tsien RW. CREB phosphorylation and dephosphorylation: a Ca2+– and stimulus duration-dependent switch for hippocampal gene expression. Cell. 1996;87:1203–14.
39. Kandel ER, Siegelbaum SA, Schwartz JH. Elementary interactions between neurons: synaptic transmission. In: Kandel ER, Schwartz JH, Jessell TM, editors. Principles of neural science. New York: McGraw-Hill; 2000. p. 175–308.
40. Kandel ER. The molecular biology of memory: cAMP, PKA, CRE, CREB-1, CREB-2, and CPEB. Mol Brain. 2012;5(1):14.
41. Josselyn SA, Nguyen PV. CREB, synapses and memory disorders: past progress and future challenges. Curr Drug Targets CNS Neurol Disord. 2005;4(5):481–97.
42. Sheng M, Thompson MA, Greenberg ME. CREB: a calcium-regulated transcription factor phosphorylated by calmodulin-dependent kinase. Science. 1991;252:1427–30.
43. Wang JH, Feng DP. Postsynaptic protein kinase C essential to induction and maintenance of long-term potentiation in the hippocampal CA1 region. Proc Natl Acad Sci U S A. 1992;89(7):2576–80.
44. Wang JH, W J. The regulation of unitary synaptic responses to multipulse inputs. Soc Neurosci Abstr. 2001. 501.1.
45. Wang J-H. Searching basic units of memory traces: associative memory cells. F1000Res. 2019;8(457):1–28.
46. Sakmann B, Neher E. Single-channel recording. New York: Plenum Publishing Corp; 1983. p. 503.
47. Hodgkin AL, Huxley AF. The dual effect of membrane potential on sodium conductance in the giant axon of Loligo. J Physiol. 1952;116(4):497–506.
48. Hodgkin AL, Huxley AF. A quantitative description of membrane current and its application to conduction and excitation in nerve. J Physiol. 1952;117(4):500–44.
49. Hodgkin A. The optimum density of sodium channels in an unmyelinated nerve. Philos Trans R Soc Lond Ser B Biol Sci. 1975;270(908):297–300.
50. Huxley AF, Stampfli R. Evidence for saltatory conduction in peripheral myelinated nerve fibres. J Physiol. 1949;108(3):315–39.
51. Johnston D, Wu SM-S. Functional diversity of voltage-gated conductances. In: Johnston D, Wu SM-S, editors. Foundations of cellular neurophysiology. Cambridge, MA/London: MIT Press; 1995. p. 183–214.
52. Linden DJ. The return of the spike: postsynaptic action potentials and the induction of LTP and LTD. Neuron. 1999;22(4):661–6.
53. Trussell LO, Fischbach GD. Glutamate receptor desensitization and its role in synaptic transmission. Neuron. 1989;3(2):209–18.
54. Feng TP. Looking back, looking forward. Annu Rev Neurosci. 1988;11:1–12.
55. Zucker RS. The calcium hypothesis and modulation of transmitter release by hyperpolarizing pulses. Biophys J. 1987;52:347–50.
56. Wang J-H, Kelly PT. Regulation of synaptic facilitation by postsynaptic Ca2+/CaM pathways in hippocampal CA1 neurons. J Neurophysiol. 1996;76(1):276–86.
57. Wang J-H, Kelly PT. Attenuation of paired-pulse facilitation associated with synaptic potentiation mediated by postsynaptic mechanisms. J Neurophysiol. 1997;78:2707–16.
58. Reyes A, et al. Target-cell-specific facilitation and depression in neocortical circuits. Nat Neurosci. 1998;1:279–85.
59. Reyes A, Sakmann B. Developmental switch in the short-term modification of unitary EPSPs evoked in layer 2/3 and layer 5 pyramidal neurons of rat neocortex. J Neurosci. 1999;15:3827–35.
60. Charlton MP, Smith SJ, Zucker RS. Role of presynaptic calcium ions and channels in synaptic facilitation and depression at the squid giant synapse. J Physiol Lond. 1982;323:173–93.

61. Katz B, Miledi R. The timing of calcium action during neuromuscular transmission. J Physiol Lond. 1967;189:535–44.
62. Muller D, et al. Phorbol ester-induced synaptic facilitation is different than long-term potentiation. Proc Natl Acad Sci U S A. 1988;85(18):6997–7000.
63. Wang JH, et al. Secondary associative memory cells and their plasticity in the prefrontal cortex. Biophys J. 2019;116(3):427a.
64. Bliss TVP, Collingridge GL. A synaptic model of memory: LTP in the hippocampus. Nature. 1993;361:31–9.
65. Malenka RC, Nicoll RA. Long-term potentiation – a decade of progress? Nat Neurosci. 1999;285:1870–4.
66. Bliss T, Gardner-Medwin A, Lømo T. Synaptic plasticity in the hippocampal formation. In: Ansell G, Bradley P, editors. Macromolecules and behaviour. London: Macmillan; 1973. p. 193–203.
67. Lynch G, Dunwiddie T, Gribkoff V. Heterosynaptic depression: a postsynaptic correlate of long-term potentiation. Nature. 1977;266:737–9.
68. Lynch G, et al. Intracellular injections of EGTA block induction of hippocampal long-term potentiation. Nature. 1983;305:719–21.
69. Hebb DO. The organization of behavior, a neuropsychological theory. New York: Wiley; 1949.
70. Hebb DO. Animal and physiological psychology. Annu Rev Psychol. 1950;1:173–88.
71. Chavez NLE, Halliwell JV, Bliss TV. A decrease in firing threshold observed after induction of the EPSP-spike (E-S) component of long-term potentiation in rat hippocampal slices. Exp Brain Res. 1990;79(3):633–41.
72. Collingridge GL, et al. Involvement of excitatory amino acid receptors in long-term potentiation in the Schaffer collateral-commissural pathway of rat hippocampal slices. Can J Physiol Pharmacol. 1991;69(7):1084–90.
73. Feng J, et al. Barrel cortical neuron integrates triple associated signals for their memory through receiving epigenetic-mediated new synapse innervations. Cereb Cortex. 2017;27(12):5858–71.
74. Wang D, et al. Neurons in the barrel cortex turn into processing whisker and odor signals: a cellular mechanism for the storage and retrieval of associative signals. Front Cell Neurosci. 2015;9:320.
75. Silva AJ, et al. Impaired spatial learning in alpha-calcium-calmodulin kinase II mutant mice. Science. 1992;257:206–11.
76. Silva AJ, et al. Deficient hippocampal long-term potentiation in alpha-calcium-calmodulin kinase II mutant mice. Science. 1992;257:201–6.
77. Wang JH, Cui S. Associative memory cells and their working principle in the brain. F1000Res. 2018;7:108.
78. Reymann KG, et al. Calcium-induced long-term potentiation in the hippocampal slice: characterization of the time course and conditions. Brain Res Bull. 1986;17(3):291–6.
79. Wang JH, Kelly PT. Postsynaptic injection of Ca2+/CaM induces synaptic potentiation requiring CaM-KII and PKC activity. Neuron. 1995;15(2):443–52.
80. Malenka RC, et al. An essential role for postsynaptic calmodulin and protein kinase activity in long-term potentiation. Nature. 1989;340(6234):554–7.
81. Nayak AS, Morre CI, Browning MD. Ca2+/calmodulin-dependent protein kinase II phosphorylation of the presynaptic protein synapsin I is persistently increased during long-term potentiation. Proc Natl Acad Sci U S A. 1996;93:15451–6.
82. Popov NS, et al. Alterations in calmodulin content in fractions of rat hippocampal slices during tetanic- and calcium-induced long-term potentiation. Brain Res Bull. 1988;21(2):201–6.
83. Reymann KG, et al. Inhibitors of calmodulin and protein kinase C block different phases of hippocampal long-term potentiation. Brain Res. 1988;461(2):388–92.

84. Wang J-H, Kelly PT. Postsynaptic calcineurin activity down-regulates synaptic transmission by weakening intracellular Ca2+ signaling mechanisms in hippocampal CA1 neurons. J Neurosci. 1997;17:4600–11.
85. Hu GY, et al. PKC injection into hippocampal pyramidal cells elicits features of long term potentiation. Nature. 1987;328(6129):426–9.
86. Chen HX, Wang JH. Potentiation of synaptic transmission induced by daicylglycerol/phosphatidylserine and its mutual occlusion with long-term potentiation in rat hippocampal slice. Chin J Physiol Sci. 1995;11:97–102.
87. Malinow R, Madison DV, Tsien RW. Persistent protein kinase activity underlying long-term potentiation. Nature. 1988;335(6193):820–4.
88. Malinow R, Schulman H, Tsien RW. Inhibition of postsynaptic PKC or CaMKII blocks induction but not expression of LTP. Science. 1989;245:862–6.
89. Malinow R, Tsien RW. Long-term potentiation: postsynaptic activation of Ca(2+)-dependent protein kinases with subsequent presynaptic enhancement. Prog Brain Res. 1991;89:271–89.
90. Nicoll RA, Kauer JA, Malenka RC. The current excitement in long-term potentiation. Neuron. 1988;1:93–103.
91. Clements MP, Bliss TV, Lynch MA. Increase in arachidonic acid concentration in a postsynaptic membrane fraction following the induction of long-term potentiation in the dentate gyrus. Neuroscience. 1991;45(2):379–89.
92. Errington ML, et al. The nitric oxide synthase inhibitor N^{ω}–nitro-L-arginine reduces the magnitude of long-term potentiation in the dentate gyrus but not in area CA1 of the hippocampus in vitro. Soc Neurosci. 1991;76:387–95.
93. Lynch MA, et al. Is arachidonic acid a retrograde messenger in long-term potentiation? Biochem Soc Trans. 1991;19(2):391–6.
94. Liao D, et al. Regulation of morphological postsynaptic silent synapses in developing hippocampal neurons. Nat Neurosci. 1999;2:37–43.
95. Petralia RS, et al. Selective acquisition of AMPA receptors over postnatal development suggests a molecular basis for silent synapses. Nat Neurosci. 1999;2:31–6.
96. Takumi Y, et al. Different modes of expression of AMPA and NMDA receptors in hippocampal synapses. Nat Neurosci. 1999;2:618–24.
97. Blank T, Nijholt I, Spiess J. Treatment strategies of age-related memory dysfunction by modulation of neuronal plasticity. Mini Rev Med Chem. 2007;7(1):55–64.
98. Byrne JH, et al. Roles of second messenger pathways in neuronal plasticity and in learning and memory. Adv Second Messenger Phosphoprotein Res. 1993;27:47–108.
99. Kemenes I, et al. Role of delayed nonsynaptic neuronal plasticity in long-term associative memory. Curr Biol. 2006;16(13):1269–79.
100. Teyler TJ, Fountain SB. Neuronal plasticity in the mammalian brain: relevance to behavioral learning and memory. Child Dev. 1987;58(3):698–712.
101. Turrigiano GG, Nelson SB. Hebb and homeostasis in neuronal plasticity. Curr Opin Neurobiol. 2000;10(3):358–64.
102. Chen N, et al. Sodium channel-mediated intrinsic mechanisms underlying the differences of spike programming among GABAergic neurons. Biochem Biophys Res Commun. 2006;346:281–7.
103. Chen N, et al. The refractory periods and threshold potentials of sequential spikes measured by whole-cell recordings. Biochem Biophys Res Commun. 2006;340:151–7.
104. Chen N, et al. After-hyperpolarization improves spike programming through lowering threshold potentials and refractory periods mediated by voltage-gated sodium channels. Biochem Biophys Res Commun. 2006;346:938–45.
105. Liu Y, et al. Piriform cortical glutamatergic and GABAergic neurons express coordinated plasticity for whisker-induced odor recall. Oncotarget. 2017;8(56):95719–40.
106. Yan F, et al. Coordinated plasticity between barrel cortical glutamatergic and GABAergic neurons during associative memory. Neural Plast. 2016;2016(ID5648390):1–20.
107. Ganguly K, Kiss L, Poo M-M. Enhancement of presynaptic neuronal excitability by correlated presynaptic and postsynaptic spiking. Nat Neurosci. 2000;3(10):1018–26.

108. Nick TA, Ribera AB. Synaptic activity modulates presynaptic excitability. Nat Neurosci. 2000;3(2):142–9.
109. Aizenmann C, Linden DJ. Rapid, synaptically driven increases in the intrinsic excitability of cerebellar nuclear neurons. Nat Neurosci. 2000;3:109–11.
110. Campanac E, Debanne D. Plasticity of neuronal excitability: Hebbian rules beyond the synapse. Arch Ital Biol. 2007;145(3–4):277–87.
111. Sourdet V, et al. Long-term enhancement of neuronal excitability and temporal fidelity mediated by metabotropic glutamate receptor subtype 5. J Neurosci. 2003;23(32):10238–48.
112. Desai NS, Rutherford L, Turrigiano GG. Plasticity in the intrinsic excitability of cortical pyramidal neurons. Nat Neurosci. 1999;2(6):515–20.
113. Burrone J, O'Byrne M, Murthy VN. Multiple forms of synaptic plasticity triggered by selective suppression of activity in individual neurons. Nature. 2002;420:414–8.
114. Ramakers GJ, Corner MA, Habers AM. Development in the absence of spontaneous bioelectric activity results in increased stereotyped burst firing in cultures of associated cerebral cortex. Exp Brain Res. 1990;79:157–66.
115. Van Den Pol AN, Obrietan K, Belousov A. Glutamate hyperexcitability and seizure-like activity throughout the brain and spinal cord upon relief from chronic glutamate receptor blockade in culture. Neuroscience. 1996;74:653–74.
116. Gao Z, et al. Associations of unilateral whisker and olfactory signals induce synapse formation and memory cell recruitment in bilateral barrel cortices: cellular mechanism for unilateral training toward bilateral memory. Front Cell Neurosci. 2016;10(285):1–16.
117. Guo R, et al. Associative memory extinction is accompanied by decayed plasticity at motor cortical neurons and persistent plasticity at sensory cortical neurons. Front Cell Neurosci. 2017;11(168):1–12.
118. Liu Y, et al. Activity strengths of cortical glutamatergic and GABAergic neurons are correlated with transgenerational inheritance of learning ability. Oncotarget. 2017;8(68):112401–16.
119. Zhao X, et al. Coordinated plasticity among glutamatergic and GABAergic neurons and synapses in the barrel cortex is correlated to learning efficiency. Front Cell Neurosci. 2017;11(221):1–12.
120. Ashby MC, Isaac JT. Maturation of a recurrent excitatory neocortical circuit by experience-dependent unsilencing of newly formed dendritic spines. Neuron. 2011;70(3):510–21.
121. Bourne J, Harris KM. Do thin spines learn to be mushroom spines that remember? Curr Opin Neurobiol. 2007;17(3):381–6.
122. Fifkova E, Van Harreveld A. Long-lasting morphological changes in dendritic spines of dentate granular cells following stimulation of the entorhinal area. J Neurocytol. 1977;6:211–30.
123. Fortin DA, Srivastava T, Soderling TR. Structural modulation of dendritic spines during synaptic plasticity. Neuroscientist. 2011;18(4):326–41.
124. Hongpaisan J, Alkon DL. A structural basis for enhancement of long-term associative memory in single dendritic spines regulated by PKC. Proc Natl Acad Sci U S A. 2007;104(49):19571–6.
125. Kasai H, et al. Structural dynamics of dendritic spines in memory and cognition. Trends Neurosci. 2010;33(3):121–9.
126. Kitanishi T, et al. Experience-dependent, rapid structural changes in hippocampal pyramidal cell spines. Cereb Cortex. 2009;19(11):2572–8.
127. Lendvai B, et al. Experience-dependent plasticity of dendritic spines in the developing rat barrel cortex in vivo. Nature. 2000;404:876–81.
128. Leuner B, Falduto J, Shors TJ. Associative memory formation increases the observation of dendritic spines in the hippocampus. J Neurosci. 2003;23(2):659–65.
129. Matsuzaki M, et al. Structural basis of long-term potentiation in single dendritic spines. Nature. 2004;429(6993):761–6.
130. Moczulska KE, et al. Dynamics of dendritic spines in the mouse auditory cortex during memory formation and memory recall. Proc Natl Acad Sci U S A. 2013;110(45):18315–20.

131. Murakoshi H, Yasuda R. Postsynaptic signaling during plasticity of dendritic spines. Trends Neurosci. 2012;35(2):135–43.
132. Zuo Y, Perkon I, Diamond ME. Whisking and whisker kinematics during a texture classification task. Philos Trans R Soc Lond Ser B Biol Sci. 2011;366(1581):3058–69.
133. Yuste R, Denk W. Dendritic spines as basic functional units of neuronal integration. Nature. 1995;375:682–4.
134. Patterson M, Yasuda R. Signalling pathways underlying structural plasticity of dendritic spines. Br J Pharmacol. 2011;163(8):1626–38.
135. Tanaka J, et al. Protein synthesis and neurotrophin-dependent structural plasticity of single dendritic spines. Science. 2008;319(5870):1683–7.
136. Harnett MT, et al. Synaptic amplification by dendritic spines enhances input cooperativity. Nature. 2012;491(7425):599–602.
137. Jung CK, Herms J. Structural dynamics of dendritic spines are influenced by an environmental enrichment: an in vivo imaging study. Cereb Cortex. 2014;24(2):377–84.
138. Schiegg A, Gerstner W, Ritz R, Leo van Hemmen J. Intracellular Ca2+ stores can account for the time course of LTP induction: a model of Ca2+ dynamics in dendritic spines. J Neurophysiol. 1995;74(3):1046–55.
139. Garin-Aguilar ME, et al. Extinction procedure induces pruning of dendritic spines in CA1 hippocampal field depending on strength of training in rats. Front Behav Neurosci. 2012;6:12.
140. Leuner B, Shors TJ. New spines, new memories. Mol Neurobiol. 2004;29(2):117–30.
141. Sanders J, et al. Elimination of dendritic spines with long-term memory is specific to active circuits. J Neurosci. 2012;32(36):12570–8.
142. Vees AM, et al. Increased number and size of dendritic spines in ipsilateral barrel field cortex following unilateral whisker trimming in postnatal rat. J Comp Neurol. 1998;400(1):110–24.
143. Ni H, et al. Upregulation of barrel GABAergic neurons is associated with cross-modal plasticity in olfactory deficit. PLoS One. 2010;5(10):e13736.
144. Wang GY, et al. Glucocorticoid induces incoordination between glutamatergic and GABAergic neurons in the amygdala. PLoS One. 2016;11(11):e0166535.
145. Ye B, et al. The functional upregulation of piriform cortex is associated with cross-modal plasticity in loss of whisker tactile inputs. PLoS One. 2012;7(8):e41986.
146. Zhu Z, et al. GABAergic neurons in nucleus accumbens are correlated to resilience and vulnerability to chronic stress for major depression. Oncotarget. 2017;8(22):35933–45.
147. Liu B, Feng J, Wang J-H. Protein kinase C is essential for kainate-induced anxiety-related behavior and glutamatergic synapse upregulation in prelimbic cortex. CNS Neurosci Ther. 2014;20:982–90.
148. Ma K, et al. Impaired GABA synthesis, uptake and release are associated with depression-like behaviors induced by chronic mild stress. Transl Psychiatry. 2016;6(e910):1–10.
149. Wen B, et al. A portion of inhibitory neurons in human temporal lobe epilepsy are functionally upregulated: an endogenous mechanism for seizure termination. CNS Neurosci Ther. 2015;21(2):204–14.
150. Xu A, Cui S, Wang J. Incoordination among subcellular compartments is associated to depression-like behavior induced by chronic mild stress. Int J Neuropsychopharmacol. 2015;19(5):pyv122.
151. Zhang F, et al. mGluR1,5 activation improves network asynchrony and GABAergic synapse attenuation in the amygdala: implication for anxiety-like behavior in DBA/2 mice. Mol Brain. 2012;5(1):20.

Chapter 5
Associative Memory Cells in Memory Trace

Abstract Associative memory is characterized as the integrative storages and the reciprocal retrievals of associated signals after their joint acquisitions. Basic units in memory traces should be morphological and functional identities to work for memory-relevant processes. Beyond the concepts of memory traces, engrams, cell assemblies, and neural plasticity, associative memory cells have been functionally and structurally identified, which encode the integrative storage and reciprocal retrieval of associated signals as well as receive synapse innervation from coactivated brain areas. Associative memory cells are identified as primary associative memory cells in sensory cortices to memorize exogenous signals and secondary associative memory cells in sensory downstream brain areas, such as the prefrontal cortex, amygdala, and hippocampus, to memorize endogenous signals generated from cognitive events and emotional reactions. Based on experimental data, these associative memory cells have the following characters. They receive synapse innervations from the coactivated brain areas in a reciprocal manner. They are able to encode the associated signals acquired in associative learning and generated in integrative cognitions. Their encoded signals and received synapse innervations come from cross-modal and intramodal sources. Their axons innervate their downstream brain regions in convergent and divergent manners. Associative memory cells and their mediated memory formation are influenced by the chain reaction including neuronal activation, epigenetic process, and the expressions of genes and proteins in relevance to axon prolongation and synapse formation. Working principles of associative memory cells are based on their reception of synapse innervations from multiple sources and mutual synaptic innervations by the coactivations, as well as neuronal encoding capability and synaptic transmission efficacy. Moreover, their functional states are modulated by the arousal system that release monoamine and acetylcholine as well as by hormones. These associative memory cells constitute the foundations of memory-related physiological and psychological processes as well as memory deficits in pathology.

Keywords Associative memory cells · Synapse · Neuron · Cognition and emotion

J.-H. Wang, *Associative Memory Cells: Basic Units of Memory Trace*,
https://doi.org/10.1007/978-981-13-9501-7_5

5.1 Introduction

Associative learning refers to the joint acquisition of multiple exogenous knowledge and experiences by sensory systems. Associative memory stands for the integrative storage of these associated signals in the brain as well as their reciprocal retrievals through recall, representation, and recollection [1–4]. Learning and memory contents include new unitary signals acquired in early life and the complicated signals organized from these unitary signals by systemic education [5]. In terms of signal sources to be learned and memorized, the associated exogenous signals come from sensory organs and reside into sensory cortices in cross-modal and intramodal styles, and the associated endogenous signals originate from cognitive events, emotional reactions, and other rehearsals that are stored in brain areas related to cognition and emotion [6, 7]. Therefore, basic units in memory trace should be able to encode these associated signals, either unitary or reorganized unitary signals, for their integrative storage and reciprocal retrievals as well as to receive synapse innervations that carry these signals for their storages [6]. Recently, a mouse model of associative learning, which is characterized by the integrative storages and reciprocal retrievals of associated signals, has been developed. In the meantime, associative memory cells in the sensory cortices and cognition-relevant brain areas are identified to fulfill these functions [3, 8, 9].

The pairings of unitary sensory signals, e.g., whisker stimulus and olfactory stimulus, lead mice to express odorant-induced whisker motion and whisking-induced olfactory responses [3, 8–10]. Some neurons in the barrel cortex and the piriform cortex become able to encode both whisker and olfactory signals, identified by electrophysiological recording and two-photon cell imaging in vivo. These barrel and piriform cortical neurons are mutually innervated, observed by neural tracing [3, 8–10]. These data indicate that the coactivations of sensory cortical neurons evoke their axon projection and mutual synapse innervations, in addition to receiving innate synaptic input. These sensory cortical neurons become able to encode the associated signals. So, primary associative memory cells are recruited in sensory cortices that are characterized to be receiving synapse innervations from the coactivated sensory cortices besides innate inputs as well as be encoding the integrative storage of associated signals. Moreover, primary associative memory cells that encode associated whisker, olfactory, and tail-tactile signals are recruited by pairing these three stimulations, which strengthen the concept of associative memory cells [11]. In terms of molecular mechanisms for the recruitment of primary associative memory cells, microRNA-324/-133, tet3, and ttbk1 are required. A chain reaction including neuronal activation, epigenetic process as well as genes and proteins relevant to axon prolongation and synapse formation is involved in the recruitment of associative memory cells [11, 12].

Once mutual synapse innervations are established between two groups of associative memory cells among sensory cortices, each group of associative memory cells is able to encode the innate signal and the new signal from cross-modal associative memory cells for their integrative storage. In other words, innate signal cue and cross-modal signal cue are potentially to activate each group of associative memory cells. Each signal cue is able to activate these two groups of associative memory cells. Based on their natural pathways for the reflexes or behavioral responses to cues, the activation of each group of associative memory cells is able to induce these two behavioral activities [3, 8–10]. This mechanism can be applied for the associative memory cells that encode multiple signals.

Meanwhile, primary associative memory cells project their axons and make synapse innervations onto their downstream neurons in brain regions relevant to cognition, emotion, and behaviors for the conversion of exogenous signals into endogenous signals for the cognitions and emotions [7, 13, 14]. The neurons that receive convergent synapse innervations from primary associative memory cells are recruited to be secondary associative memory cells. These secondary associative memory cells are responsible for conducting logical reasoning, associative thinking, and imagination as well as for memorizing associated endogenous signals generated during cognitive events, emotion reactions, and other rehearsals [5]. Moreover, the divergent synapse innervations from primary associative memory cells onto secondary associative memory cells confer the storage of three associated signals in multiple places to long-term maintenance as well as efficient retrieval [5].

5.2 Associative Memory Cells in Theory and Experiments

The stores to memorize the learnt information and knowledge have been proposed to be various terms, such as memory traces, engrams, cell assemblies, and neural plasticity [4, 15–20]. Numerous studies have been done in order to reveal memory units and their working principle. However, the clear views about the structure and function of these basic units in memory traces for the integrative storage and reciprocal retrieval of associated signals remain needed. The reliable criteria to define memory cells in engrams and to identify molecules in memory cells specifically for labeling and indicating the recruitment of neurons into memory cells have not been well established. These are no clear approaches to validate behaviors specifically and directly initiated by memory cells [5]. In order to set up reliable standards for judging whether cell assembles being recruited are correlated to memory formation and retrievals, the alternations at the levels of molecules, neurons, and behaviors relevant to associative learning and memory should be precisely evaluated and validated.

Memory formation has been routinely examined by memory retrieval, in which the cue-induced expression of specific behaviors that have been represented in learning and previous memory recalls is better applied to indicate the maintenance of memory traces. In terms of pitfalls of this strategy, a few of thoughts are given below. Behaviors, perception, and cognitions are quickly developed [21, 22]; however, their postnatal developments are not parallel in behavioral patterns and memory-relevant cognition contents [5, 23–25]. The number of verbal language patterns, such as words and their organizations, is larger than the number of memory contents. On the other hand, the number of arm-body language patterns is much lower than the number of memory contents. In these regard, memory contents can be represented well by verbal language patterns. However, one pattern of arm-body language may represent several memory contents. For example, a thumb-up gesture often signifies memory contents in relevance to positive events. Moreover, the limited number of behavior patterns cannot well present the enriched patterns and varieties in sensory input signals, memory contents, cognitive events, or emotional reactions. As control mechanisms for simple behavior may be common output pathways for the representation of many memory contents and memory-relevant cognitions, a behavior pattern can signify numerous perception signals, memory contents, and cognition processes. For instance, "OK" gesture is used to express appropriate sensory stimulations, good perception, successful memory retrieval, and other positive cognitions. In other words, behaviors as arm-body language or gestures may not signify the retrievals of specific memory contents, except for verbal language [5]. This limitation of behavior patterns to represent memories and cognitions has to be seriously placed in mind in the study of memory retrievals based on animal behaviors. The simply use of behavioral responses as signs to indicate memory retrieval should be in caution. For example, body freeze and involuntary/voluntary shaking used to signify fear memory do not specifically denote phobia to contexture associated with electrical shock. These behaviors can also be induced by intensive fear in panic, anxiety, emotion reactions (anger and fighting), and some physiological responses (hypoglycemia and hypothermia). In addition, the brain is highly connected, and its distinct areas are mutually innervated after the maturation in human being and animals [26]. Stimulations to the pretended engram cells through the electrical or optogenetic methods in a brain area may indirectly trigger other areas, which interconnect with the stimulus area via anterograde and retrograde manners. With this indirect activation of brain areas highly rewired, memory-relevant behaviors may be indirectly activated, and even behaviors across or similarly to memory retrieval may be activated [5]. If it is a case, the replays of so-called memory-relevant behaviors may not be realistically or directly controlled by particular memory traces.

Associative learning includes the joint acquisitions of unitary signals and complicated signals reorganized from unitary signals. After unitary signals are memorized in young age, such as letters, words, and simple concepts, the advanced learning moves forward to acquire the complicated concepts from various organizations of those unitary signals. For instance, sentences and articles are learnt by the association of letters and words in language as well as the theories and principles are learnt by the association of various concepts. Unlikely verbal presentation, arm-body behaviors are not obviously upgraded to more complicated versions to express advanced language and knowledge during postnatal development, such that the similar behaviors may represent different contents and knowledge. The retrievals of various memorized signals by the replay of similar behaviors cannot indicate the retrieval of particular memory contents, i.e., behavior replays are unreliable, except for the reemergence of associated behaviors induced by learning cues specifically [5].

The brain consists of neurons and glia cells, which may be the basic units in memory traces. What kinds of nerve cells are ready to be recruited as memory cells for storing particular new signals is unclear. What new features have been added to those ready-recruited neurons to convert them into memory cells is unknown. The elucidation of these questions is critically important to figure out the working principles of memory cells in engrams for memory formation and retrieval. As memory cells are recruited from neurons, they have conventional neuronal features, such as the innate synapse input, synapse transmission, neuron excitability, and excitation propagation output. In addition, these memory cells theoretically encode the newly learned signals and receive new synapse innervations that carry these learned signals [6]. The most common pattern of learning and memory in life is associative in nature, in which the integrative storage and reciprocal retrieval of these associated signals include the new signal to be learned and innate signals or previously learnt signals. Associative memory cells recruited from neurons during associative learning should be featured to encode this new signal and innate signal (or previously learned signals) as well as to receive new axon projection and synapse innervations from coactivated brain regions in addition to innate inputs [6, 7]. The detection and identification of new synapse innervations and multi-signal encodings through recording approaches (cell electrophysiology and imaging) is important in searching basic units in memory trace. Moreover, learning and memory are characterized as memories to unitary signals in the young and memories to the complicated signals reorganized from these unitary signals in later periods. Associative memory cells presumably encode multiple unitary signals, and their assemblies or groups work together to integrate unitary signals into the complicated signals through various reorganizations of the unitary signals. In other words, the ensembles of associative memory cells store advanced knowledge contents in specific spatial and temporal patterns [7].

Neuronal excitation is driven by synapse inputs, and neuronal excitability is controlled by spiking threshold [27–34]. Neuronal spike-encoding capability including spike number and pattern is affected by the intrinsic spiking threshold as well as the driven force from the interactions between excitatory and inhibitory synaptic transmission [34, 35]. Therefore, the strength of neuronal activities is proportional to the weight of synapse inputs, but not the nature of synapse input contents. The spiking pattern and frequency of memory cells in neural circuit signify their activity strength and memory depth, but not synapse input specificity and memory contents. Based on this consideration, the replay of some neuronal activity patterns, e.g., spontaneous sharp-wave ripple, may not be relevant to memory characters and contents. The replay of neuronal activity patterns is likely implying the reemergence of neuronal activity strength for memory depth [5]. Efforts to correlate memory cell activity patterns to particular memory contents (memory specificity) [36, 37] may not be strongly practices. Neuronal activities in response to learning cues may reflect the retrieval of memory contents. The identification of memory cells by using learning cues to evoke their activity may be better used to track associative memory cells about their location and distribution in engram circuits [5].

To identify memory cells as well as to reveal their characters and work principle during memory formation and retrieval, researchers expect to find out specific molecules in associative memory cells that can be used to label these cells for studies to answer memory-related questions. In the formation of new memory cells recruited from the ready-recruited neurons, the interconnections among these neurons based on axon prolongation for its projection and synapse innervations are critical. These two processes are nonspecific for the neurons in the central nervous system. Therefore, efforts to identify molecular markers for specifically labeling memory cells are challenge. Recently, some immediate early genes in response to intensive cell activities have been introduced to label memory cells [38, 39]. This practice is based on proposal that active memory cells express immediate early genes [40, 41]. However, the expression of these immediate early genes is proportional to the activity strengths of cerebral neurons, such as the seizure discharge in epilepsy [42–45] and neuron toxicity in brain ischemia [46–48]. In this regard, immediate early genes may not be nonspecific for labeling memory cells.

Taken these messages thoughts together, the criteria for neurons to be defined as memory cells include the cue-induced replay of neuronal activity, new synapse innervations, and active molecule labeling and the emergence with cue-induced behavior. Moreover, associative memory is characterized as the integrative storage and reciprocal retrieval of associated signals. Basic units in memory trace should be able to encode the associated signals for their integrative storage and reciprocal retrievals as well as to receive synapse innervations that carry in these signals for their storages [6]. Therefore, the proposed associative memory cells should possess all of these characteristics.

The encoding of associated signals by the memory cells in engrams can be detected by electrophysiological recording for spiking patterns and two-photon cell imaging for their intracellular calcium level that is proportional to spike frequency in response to learning cues. The electrophysiological method to identify memory cells includes the recordings of local field potentials and intracellular spikes [49–62]. Local field potential recording is to detect multiple units, or a population of neurons, in which some neurons are full associative memory cells to encode all of the associated signals, and other neurons are incomplete associative memory cells to encode parts of these associated signals [3, 11]. In other words, the neurons recorded by local field potentials may not be all of full associative memory cells. However, intracellular recording can show whether those recorded neurons are associative memory cells or not, based on the encoding patterns of associative memory cells in response to the different learning cues or associated signals [3].

Local field potentials are recorded in sensory cortices to identify primary associative memory cells as well as the prefrontal cortex or hippocampus to identify secondary associative memory cells by glass electrodes. These pipettes to recording neuronal electrical signals include standard pipette solution (150 mM NaCl, 3.5 mM KCl, and 5 mM HEPES). The pipettes filled with this standard solution possess the resistance around 5–7 MΩ. The electrical signals recorded by these pipettes are inputted to an AxoClamp-2B amplifier, which connects with the pClamp 10 system for data acquisition and analysis. In this system, the electrical signals are digitized at 10 kHz and filtered by low-pass at 0.5 KHz. In the data analysis, another band-pass filter and the second-order "Savitzky-Golay" filter are applied to isolate the signals from local field potentials. The signals are complex and variable. Individual events of local field potentials induced by learning cues last for 10 ms with a sharp negative response. Differences between negative peaks and baseline in individual signals are measured and averaged to show cue-evoked signal amplitudes. Spike frequency is calculated as one over inter-spike intervals. The intracellular recording of synaptic activity and neuron spikes in memory cells is conducted in sensory cortices to identify primary associative memory cells and the hippocampus to identify secondary associative memory cells by sharp electrodes that contain the standard pipette solution (2 M KAc). The pipettes filled with this solution possess the resistance around 50–60 MΩ. Electrical signals are inputted to an AxoClamp-2B amplifier and a pClamp 10 system for data acquisition and analyses. In this system, the electrical signals are digitized at 20 kHz and filtered by low-pass (3 KHz). In the data analyses, the fluctuations of membrane potential signals including integrated synapse events and neuronal spikes are isolated, and inter-event intervals are measured to indicate the frequencies of these events that equal to one over inter-event intervals [3, 11]. The recordings of local field potential and intracellular signals are conducted in the identical areas of sensory cortices [63].

Two-photon cell calcium imaging can monitor the activities from a population of neurons to identify the distribution and the correlation of associative memory cells in vivo [63–66]. Three steps are applied to detect neuronal activities. (1) Calcium-sensitive dyes or calcium-sensitive fluorescent proteins by adeno-associated viruses are loaded into neurons in the interested areas, such as sensory cortices and the prefrontal cortex. (2) Two-photon cell calcium imaging is conducted in these areas, in which two-photon laser beam (e.g., 810 nm) is used to activate these fluorescents with specific wave length and to detect emission waves from these fluorescents in the labeled cells with the photomultiplier tubes of specific wavelength zone (e.g., 523 nm for calcium-binding Oregon-BAPTA-AM and 610 nm for sulforhodamine-101 to show astrocytes). All of the parameters set for the two-photon laser beam and photomultiplier tube are locked for the measurements throughout all experiments to maintain consistent conditions in data comparisons among groups. (3) Cellular Ca^{2+} fluorescence signals in response to learning cues are acquired by commercial softwares from different vendors. These acquired signals are analyzed in cell soma by public softwares, such as NIH ImageJ and MATLAB (MathWorks). In the analyses of Ca^{2+} signals, the regions of interest (ROI) from each of the cells with fluorescent inclusion are marked by circles around cell somata for measuring the spatial and temporal Ca^{2+} signals. The median filter (radius, 1 pixel) is used to all of the images, in order to reduce photon and PMT noises. Cell Ca^{2+} fluorescence signals in response to learning cues are digitized to be the traces of the neuronal signals. These digitized signals are normalized and represented as the relative fluorescence alternations ($\Delta F/F$). Baseline fluorescence (F) is an averaged value in the ROI before stimulations. ΔF values refer to the differences between the evoked cell Ca^{2+} signals and the baseline. Fluorescence signals are also subtracted from the noise signals of unstained blood vessels. The normalized Ca^{2+} signals are smoothed by a low-pass Butterworth filter to remove low-level fluctuation and minimize distortions from fast Ca^{2+} transients. The effective Ca^{2+} signals from active cells are judged based on the criterion that ΔF/F was greater than 2.5 times of the standard deviation of baseline values lasting for 500 ms [3, 11, 63].

In order to detect the responses of astrocytes, sulforhodamine-101 is thought to be a specific fluorescent dye to label them [63]. If activities of glutamatergic and GABAergic neurons need to be isolated, transgenic mice with labeling glutamatergic neurons by yellow fluorescent protein (YFP) and GABAergic neurons by green fluorescent protein (GFP) are used [8, 9, 67, 68]. For instance, Oregon-BAPTA is used to detect whether glutamatergic and GABAergic neurons have been recruited as associative memory cells. The wave of two-photon laser beam at 810 nm is given for its excitation to examine neuronal responses to learning cues. The emission wave is 525 nm for Ca^{2+}-binding Oregon-BAPTA. Then, neuron identities are tested by giving a wave of 920 nm to excite GFP that labels GABAergic neurons and YFP that labels glutamatergic neurons. Although the peaks of optical emission waves are closely for GFP (510 nm) and YFP (525 nm), the separation of their images is conducted by setting optical gratings with different windows for their unmixing. Average power delivered to the cortices should be as low as possible to minimize photo bleach [3, 11, 63].

In the studies with electrophysiology and two-photon cells imaging in vivo, the activity levels of associative memory cells can be read from the frequency and amplitude of field potentials, the frequency of neuronal spikes and synaptic potentials in intracellular recording, as well as the response strength and duration of calcium signals in two-photon calcium imaging. Activity strength and synchrony are used to represent the spatial and temporal patterns of neuronal activities, respectively. The cross-correlations among the recorded cells may be negative, but their absolute values are applied for data presentation. In general, the larger the cross-correlation value is, the better the synchrony among the associative memory cells is, or vice versa. In the analyses of the neurons to encode learning cues by two-photon imaging and intracellular recording, if the recorded neurons become encoding the associated signals, they are associative memory cells to memorize these associated signals. If the spatial and temporal activities of these neurons are distinct in response to the associated signals, these associative memory cells are able to distinguish the difference of these signals [3, 11].

Neural tracing by adeno-associated viruses (AAV) to carry the genes of fluorescent proteins (FP) is used to detect axon projection and synapse innervation [11, 12, 69], in order to morphologically identify associative memory cells that receive synapses innervations from the coactivated brain areas. In this study, AAV-FP is injected into brain areas that are thought to be the source of neuronal axon projection, and distributions of axonal boutons from AAV-FP-labeled neurons in potential brain areas are detected by confocal microscope. For instance, to study the convergence from two sensory cortices onto neurons in a cortical area, AAV-CMV-tdTomato and AAV-CMV-ZsGreen are injected into these sensory cortices, respectively, and the presence of fluorescence labeled axon boutons and synapse contacts is detected onto neurons in this cortical area. Glutamatergic neurons in C57 mice were genetically labeled by YFP, and GABAergic neurons were labeled by GFP [68]. The working principle of AAV-CMV-tdTomato is that CMV promoter initiates tdTomato expression in presynaptic boutons. That GFP-labeled boutons innervated onto the dendritic spines of YFP-labeled glutamatergic neurons are presumably synaptic contacts, and that RFP-labeled boutons innervated onto GFP-labeled GABAergic neurons are also synaptic contacts. In AAV injection, glass pipettes were positioned in the source areas of sensory cortices, based on the map of mouse brain [70]. After AAVs are injected into cortices for 2 weeks, axon projection and synapse formation were analyzed. In order to clearly show three-dimensional images for new synapses in the cortex, brain slices are placed into Sca/eA2 solution for a few hours to make them transparency [71].

Putative new synapses are identified and quantified based on those images that illustrate synaptic contacts between GFP-labeled axon boutons and YFP-labeled glutamatergic neuron spines or between RFP-labeled axon boutons and GFP-labeled GABAergic neurons. The images of showing synaptic contacts in cortical layers II~III are photographed under this confocal microscopy with oil lens (Plan Apo VC 60X and 1.4NA). The excite wavelength for tdTomato is 561 nm, and the emission wavelength for tdTomato is 596–644 nm. The excite wavelength is 488 nm for GFP and YFP. Although peaks of GFP and YFP emission wavelengths are 510 and 525 nm, respectively, the images of these neurons are scanned by setting two optical gratings in the band of 505–515 nm for GFP and in the band of 545–555 nm for YFP, respectively, in order to separate their fluorescent images [72]. These separated fluorescence images are merged to reconstruct the newly formed synapses. In the study of cellular images with a confocal microscope, the spatial resolution is set at 0.05 μm per pixel, such that minimal pixels for the imaged spines are about 9–10 in the scanning lines for their high spatial resolution. The morphology and density of new synapses formed during associative learning are analyzed by applying a public software ImageJ (version. 1.47; the National Institutes of Health, USA) and commercialized software Imaris (version 7.2.3; Bitplane, England). In the analysis of axon boutons and synapses, the contacts between presynaptic and postsynaptic identities are defined as synapses based on their separations less than 0.1 μm. Synaptic contacts per mm dendrites are calculated. As the YFP does not label all glutamatergic neurons and the GFP does not label all GABAergic neurons due to the weaker Thy1 promoter and GAD promoter, synapsin-GFP-labeled and tdTomato-labeled presynaptic boutons may innervate the spines on non-YFP glutamatergic neurons and the processes on non-GFP GABAergic neurons, so that the densities of GFP-labeled and tdTomato-labeled boutons per mm^3 are also calculated to assess the number of new synapses.

Associative memory cells are basic units to encode the integrative storage of multiple associated signals by receiving the innate and new synapse innervations [5–7]. Therefore, activity-dependent synaptic plasticity (long-term potentiation and depression of synaptic transmission) in single neural pathway [73–75] and activity-dependent neural plasticity [29, 76–80] may not be directly involved in associative memory, though these types of neural and synaptic plasticity may influence memory retrieval [6, 7]. In terms of the location of information storages, memory traces appear widely distributed in the brain, such as the prefrontal cortex, hippocampus, amygdala, motor cortex, sensory cortices, and associative cortices [1, 13, 61, 81–91]. The information has been hypothetically memorized in cell assemblies by strengthening their interconnections that are instigated by their correlated activities in learning [15]. These studies have not clearly interpreted when these cell assemblies in relevance to memory-relevant activities are formed, why these cell assemblies are widely distributed in the brain, as well as how the strengthening at synapses and neurons is able to coordinately integrate associated signals for their integrative storage in primary and secondary manners [6, 7]. In other words, the characters and working principles of these memory-relevant neurons that coordinately encode associative memory remain unknown. Except for the activation efficiency, neuronal plasticity cannot be used to interpret memory patterns, such as explicit versus implicit memory, episodic versus semantic memory, and the transformation memory patterns [92–96], the temporal features of associative memory as well as the impacts of associative memory onto associative thinking and logical reasoning [97–102]. The basic units in memory traces need to be figured out to reveal how endogenous signals generated in associative thinking and logical reasoning are memorized for future representation, how memory is encoded under different consciousness states, as well as what are comprehensive characters of these cell assemblies, the patterns of their connection strengthening, and the coordination of their encoded memory [5].

Associative memory cells that encode multiple associated signals as well as receive innate and new synapse inputs have been detected to be recruited by the coactivation of cortical neurons [3, 6, 7, 11, 103]. The coactivation of sensory cortices evokes their mutual synapse innervations and recruits associative memory cells to integrate and encode associative signals [3, 10, 104]. Based on mutual innervations among associative memory cells in sensory cortices [3, 8, 11], sensory signals can be associatively inputted and integrated among the sensory cortical neurons. The integrative storage of these sensory signals in primary associative memory cells enables each signal to induce the retrieval of its associated signals or the other around. In the meantime, these primary associative memory cells in sensory cortices prolong their axons and project toward brain regions in relevance to cognitions, emotions, and behaviors, as long as these areas are coactive during associative learning. These primary associative memory cells are synaptically convergent onto the individual neurons in these cognitive areas and association cortices during logical reasoning and associative thinking to recruit their downstream neurons as secondary associative memory cells [6, 7, 13]. With mutual synapse innervations among associative memory cells as well as convergent innervation, such primary associative memory cells in sensory cortices and their downstream secondary associative memory cells constitute basic cellular architectures for the reciprocal retrieval of associated signals, the automatic conversion of associated signals among cross-modalities during their retrieval as well as the cognitions and emotions at the higher orders [6, 7] (Figs. 1.2 and 5.1).

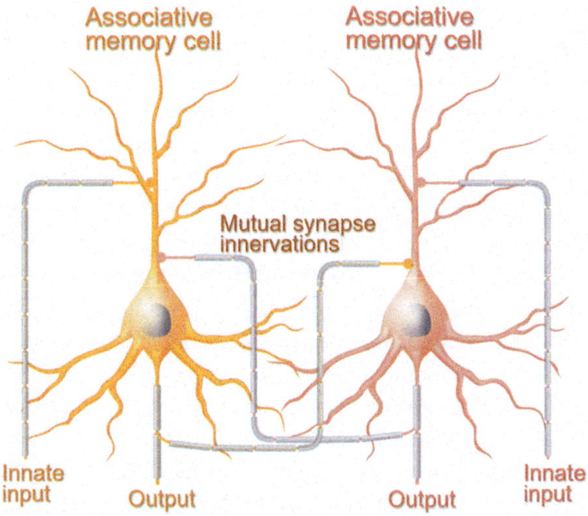

Fig. 5.1 Associative memory cells and their mutual innervations. Illustration shows two neurons (orange and red). These two associative memory cells interconnect each other through their axonal branches alongside their innate inputs, respectively. Innate synapse inputs are labeled with identical color with associative memory cells

5.3 Associative Memory Cells in Sensory Cortices

In the retrieval of memorized signals, individuals can tell the sources of signals, i.e., which sensory system takes in these signals for their memory in the brain. For instance, when image signals are retrieved, individuals can tell these signals from previous sight. When a story is retrieved, individuals can tell this story from reading book or through listening. In this regard, these signals are likely stored primarily in sensory cortices [5]. The identification of associative memory cells is initially performed by author's lab in a mouse model of associative learning and memory with the integrative storages and reciprocal retrievals of associated signals [3, 8, 9]. Associative learning in mice by simultaneously paring whisker and odor stimulations leads to the reciprocal retrievals of these associated signals induced by each of these signals, including odorant-induced whisker motion and whisking-induced olfaction responses [3, 8, 69]. After the formation of this associative memory, associative memory cells are searched by the electrophysiology and two-photon cell imaging for their functional identification, as well as the neuronal tracing for their morphological identification.

In the in vivo recording of local field potentials at the barrel cortex from mice that express odorant-induced whisker motion and whisker-induced olfaction responses, a population of barrel cortical neurons are able to encode new odor signal alongside innate whisker signal [3], in comparison with the mice without paired odor and whisker stimulations. In the meantime, a population of piriform cortical neurons becomes able to encode the newly learnt odor signal alongside the innate whisker signal [8]. Moreover, mice that have been trained by pairing whisker, odor and tail stimulations can express odorant-induced whisker motion and tail-induced whisker motion [11]. In mice of showing associative memory for these triple signals, a population of barrel cortical neurons become to encode the newly learned odor and tails signals alongside innate whisker signal. These results from the electrophysiological recording of local field potential in vivo indicate that a portion of sensory cortical neurons are recruited to memory cells of encoding associatively learned signals or associative memory cells. In order to ensure this indication, the intracellular recording of individual neurons in vivo is used, in which the single barrel cortical neurons become able to encode these associated signals in response to the learning cues [3].

Moreover, the functional identification of associative memory cells is conducted by using two-photon cell imaging in vivo. Certain barrel cortical neurons are able to encode whisker and odor signals in mice with behaviors relevant to associative memory of these two signals [3] and to encode whisker, odor, and tail signals in mice with behaviors relevant to associative memory of these three signals [11]. Studies by two-photon cell imaging support the conclusion about the recruitment of associative memory cells in the sensory cortices after associative learning [3, 11]. It is noteworthy that certain neurons encode triple associated signals, some neurons encode two associated signals, and others encode one signal. These neurons that encode an innate signal are called as innate memory cells. The neurons that encode the newly learned signal only are called as new memory cells. The neurons that encode two or more signals including new signals and an innate signal are called as associative memory cells. If neurons encode all of the associated signals, they are full associative memory cells. If the neurons encode parts of associated signals are incomplete associative memory cells [3, 5, 11]. The mechanism underlying the recruitments of full associative memory cells, incomplete associative memory cells, and new memory cells may be caused by a fact that the activation strength of ready-recruited neurons is different. The reasons for their different activations may be due to their various excitabilities [34, 105] or their activations by the ascending arousal system differently [5–7]. The physiological impact for a presence of these associative memory cells with different encoding capability and capacity may be that the different populations of these associative memory cells have made all of these associated signals to be dissected into unitary signals being memorized in sensory cortices with different group patterns [6]. These associative memory cells in the storage of these unitary signals can be reorganized in their subsequent coactivations during new learning activities and cognitive processes in order to generate advanced mutual synapse innervation and associative memory cells for encoding certain inspired and complicated signals [5].

There are a few discoveries from such studies. (1) In addition to sensory cortical neurons, astrocytes become able to encode two or three associated signals in associative memory formation. The results indicate that astrocytes may coordinate neurons in their recruitment and functional upregulation [3, 11]. (2) Both glutamatergic and GABAergic neurons become able to encode two or three associated signals [9, 10, 104]. These glutamatergic associative memory cells are upregulated in their functions including excitability and excitatory synaptic transmission. These GABAergic associative memory cells are functionally downregulated [3, 8, 9]. The recruitment of both glutamatergic and GABAergic neurons as associative memory cells may facilitate their coordination to shift the excitatory and inhibitory balance toward more excitation for recruiting more associative memory cells [6, 7].

In the morphological identification of associative memory cells in sensory cortices, neural tracing by AAV-carried fluorescent proteins is used to search mutually axon projections and synapse innervations among associative memory cells and among coactivated sensory cortices [3, 8–11, 69, 104]. To associative memory trained by whisker and odor signals, AAVs that carry genes to express fluorescent proteins, AAV-CMV-ZsGreen and AAV-CMV-tdTomato, are injected into the thalamus and the piriform cortex, respectively. YFP-labeled glutamatergic neurons in the barrel cortex are scanned under confocal microscope to examine synapse contacts made by GFP and RFP-labeled axonal boutons. In another group of studies, AAVs that carried genes to express fluorescent proteins, AAV-CMV-ZsGreen and AAV-CMV-tdTomato, are injected into the thalamus and the barrel cortex, respectively. YFP-labeled glutamatergic neurons in the piriform cortex are scanned under confocal microscope to track synapse contacts made by GFP and RFP-labeled axonal boutons. In both cases, YFP-labeled neurons receive the convergent synapses innervations from two sources where AAV-CMV-carried genes are injected. These results indicate the morphological identification of associative memory cells in sensory cortices that can encode two associated signals.

The morphological identification of associative memory cells is also conducted in mice that demonstrate associative memory trained by triple signals, i.e., whisker, odor, and tail signals [11]. In this study, YFP-labeled glutamatergic neurons receive synapse innervations from piriform cortical and S1-tail cortical neurons in addition to innate innervation from the thalamus.

Based on the studies of associative memory cells by functional and morphological approaches, a portion of sensory cortical neurons in mice after associative learning become able to encode associated signals and receive new synapse innervations from the coactivated brain areas to establish their mutual innervations alongside innate inputs. These sensory cortical neurons that encode associated signals are named as primary associative memory cells [3, 10, 11, 69, 104]. These associative memory cells have been assured to include glutamatergic neurons, GABAergic neurons, and astrocytes [3, 8–11, 69, 104]. The coactivations or simultaneous activities of sensory cortices can trigger the new synaptogenesis for mutual synapse innervations and the recruitment of associative memory cells that encode the integrative storage and reciprocal retrieval of associated signals. Based on the studies, results and indication in barrel, piriform, and S1-tail cortices, the concept about a recruitment of associative memory cells may be extended to other sensory cortices. That is, the association of cross-modal sensory signals may occur in the coactivation or simultaneous activity among sensory cortical neurons in all of the cross-modalities. For instance, the visual signal is associated with the auditory, olfactory, taste, and somatosensory signals. The auditory signal is associated with somatosensory, visual, olfactory, and taste signals. In this regard, primary associative memory cells are presumably recruited in all sensory cortices by their mutual synapse innervations, provided these auditory, visual, olfactory, gustatory, and somatosensory cortices are coactivated [6, 7] (Fig. 1.2 as well as Figs. 5.1, 5.2, and 5.3).

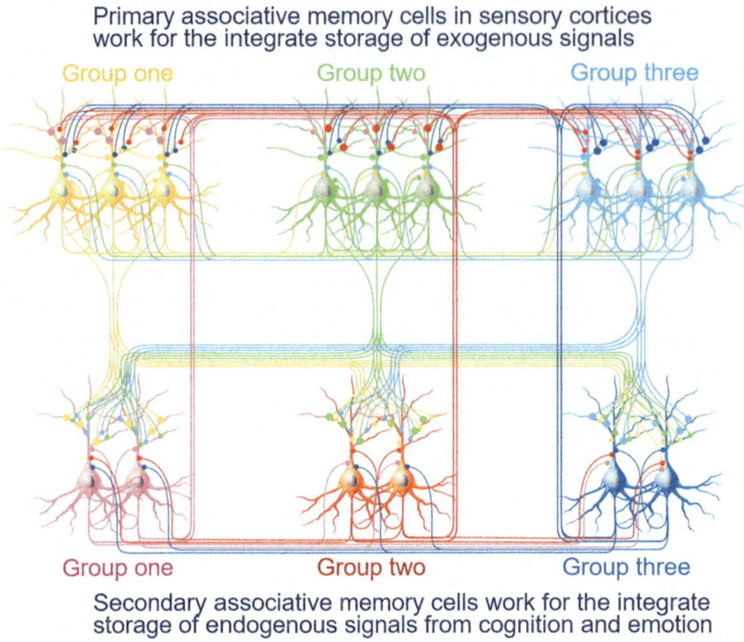

Fig. 5.2 Associative memory cells and their interconnections. Three clusters of primary associative memory cells (yellow, green, and blue) in the sensory cortices possess mutual synapse innervations in intramodal and cross-modal manners. Three clusters of secondary associative memory cells (pink, red, and blue) in brain regions related to the cognition, emotion, and behaviors synaptically interconnect one another in intramodal and cross-modal manners. The axons of these primary associative memory cells can convergently and divergently innervate secondary associative memory cells (feedforward). The axons of these secondary associative memory cells can innervate back to primary associative memory cells (feedback). All of these associative memory cells receive innate synapse innervations that are labeled by identical color with their somata. Synapse innervations among the functionally correspondent groups of primary and secondary associative memory cells in the same vertical lines are labeled by the bigger presynaptic boutons. (This illustration is modified from a reference [5])

Fig. 5.3 (continued) acetylcholine (ACh), respectively. These neurotransmitters can maintain wakefulness, permit normal consciousness, and support specific attentions. In addition, associative memory cells are regulated by the hormones that are released from the hypothalamus-pituitary glands. The upregulations of AMC number and encoding capability facilitate memory to be impression and maintenance, or vice versa. The functional downregulation of memory output cells in motor cortex leads to the incapability of memory retrieval and presentation. Each of associative memory cells receives the innate synapse input, and colors for associative memory cells and innate input are labeled to be identical. (This illustration is modified from a reference [5])

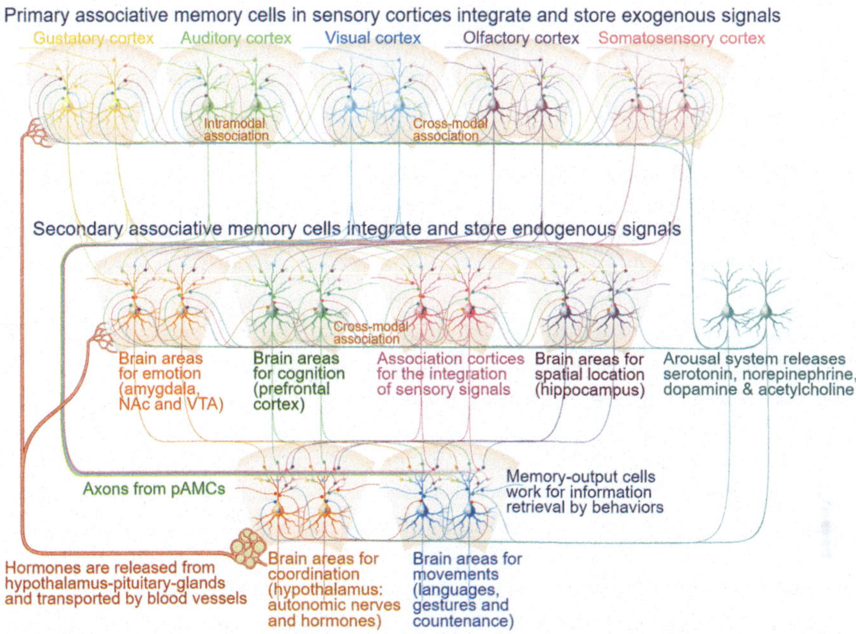

Fig. 5.3 Associative memory cells, memory output cells, and their interconnections for memories, cognitions, and emotions. Associative learning and memory include the acquisition of associated signals, the integration and storage of exogenous signals, the integration and storage of endogenous signals as well as the memory retrievals through behavioral representation. Associative memory cells (AMC) are classified into primary AMC and secondary AMC. In top horizontal panels, primary AMCs are distributed in auditory, visual, gustatory, olfactory, and somatosensory cortices for the integrative storages and reciprocal retrievals of exogenous associated signals. These primary AMCs are synaptically interconnected in cross-modal and intramodal manners. The axons of primary AMCs convergently and divergently innervate to secondary AMCs for cognitions, emotion, sensory integrations, and spatial localization. In middle horizontal panels, secondary AMCs are distributed in brain regions relevant to cognitive events (logical reasoning, associative thinking, imagination, computation, concept, judgment, conclusion, and decision-making in the prefrontal cortex), emotional reactions (happiness, fear, aversion, and angry in the ventral tegmental area (VTA), nucleus accumbens (NAc), and amygdala), the integration of multiple sensations for understanding and perception in the association cortices, as well as the spatial localization in the hippocampus. Secondary AMCs are interconnected via their synapse innervations for the integrations of cognitions, emotion, perception, and localization. The axons of these primary and secondary associative memory cells synaptically innervate onto memory output cells in the brain regions relevant to behavioral outputs, such as language, writing, gesture and countenance, and their coordination. This coordination is fulfilled by the systems that maintain the homeostasis of the internal environment, such as the hypothalamus to control autonomic nerves and to release hormones. The cross-modal associative memory cells are recruited through mutual synapse innervation among sensory cortices or among brain areas in relevance to cognition, emotion, and localization. Intramodal associative memory cells are recruited through mutual innervations among the neurons within single-modality cortex. Associative memory cells can be activated by the innate synapse input and new synapse innervations from the coactivated brain regions. Associative memory cells can also be activated by the arousal system from the brain stem and the thalamus including the midbrain raphe nuclei, locus coeruleus, substance nigra, and cholinergic nuclei that release serotonin (5-HT), norepinephrine (NE), dopamine (DA), and

The coactivation of ready-recruited neurons in sensory cortices recruits primary associative memory cells; however, the encoding capabilities in strength and contents at these associative memory cells are diversified. Some associative memory cells are able to encode all of the associated signals, which are called as full associative memory cells. Some associative memory cells can encode one or more new signals, such as two or one of olfactory, whisker, and tail signals, which are called as incomplete associative memory cells [11]. The reason for this feature may be based on coactivity together and connection together, in which the stronger coactivity has the higher ability to attract more synapse innervations among sensory cortical neurons [5]. The coactivity strength in these ready-recruited neurons is likely different based on their variable excitability [34]. Neurons that encode one signal are called as new memory cells or innate memory cells [3, 11]. The recruitment of different populations of associative memory cells in their capability of encoding strength and contents may lead to two situations. The repeated coactivity may convert those incomplete associative memory cells into full associative memory cells. The diversified groups of associative memory cells have dissected those complicated associations of processes, objects, or images into simple unitary signals. The dissection of those complicated associations by recruiting different groups of associative memory cells in their encoding capability may facilitate their storages in unitary manners. This incomplete and full associative memory may further reorganize by their coactivity and mutual synapse innervation to recruit associative memory cells that distort original associations as well as encode new inspired signals or weird signals. Their further retrievals in differently organized patterns and various organizations of unitary signals make new in future associative learning [5, 7]. The repetitive coactivations of sensory cortices may facilitate the recruitment of full associative memory cells from incomplete associative memory cells and the formation of more en passant synapses among their mutual innervation. The number and the activity strength of associative memory cells are upregulated [5] (Fig. 5.4). Therefore, there is proportional relationship among associative memory efficiency, associative memory cells, and their upregulations [3, 11, 12, 69, 106–108], also called as activity-dependent positive cycle in the recruitment and refinement of associative memory cells [7].

Associative memory cells are characterized as mutual axon projections and synapse innervations to encode multiple associated signals. The molecules potentially for building axonal growth and synapse formation as well as signaling pathways for regulating axons and synapses [109–111] are likely substrates underlying the recruitment of associative memory cells. Recent studies indicate that antagomirs for microRNA-324 and microRNA-133a by influencing *Ttbk1* and *Tet3* expressions attenuate associative memory, new synapse innervations, and associative memory cell formation [11, 12]. Moreover, the downregulation of miRNA-342 expression and the upregulation of *Nlgn3* and *Nrxn1* expression are linked to the recruitments and refinements of associative memory cells [8, 9]. These genes and proteins have been found to be involved in the axon prolongation and synapse formation [112]. The formation of synapse innervations and the recruitment of associative memory cells may be based on a chain reaction from intensive neuronal

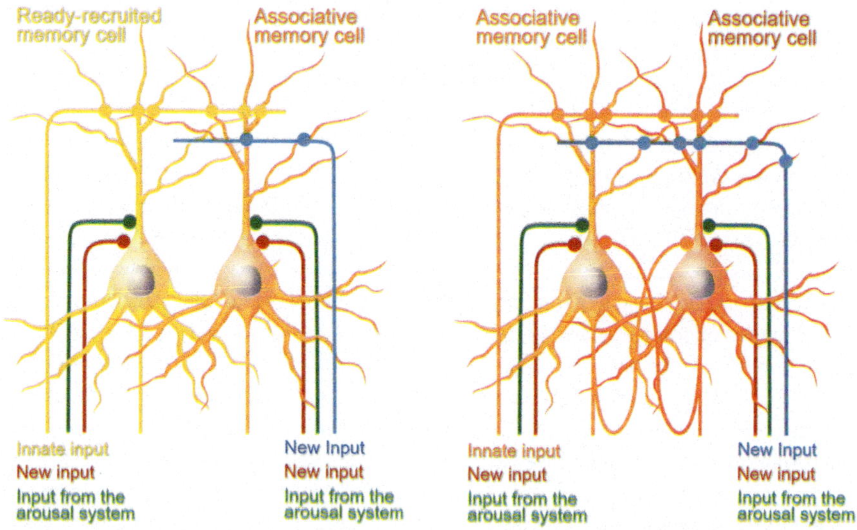

Fig. 5.4 The activity-dependent positive cycle in the recruitment and upregulation of associative memory cells is based on local axon growth and en passant synapse formation by their coactivation. Left panel illustrates a pair of associative memory cells that receive the synapse innervations from the innate input with three en passant synapses (yellow), the arousal system input (green), and the new input (red). One of these two associative memory cells on the right side also receives another new input with two en passant synapses (blue). After this pair of associative memory cells is coactivated with other associative memory cells that axons are labeled by blue color, the new axonal input (blue) will be locally grown and make synapse contacts on its targets. Right panel illustrates this pair of associative memory cells after coactivation, where en passant synapses on the right side of associative memory cells are raised from two to four and on left side of associative memory cells are raised from zero to two. In addition, the increased encoding capability after their coactivation is showed by making color to be deep in these associative memory cells. Each of associative memory cells receives innate synapse input, and the colors for associative memory cells and innate input are labeled to be identical

spikes to microRNA-regulated genes and proteins that can specifically manage the axon prolongation and synapse formation [11, 12, 69]. In addition, the inhibition of sensory cortices by pharmacological approaches blocks associative memory [3, 69]. The injections of the mixture of microRNA-324 and microRNA-133a antagomirs into barrel cortex reduce the depth of associative memory as well as lower the recruitment of new synapse innervations and primary associative memory cells [11, 12]. These results also indicate that the location to encode associative memory is primarily in sensory cortices, where mutual the synapse innervations are formed and the primary associative memory cells are recruited [6, 7].

Pair-coding neurons that encode two visual pictures have been recorded in the animal visual cortex in vivo [60, 113], similar to the encoding characters of primary associative memory cells in cross-modal sensory cortices. These pair-coding neurons within intramodal sensory cortices may encode the integrative storage of the associated signals inputted from single sensory modality, such as numerous photon beams in video pictures associatively to stimulate the visual system, odorant signals associatively to activate the olfactory system, letters, and words in the verbal language associatively to trigger the auditory system, and so on (Fig. 5.3). However, morphological evidences about mutual synapse innervations among these cortical pair-coding neurons in single modality remain to be provided because of the challenge in tracing neuron-by-neuron.

Because primary associative memory cells are recruited from sensory cortical neurons, these cells should possess the characters of associative memory cells that receive synapse innervations from coactivated cortices and encode associated signals carried by these synapse inputs, as well as the general characters of neurons [5]. Specific characters of associative memory cells can be applied as criteria to identify whether the neurons detected in experiments are associative memory cells. General characters expressed in neurons can be used to assess that associative memory cells may have events occurred in neurons, such as morphological and functional plasticity induced by activities. As the coactivation via the synchronous activities of cortical neurons that are readily recruited to be memory units triggers their mutual synapse innervations and recruits them as associative memory cells, these specific characters of associative memory cell are presented below [5–7].

In terms of morphology, primary associative memory cells receive the new synapse innervations from those coactivated neurons in sensory cortices for their interconnections alongside innate sensory inputs. In their functions, primary associative memory cells are able to encode the innate signal and newly learned signals for their integrative storage and reciprocal retrieval. Their axons prolong and project toward active downstream neurons in convergent and divergent manners to make synapses onto these neurons in brain regions relevant to cognitive events, emotional responses, and memory-relevant behaviors. The recruitment of the primary associative memory cells is influenced by microRNA-regulated genes and proteins for the axon prolongation and synapse formation [11, 12, 69]. Mutual synapse innervations among these cross-modal associative memory cells confer the reciprocal retrieval of associated signals by one of learning cues as well as the automatic conversion of signal retrievals among different modalities, such as visual signals in images are presented by verbal language and verbal signals in stories are presented by visual diagrams. The synapse convergences onto downstream neurons and the activation of associative memory cells confer associative thinking, logical reasoning, imagination, computation, and so on. Thus, the synapse innervations determine the specificity of memory contents [5–7].

To general characters for neurons and their functional consequences, the number and the functional state of primary associative memory cells influence memory strength and maintenance. The recruited number of associative memory cells is proportional to the number of their mutual synapse innervations produced by neuronal coactivation strength and repetitive coactivations during learning as well as is influenced by developmental stages [3, 11]. The function state of associative memory cells is regulated by the strengths of innate and new synapse inputs as well as synapse inputs from the arousal system, their ability to convert synaptic analogue signals into digital spikes, and their capability to output spikes [34, 114–116]. In addition, glutamatergic associative memory cells may suppress the activity of other neurons through GABAergic associative memory cells and lateral inhibition to have themselves to be dominantly active for memory contrast [9, 104].

Once mutual synapse innervations are established between two groups of associative memory cells among sensory cortices, each group of associative memory cells is able to encode the innate signal and the new signal from cross-modal associative memory cells for their integrative storage. In other words, innate signal cue and cross-modal signal cue are potentially to activate each group of associative memory cells. Each signal cue is able to activate these two groups of associative memory cells. Based on their natural pathways for the reflexes or behavioral responses to cues, the activation of each group of associative memory cells is able to induce these two behavioral activities [3, 8–10]. For instance, stimulations to one group of associative memory cells by electrical or optogenetic approach should induce two behavioral responses naturally to their innate cues. The activation to one group of associative memory cells by one of these two cues should induce these two behavioral responses. Spontaneous activation in one group of associative memory cells will lead to these two behavior activities. Therefore, the identification of associative memory cells, which are recruited to encode two signals, by the stimulation or activation protocol, should see two behavioral activities naturally induced by these two cues [5]. This principle can be used for the associative memory cells that encode multiple signals.

Synapse innervations to associative memory cells determine memory specificity. The number and function state of associative memory cells as well as the connection and activity strengths in their synapse inputs and axon output partners influence memory power and persistence [11, 12, 106, 108]. For instance, associative memory cells in the barrel cortex receive new synapse innervations from associative memory cells in the piriform cortex after associative learning alongside the innate inputs from the thalamus. Synapse activity in the pathway of odor signal will drive barrel cortical associative memory cells toward spiking threshold under the basal activity driven by thalamic inputs. Once spike threshold reaches, the spikes trigger downstream motor cortical neurons for odorant-induced whisker motion. With associative memory cells in sensory cortices [3, 11, 12], their axon-innervated downstream neurons are able to encode these associated signals [13, 67, 117–121]. Stimulations to any of these areas in neural circuits from sensory cortices to behavior- and emotion-related brain nuclei induce memory retrievals [61, 84, 85, 87, 88, 91]. In terms of the determination of proteins or memory cells for particular memory contents, our thought is given below. There are approximately 10,000 types of proteins in living cells [122]. The kinds of proteins are much less than the quantity of unitary signals remembered in the life, such as words, unitary visual signals, odorants, and so on. On the other hand, more than ten billion of neurons are included in the central nervous system. Those neurons plus synapse interconnections patterns, i.e., associative memory cells and their circuits, may constitute the enriched number of basic units in memory traces that are responsible for storing various learnt signals. Therefore, it is less possible for a specific protein memorizing a specific content of memory [5].

Each of sensory cortical neurons receives one type of innate sensory inputs via sensory afferent axons and their synapse innervations as well as encodes this particular innate sensory signal. Once this group of sensory cortical neurons is coactivated with sensory cortical neurons that encode other sensory signals in cross-modal or intramodal manner during learning processes, their mutual synapse innervations recruit them to be primary associative memory cells that encode the integrative storage and the reciprocal retrieval of innate signal and associatively learned signals [5]. The types of associatively memorized signals in each sensory cortical neuron are dependent on how many kinds of sensory cortical neurons have been coactivated during various associative learning processes. In other words, when individuals have experienced more associative learnings, each of the associative memory cells in their brains will store more types of associated signals for them to efficiently perform the associations relevant to future associative learning, new associative memory, reciprocal retrieval, logical reasoning, associative thinking, and so on [6, 7].

5.4 Associative Memory Cells in Cognition- and Emotion-Related Brain Areas

The axons of primary associative memory cells innervate their downstream neurons in brain areas related to cognition, emotion, and behaviors, in which convergent and divergent innervations recruit their downstream neurons to be secondary associative memory cells [5–7]. In other words, besides primary associative memory cells recruited in sensory cortices to encode the integrative storage and reciprocal retrieval of exogenous associated signals, the secondary associative memory cells that integrate and store endogenous signals may be recruited during cognition, emotion, and behaviors [6]. As we know, contents, events, and consequences produced in associative thinking, imagination, logical reasoning, and computation can be remembered. In addition, emotional responses to different natural stimulations and social interactions can be represented and recollected. Based on the associative storage of learned exogenous signals in sensory cortices, such as pictures, stories, tastes, and odors, specific contents in cognition and emotion in mind may be generated and stored. These exogenous signals are transmitted to neurons in cognition-, emotion-, and behavior-relevant brain areas by the axons of primary associative memory cells in sensory cortices and can be memorized in the downstream brain areas of sensory cortices in an integrative manner for subsequent retrievals. In terms of cellular substrates, the reorganized associations of signals stored in sensory cortices may enable primary associative memory cells to yield more mutual synapse innervations among them and to strengthen their mutual synapse innervation. Their activity also strengthens convergent innervation onto downstream neurons and receives the feedback synapse innervations during cognitive events and emotional activity. The downstream neurons of these primary associative memory cells become able to encode those associated signals and are recruited to be secondary associative memory cells that memorize specific contents generated during logical reasoning and associative thinking [13, 14]. The feedback and feedforward interactions among these primary and secondary associative memory cells can enable sensory sources be included in logical reasoning and associative thinking. Their interconnections constitute the bases for memory to be essential for cognition and emotion as well as for the cognition and emotion to influence memory formation and efficiency [6, 7] (Figs. 5.2 and 5.3).

In terms of brain areas that may produce secondary associative memory [7], prefrontal cortical neurons show a sustained activity after pair stimulation [119, 120]. The inferotemporal cortical neurons become responding to learning cues after associative learning [118]. The amygdala neurons become responding to conditioned and unconditioned stimulations and their response transformation [123]. Hippocampal and amygdala neurons have been found to work for the contextual fear memory [124]. The memory cell assemblies in response to the temporal signals are overlapped and recorded in the hippocampus [117]. The stimulations to engram cells in the amygdala or the hippocampus are able to induce fear response [84, 85]. These experiment results indicate that memory cells can be recruited in the hippocampus, amygdala, prefrontal cortex, and associative cortices for information retrievals [88, 125]. It remains unclear whether these memory cells receive synapse innervations from primary associative memory cells in the sensory cortices. Whether these memory-relevant neurons in the prefrontal cortex, amygdala, hippocampus, and temporal cortex are able to encode associated signals, presumably secondary associative memory cells need to be examined.

In the mouse model of associative learning by pairing whisker and odor stimulations, neuronal activities are in vivo recorded by unitary discharges of local field potential in the hippocampus [13, 14]. In mice showing odorant-induced whisker motion and whisker-induced olfactory responses, a portion of unitary neurons in the dentate gyrus of hippocampal formation are able to encode both whisker and odor signals. As their activities are blocked by pharmacologically inhibiting the barrel and piriform cortices with CNQX and D-AP5 (the antagonists for AMPA and NMDA types of glutamatergic receptors, respectively), the encoding of these dentate gyrus neurons to whisker and olfactory is secondarily to sensory cortices. The hippocampal neurons that encode the storage of associated signals are called as secondary associative memory cells. In addition, AAV-CMV-ZsGreen and AAV-CMV-tdTomato are injected into barrel and piriform cortices, respectively, and some neurons in hippocampal dentate gyrus receive the convergent synapse innervation from barrel and piriform cortices. These morphological and functional studies indicate the recruitment of secondary associative memory cells in the hippocampal dentate gyrus [13, 14].

Furthermore, the activities of prefrontal cortical neurons are in vivo recorded by two-photon cell imaging in the mouse model of associative learning by pairing whisker, tail, and odor stimulations in order to identify secondary associative memory cells [13, 14]. In mice showing associative memory to all of these three signals, a proportion of unitary neurons in the prefrontal cortex becomes able to encode whisker, tail, and odor signals. As their activities are blocked by pharmacologically inhibiting barrel, piriform, and S1-tail cortices with CNQX and D-AP5, the encoding of whisker, tail, and olfactory in these prefrontal cortical neurons is secondarily to sensory cortices. The prefrontal cortical neurons that encode the storage of associated signals are called as secondary associative memory cells. Moreover, AAV-CMV-ZsGreen, AAV-CMV-tdTomato, and AAV-CMV-BFP are injected into barrel, S1-tail, and piriform cortices, respectively, and some neurons in the prefrontal cortex receive convergent synapse innervations from barrel, S1-tail, and piriform cortices. These functional and morphological studies indicate the recruitment of secondary associative memory cells in the prefrontal cortex [13, 14].

After associative learning by simultaneously pairing sensory signals, neurons that encode these associated signals are identified in the prefrontal cortex and hippocampus [13, 14], besides barrel and piriform cortices [3, 8, 11]. Responses of the neurons in the prefrontal cortex and hippocampus to associated signals are weakened by blocking sensory cortical functions. Individual neurons in the prefrontal cortex and hippocampus receive synapse innervations from coactivated sensory cortices after paired stimulations [13, 14]. These results provide functional and morphological evidences for the recruitment of secondary associative memory cells in the prefrontal cortex and the hippocampus. A recruitment of secondary associative memory cells needs their coactivity with primary associative memory cells in sensory cortices [6].

Recent studies have indicated that memory cells in the downstream of sensory cortical associative memory cells undergo cross-modal connections [6]. Social memory needs the pathway from the ventral hippocampus to the nucleus accumbens [126]. Contextual fear memory is accompanied by the emergence of engram cells in the prefrontal cortex with receiving innervations from the hippocampus and amygdala [127]. Fear memory is also accompanied by axon projection from the prefrontal cortex and the hippocampus to the amygdala [128]. Moreover, reward memory requires the neural pathway from the prefrontal cortex to the striatum [87].

The characters of secondary associative memory cells in the brain regions related to cognition and emotion as well as association cortices include the natures for associative memory cells and neurons. Secondary associative memory cells receive new axon projection and synapse innervations from primary associative memory cells in those coactivated sensory cortices in convergent and divergent manners during cognitive events and emotional reactions. Secondary associative memory cells are able to encode endogenous associated signals propagated through the axons of primary associative memory cells in sensory cortices for their integrations and storages. The associations of those cognitive activities and emotional reactions induce synapse interconnections among secondary associative memory cells. The axons of these secondary associative memory cells project to the memory output cells in behavior-relevant brain regions for executing memory retrieval by language, writing, gesture, and countenance. There is a proportional relationship between the recruited number of secondary associative memory cells and their synapse interconnections induced by coactivation strength and repetitive coactivations during the cognition. These processes are influenced by development stage [5] (Figs. 5.2 and 5.3).

In addition, secondary associative memory cells recruited in the association cortices as well as brain areas in relevance to cognition and emotion possess the general characters of neurons in the brain. The encoding ability of secondary associative memory cells is controlled by the driving force from various synapse inputs, their capability to convert synaptic signals into neuronal spikes as well as their capability to propagate spikes out that trigger memory output cells [6]. The synapse inputs onto secondary associative memory cells can determine the specificity of their memory contents generated during cognitive activities and emotional reactions. The number and encoding capability of secondary associative memory cells as well as their interconnections align memory strength and maintenance. The stimulation and activity of secondary associative memory cells will lead to the rehearsals of logical reasoning, associative thinking, and emotional reactions. In addition, axonal outputs of secondary associative memory cells innervate the hypothalamus and extrapyramidal system. These synapse innervations will regulate the sympathetic and parasympathetic nerve balance through interaction between the hypothalamus and brain stem. The hypothalamus will also set up body temperature, regulate flight-fight, balance food ingestion, and regulate circadian and hormones that are involved in emotion reactions and behaviors [5, 7] (Fig. 5.3).

Associative memory cells detected in cerebral cortices include glutamatergic neurons, GABAergic neurons, and astrocytes [3, 8–11, 69, 104]. The interconnections between glutamatergic associative memory cells and GABAergic associative memory cells are mutually upregulated after memory formation [8, 9]. It is interested to know how these associative memory cells as basic units in memory traces work together to imprint those specific associated signals in the brain. When glutamatergic associative memory cells are activated by synapse inputs carrying in associated signals, their excitation activates GABAergic associative memory cells and in turn limits the activity of their neighboring associative memory cells by the lateral inhibition mediated through these GABAergic associative memory cells. This microcolumn of alternative excitation and inhibition from associative memory cells may lead to the memory of associated signals in a contrast manner. In the meantime, these glutamatergic associative memory cells also activate GABAergic associative memory cells and regulate themselves not to be overexcited by the recurrent inhibition [7] (Fig. 5.5). In terms of interactions among associative memory neurons and astrocytes, the working load of associative memory neurons can be supported by associative memory astrocytes that transfer the nutrients and waste products between the neurons and blood vessels [3, 7, 11].

Associative memory cells can be recruited among cross-modal sensory cortices, association cortices as well as brain areas in relevance to cognitive events and emotions. As the pair-coding neurons in response to two pictures have been recorded in the visual cortex [60, 113], associative memory cells can be recruited in intramodal cortices, such as associated photon beams in pictures to the visual system, associated odor signals to the olfactory system, associated letters and words to the auditory system, and so on (Figs. 5.2 and 5.3). Neural afferent pathways bringing the associated signals in a sensory modality may synaptically connect to several groups of neurons in the genetically and developmentally assigned sensory cortex. The

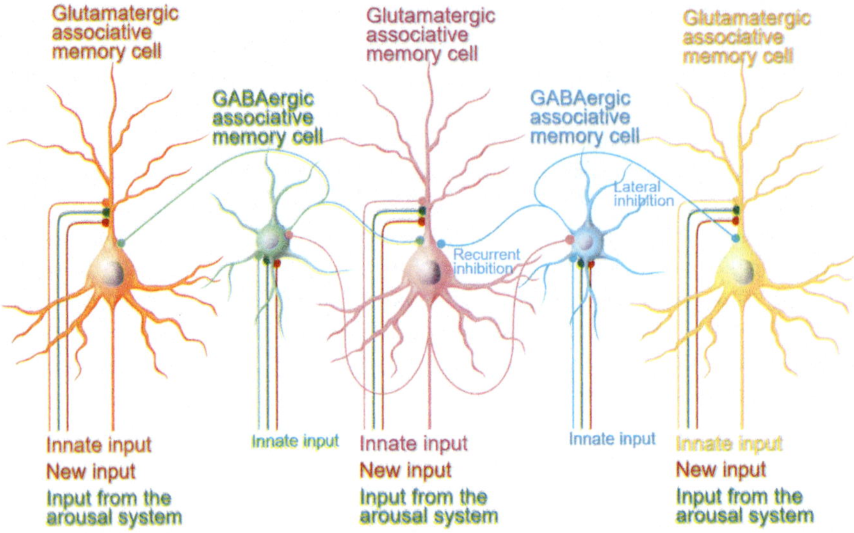

Fig. 5.5 Glutamatergic and GABAergic associative memory cells and their interconnections are illustrated. Three glutamatergic associative memory cells and two GABAergic associative memory cells receive their innate synapse input (identical color with soma), the arousal system input (green) and the new synapse input (red). Glutamatergic associative memory cells in the middle place send its axonal branches to innervate GABAergic associative memory cells. If the GABAergic associative memory cell innervates neighboring glutamatergic associative memory cells, this pathway constitutes the lateral inhibition to suppress the activity of these neighbors. If the GABAergic associative memory cell innervates back to this glutamatergic associative memory cell, this pathway constitutes the recurrent inhibition to suppress itself not to be overly excited

neurons within each group encode one of these associated signals. For example, the different groups of neurons in the auditory cortex receive neural afferents that carry in neural signals about different frequency sounds in a point-by-point manner. Each group of auditory cortical neurons encodes one of specific frequency sounds. Different neurons in the visual cortex receive neural afferents that carry in neural signals about different photon beams in an image that are dissected by different retina cone cells through point-by-point way. The coactivations of the sensory cortical neurons that encode different intramodal signals are able to induce their synapse interconnection. Associative memory cells in a single modality of sensory cortices are recruited. These different groups of associative memory cells in a sensory cortex can encode the integration and the storage of the intramodal signals about their different features, strengths, and locations. Therefore, memories to intramodal associated signals can be retrieved in the future. For example, object one can induce the recall of object two in a picture, an odorant induces the recall of another odorant, and word one induces the recall of word two, or the other way around [7]. The time delays often occur among intramodal signals, such as picture-by-picture in videos and letter-by-letter in language. The activity persistence among different sets of sensory cortical neurons in response to these sequential intramodal signals may play an important role in the overlap of their activities. For example, the afterdischarge from group one of neurons in response to intramodal signal one coincides the discharger from group two of neurons in response to intramodal signal two. This neuronal feature may allow the partially temporal overlap in the coactivity among sensory cortical neurons. Their coactivations induce their mutual synapse innervations and in turn recruit them as intramodal associative memory cells. The different portions, activity strengths, and connections of these intramodal associative memory cells work for the integrative storage and retrieval of intramodal signals with different features [5, 129]. The intramodal associative memory cells may also be recruited within one of cognition-/emotion-relevant brain regions as the secondary grade in nature, which work for cognitive activities and emotional reactions [6] (Figs. 5.2 and 5.3).

The relationship between primary and secondary associative memory cells and their role in memory-relevant events are summarized below. Basic architectures for their interactive working together are their connection patterns. For instance, mutual synapse innervations are present among primary associative memory cells in sensory cortices [3, 8, 9, 11]. The axons of the primary associative memory cells innervate convergently onto the secondary associative memory cells in association cortices as well as in brain regions related to cognitive activities and emotional reactions [13, 14, 121]. Each set of primary associative memory cells reciprocally connects with one group of secondary associative memory cells, whose functional roles are closely relevant (Fig. 5.2). The axons from all these primary and secondary associative memory cells terminate onto memory output cells in motor-related cortices for the memory output and innate reflex (Fig. 5.3). These primary associative memory cells based on their mutual synapse innervations found interactional circuits for the integrative storage of associated signals and their reciprocal retrievals. The reciprocal retrievals of associated signals include the learning cue-induced retrieval and the automatic conversion retrieval among different modalities. The reciprocal retrieval induced by learning cue refers to a situation that a signal induces the recall of its associated signal, or the other way around. The reciprocal retrieval in automatic conversion among different sensory modalities refers to a situation that the learnt and stored signals from a sensory modality can be automatically represented by another modality, e.g., those signals in an image are recalled by verbal language spontaneously [5]. Convergent synapse innervations from primary associative memory cells to secondary associative memory cells (Fig. 5.3) will confer logical reasoning, associative thinking, and other integrative cognitions (such as abstraction) induced by one of cues [13]. As showed in Fig. 5.2, one of secondary associative memory cells receives convergent synapse innervations from three groups of primary associative memory cells. The convergence of three signals from primary associative memory cells maintains basal activity in this secondary associative memory cell. When the learning cue activates three sets of primary associative memory cells through their synapse interactions, these primary associative memory cells will activate this secondary associative memory cell through spatial summation. In the meantime, this secondary associative memory cell will be dominantly activated by its innate chain from one group of primary associative memory cells. Therefore, three kinds of signals triggered by one of these learning cues can drive this secondary associative memory cell to integrate three associated signals for integrative cognitions, such as logical reasoning, associative thinking, and summary. Moreover, this integration may be facilitated by mutual synapse innervations among secondary associative cells that contribute to the interactions of the higher order cognition and emotion (Fig. 5.2). On the other hand, the divergent synapse innervations from primary associative memory cells in sensory cortices to secondary associative memory cells in the prefrontal cortex, the hippocampus, and other areas [13, 14, 121] (Figs. 5.2 and 5.3) enable the associated signals be stored in several brain regions for persistent maintenance with less chance being loss, to be used for different cognitive events and emotional reactions as well as to be generalized in concepts and knowledge [7, 13]. Furthermore, the functional compatibility between presynaptic axon branches and postsynaptic partners [114] plays critically important role in the coordination of these associative memory cells for memory formation and decay in multiple uses (Fig. 5.6). Besides this feedforward connection from primary associative memory cells to secondary associative memory cells, there may be feedback connections from secondary associative memory cells to primary associative memory cells. These feedforward and feedback connections grant consequences that the learned exogenous signals automatically initiate cognitions and emotions as well as those endogenous signals generated from cognitive events and emotional reactions contain sensory signal origins [7] (Fig. 5.2).

In addition to the interconnections among the associative memory cells in the linear manners of divergence, convergence, feedforward, and feedback, the mutual synapse innervations can be crossed among associative memory cells and/or their groups. (1) One associative memory cell or one group of associative memory cells can become the core connected with all other associative memory cells, which may fall in intramodal and cross-modal manners. The mechanism for this core of associative memory cells is proposed below. In addition to innate synapse innervations, new and mutual synapse innervations are formed during the coactivation of ready-recruited neurons with other cells. Once the new and mutual synapse innervations are formed among two or more groups of ready-recruited neurons, associative memory cells are recruited from these neurons. The specificity of associated signals stored in associative memory cells or their group is based on their synapse innervations that carry in these signals. In future learnings, if one group of associative memory cells is associatively activated with other groups of ready-recruited neurons or of associative memory cells that memorize different types of associated signals, this group of associative memory cells may form mutual synapse innervations with other groups of associative memory cells. With convergent innervations from other groups of associative memory cells, this group of associative memory cells may take a position to be the core of other groups of associative memory cells (Fig. 5.7a). This type of interconnections among the groups of associative memory cells may work for the integrative storage of those associated signals that share one of common characters. In addition, all these groups of

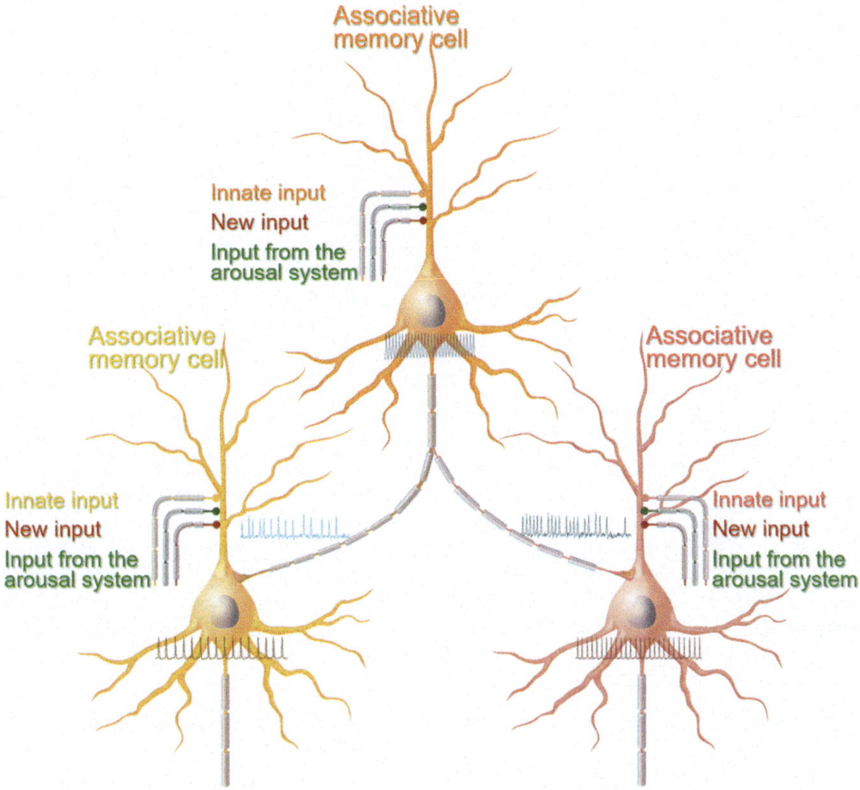

Fig. 5.6 The functional compatibility between the axon branches of presynaptic associative memory cell and their postsynaptic associative memory cell partners. Presynaptic associative memory cells have two axonal branches that synaptically innervate onto two partners of postsynaptic associative memory cells, respectively. These associative memory cells receive their innate synapse input (identical color with soma), the arousal system input (green) and the new synapse input (red). Sequential spikes induced in presynaptic associative memory cells by depolarization pulse. These spikes are propagated to axonal branches and further toward presynaptic terminals to trigger neurotransmitter release. These neurotransmitters act on postsynaptic receptors to drive postsynaptic associative memory cells to fire sequential spikes. Two presynaptic axon branches show spike patterns with high and low frequencies, respectively. Two postsynaptic associative memory cells show spike patterns high and low frequencies, respectively. Presynaptic axon branches with high spike frequency connect postsynaptic cells with high spike frequency, or vice versa. Thus, there is the compatibility between presynaptic axon branches and their postsynaptic partner cells in spike-encoding capability that is denoted by spike frequency

associative memory cells can be mutually innervated once they have been coactivated (Fig. 5.7b), which may fulfill the integrative storage of associated signals that are grouped in similar concepts or knowledge. The functions of these types of interconnections among associative memory cells are given in Chaps. 7 and 8.

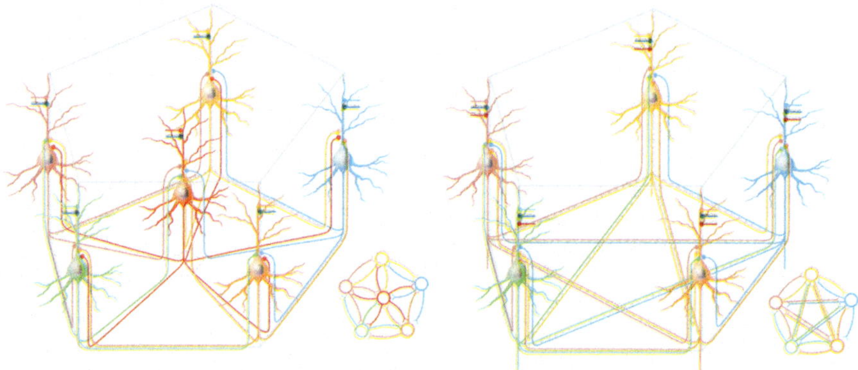

Fig. 5.7 Associative memory cells and their connection patterns. Left panel shows an associative memory cell in the center area (red) interconnects with five associative memory cells in surround by mutual synapse innervations. Their interconnection pattern is showed as a horizontal section in bottom view. This connection pattern of primary and secondary associative memory cells is presumably relevant to a core signal for other associated signals among associative memory cells in sensory cortices and in cognition-relevant brain areas. Right panel shows all associative memory cells interconnected one another by their mutual synapse innervations. Their interconnection pattern is also showed as a horizontal section in bottom view. These associative memory cells receive their innate synapse input (identical color with soma), the arousal system input (green) and the new synapse input (red)

5.5 Memory Output Cells in Behavior-Related Brain Areas

In addition to primary associative memory cells to encode the integrative storage of exogenous associated signals and secondary associative memory cells to encode integrative storage of endogenous signals from cognition and emotion, there is a group of the neurons called as memory output cells in motor cortices [6]. These memory output cells receive synapse innervations from primary associative memory cells in sensory cortices, which work for reflexes induced by innate signals, i.e., innate reflexes, and responses induced by one of associated signals, i.e., the memory retrieval of primary associated signals [121]. These memory output cells also receive synapse innervations from secondary associative memory cells in cognition- and emotion-relevant brain areas, which execute the retrieval of signals generated from cognitive events and emotional reactions [5–7] (Fig. 5.3).

In terms of experimental evidences for these memory output cells, the electrophysiological recording of unitary discharges in local field potentials is performed in the motor cortex and sensory cortex in vivo. In mice that show associative memory trained by pairing whisker and odor signals, a portion of barrel cortical neurons becomes able to encode whisker and olfactory signals, i.e., primary associative memory cells. In the meantime, some motor cortical neurons are able to encode these associated signals, i.e., memory output cells. The simultaneous presences of primary associative memory cells in sensory cortices and memory output cells in the motor cortex are identified by two-photon cell imaging in vivo

[121]. Their primary and secondary natures are confirmed by pharmacological approach. The inhibition of sensory cortices by using antagonists of NMDA and AMPA types of glutamatergic receptors completely blocks the functions of primary associative memory cells and memory output cells. On the other hand, the inhibition of the motor cortex by using the antagonist of NMDA and AMPA receptors completely blocks the functions of memory output cells, but primary associative memory cells are still functional. This result indicates that memory output cells are located after primary associative memory cells and are influenced by primary associative memory cells. This indication has been proved by morphological study by using AAV-CMV-fluorescent proteins, in which motor cortical neurons receive synapse innervations from barrel cortical neurons [121].

The persistence of encoding associated signals appears to be different in primary associative memory cells and memory output cells. After odorant-induced whisker motion and whisker-induced olfactory responses are onset trained by pairing whisker and odor stimulations in mice, the discontinued training about 1–2 weeks leads to the significant decay of associative memory and even disappearance [106, 121]. This phenomenon seems to indicate memory extinction without learning practices. When electrophysiological recording in vivo is performed in the barrel cortex and the motor cortex in mice, the onset of associative memory is followed by detecting primary associative memory cells that encode two associated signals in the barrel cortex and memory output cells that encode these two associated signals in the motor cortex, whereas the disappearance of associative memory is accompanied by the failure detection of memory output cells in the motor cortex but the presence of the primary associative memory cells in the barrel cortex. This result indicates that the decay or extinction of associative memory is likely based on the functional decay of memory output cells [106, 121]. Since primary associative memory cells are still encoding associated signals, the rehearsal of associated signals will convert silent or inactive memory output cells to be functional or active, which has been confirmed by experiments that an additional training evokes memory output cells to be reemergence [106, 121]. In real life, learning cues are much helpful in the retrieval of previously learned information that has been thought of extinction [5].

In addition, the functional plasticity including neuronal excitability and synaptic transmission in memory output cells appears upregulated in the onset of associative memory, decayed after the disappearance of associative memory and upregulated after retaining for the reemergence of associative memory again [106]. The coordinated changes in the conversion between inactive and active states as well as between functional upregulation and decay may well explain memory formation, extinction, and reemergence [5].

Therefore, the characters of memory output cells are described below. Memory output cells receive synapse innervations from primary and secondary associative memory cells. Memory output cells encode associated signals based on the activities of primary and secondary associative memory cells. Memory output cells express functional fluctuation in their encoding associated signals and functional plasticity, which is largely based on giving the cues of associated signals. The more active the memory output cells are, the more persistent the memory output cells encode associated signals. This feature indicates an approach to enhance memory retrieval through more practices that can maintain memory output cells to be persistently active [5]. It is noteworthy that memory output cells receive the direct synapse innervations from primary associative memory cells; they may be thought to fall into the category of secondary associative memory cells [5]. The characters of associative memory cells may also suitable for memory output cells. For instance, the function state of memory output cells is influenced by synapse inputs, their capability to convert synaptic analogue signals into digital spikes as well as their capability to propagate spikes output, which activate the motor neurons to guide behavior [6]. The synapse innervations onto memory output cells determine the specificity of behavior output in relevance to memory contents. The number and excitability of memory output cells as well as their connection and activity strengths set up memory persistence and strength dominantly relevant to memory contents.

References

1. Kandel ER, Pittenger C. The past, the future and the biology of memory storage. Philos Trans R Soc Lond Ser B Biol Sci. 1999;354(1392):2027–52.
2. Suzuki WA. Associative learning signals in the brain. Prog Brain Res. 2008;169:305–20.
3. Wang D, et al. Neurons in the barrel cortex turn into processing whisker and odor signals: a cellular mechanism for the storage and retrieval of associative signals. Front Cell Neurosci. 2015;9:320.
4. Wasserman EA, Miller RR. What's elementary about associative learning? Annu Rev Psychol. 1997;48:573–607.
5. Wang J-H. Searching basic units of memory traces: associative memory cells. F1000Research. 2019;8(457):1–28.
6. Wang JH, Cui S. Associative memory cells: formation, function and perspective. F1000Res. 2017;6:283.
7. Wang JH, Cui S. Associative memory cells and their working principle in the brain. F1000Res. 2018;7:108.
8. Liu Y, et al. Piriform cortical glutamatergic and GABAergic neurons express coordinated plasticity for whisker-induced odor recall. Oncotarget. 2017;8(56):95719–40.
9. Yan F, et al. Coordinated plasticity between barrel cortical glutamatergic and GABAergic neurons during associative memory. Neural Plast. 2016;2016(ID5648390):1–20.
10. Wang JH, et al. Upregulation of glutamatergic receptor-channels is associated with cross-modal reflexes encoded in barrel cortex and piriform cortex. Biophys J. 2014;106(2):supplement 191a.
11. Feng J, et al. Barrel cortical neuron integrates triple associated signals for their memory through receiving epigenetic-mediated new synapse innervations. Cereb Cortex. 2017;27(12):5858–71.
12. Lei Z, et al. Synapse innervation and associative memory cell are recruited for integrative storage of whisker and odor signals in the barrel cortex through miRNA-mediated processes. Front Cell Neurosci. 2017;11(316):1–11.
13. Wang J-H, et al. Prefrontal cortical neurons are recruited as secondary associative memory cells for associative memory and cognition. Biophys J. 2018;114(3):155a.
14. Wang JH, et al. Secondary associative memory cells and their plasticity in the prefrontal cortex. Biophys J. 2019;116(3):427a.

15. Hebb DO. The organization of behavior, a neuropsychological theory. New York: Wiley; 1949.
16. Hebb DO. Animal and physiological psychology. Annu Rev Psychol. 1950;1:173–88.
17. Poo MM, et al. What is memory? The present state of the engram. BMC Biol. 2016;14:40.
18. Kandel ER, Dudai Y, Mayford MR. The molecular and systems biology of memory. Cell. 2014;157(1):163–86.
19. Semon RW. In: Semon RW, editor. Mnemic psychology. London: Allen, Unwinis; 1923.
20. Tonegawa S, et al. Memory engram storage and retrieval. Curr Opin Neurobiol. 2015;35:101–9.
21. Black J, Greenough W. Developmental approaches to the memory process. In: Martinez J, Kesner R, editors. Learning and memory. San Diego: Academic; 1991. p. 61–91.
22. Brown M, Keynes R, Lumsden A. Development of cerebral cortex and cerebellar cortex. In: Brown MEA, editor. The development of brain. New York: Oxford University Press Inc; 2001. p. 169–93.
23. Chang H, et al. The protocadherin 17 gene affects cognition, personality, amygdala structure and function, synapse development and risk of major mood disorders. Mol Psychiatry. 2017;231:1–13.
24. Dajani DR, Uddin LQ. Demystifying cognitive flexibility: implications for clinical and developmental neuroscience. Trends Neurosci. 2015;38(9):571–8.
25. Dumas TC. Developmental regulation of cognitive abilities: modified composition of a molecular switch turns on associative learning. Prog Neurobiol. 2005;76(3):189–211.
26. Cohen Kadosh R, Walsh V. Cognitive neuroscience: rewired or crosswired brains? Curr Biol. 2006;16(22):R962–3.
27. Chen N, et al. The refractory periods and threshold potentials of sequential spikes measured by whole-cell recordings. Biochem Biophys Res Commun. 2006;340:151–7.
28. Chen N, et al. Sodium channel-mediated intrinsic mechanisms underlying the differences of spike programming among GABAergic neurons. Biochem Biophys Res Commun. 2006;346:281–7.
29. Chen N, Chen X, Wang J-H. Homeostasis established by coordination of subcellular compartment plasticity improves spike encoding. J Cell Sci. 2008;121(17):2961–71.
30. Hodgkin AL, Huxley AF. Resting and action potentials in single nerve fibres. J Physiol. 1945;104(2):176–95.
31. Hodgkin AL, Huxley AF. A quantitative description of membrane current and its application to conduction and excitation in nerve. J Physiol. 1952;117(4):500–44.
32. Shepherd GM. Synaptic potentials and synaptic integration. In: Shepherd GM, editor. Neurobiology. New York: Oxford University Press; 1998. p. 97–117.
33. Stevens CF, Wang Y-Y. Facilitation and depression at single central synapses. Neuron. 1995;14:795–802.
34. Wang JH, et al. The gain and fidelity of transmission patterns at cortical excitatory unitary synapses improve spike encoding. J Cell Sci. 2008;121(17):2951–60.
35. Kandel ER. Nerve cells and behavior. In: Kandel ER, Schwartz JH, Jessell TM, editors. Principles of neural science. 4th ed. New York: McGraw-Hill; 2000. p. 19–35.
36. Thompson RF. In search of memory traces. Annu Rev Psychol. 2005;56:1–23.
37. Thompson RF. An essential memory trace found. Behav Neurosci. 2013;127(5):669–75.
38. Reijmers LG, et al. Localization of a stable neural correlate of associative memory. Science. 2007;317(5842):1230–3.
39. Tayler KK, et al. Reactivation of neural ensembles during the retrieval of recent and remote memory. Curr Biol. 2013;23(2):99–106.
40. Link W, et al. Somatodendritic expression of an immediate early gene is regulated by synaptic activity. Proc Natl Acad Sci U S A. 1995;92(12):5734–8.
41. Morgan JI, Curran T. Stimulus-transcription coupling in neurons: role of cellular immediate-early genes. Trends Neurosci. 1989;12(11):459–62.
42. Kiessling M, Gass P. Immediate early gene expression in experimental epilepsy. Brain Pathol. 1993;3(4):381–93.

43. Meldrum BS. Concept of activity-induced cell death in epilepsy: historical and contemporary perspectives. Prog Brain Res. 2002;135:3–11.
44. Wang X, et al. Persistent hyperactivity of hippocampal dentate interneurons after a silent period in the rat pilocarpine model of epilepsy. Front Cell Neurosci. 2016;10:94.
45. Simonato M, et al. Differential expression of immediate early genes in the hippocampus in the kindling model of epilepsy. Brain Res Mol Brain Res. 1991;11(2):115–24.
46. Abe H, Nowak TS Jr. Induced hippocampal neuron protection in an optimized gerbil ischemia model: insult thresholds for tolerance induction and altered gene expression defined by ischemic depolarization. J Cereb Blood Flow Metab. 2004;24(1):84–97.
47. Bokesch PM, et al. Dextromethorphan inhibits ischemia-induced c-fos expression and delayed neuronal death in hippocampal neurons. Anesthesiology. 1994;81(2):470–7.
48. Kiessling M, et al. Differential transcription and translation of immediate early genes in the gerbil hippocampus after transient global ischemia. J Cereb Blood Flow Metab. 1993;13(6):914–24.
49. Carr MF, Jadhav SP, Frank LM. Hippocampal replay in the awake state: a potential substrate for memory consolidation and retrieval. Nat Neurosci. 2011;14(2):147–53.
50. Harris KD, et al. Organization of cell assemblies in the hippocampus. Nature. 2003;424(6948):552–6.
51. Ji D, Wilson MA. Coordinated memory replay in the visual cortex and hippocampus during sleep. Nat Neurosci. 2007;10(1):100–7.
52. Jadhav SP, et al. Awake hippocampal sharp-wave ripples support spatial memory. Science. 2012;336(6087):1454–8.
53. Kay K, et al. A hippocampal network for spatial coding during immobility and sleep. Nature. 2016;531(7593):185–90.
54. Kudrimoti HS, Barnes CA, McNaughton BL. Reactivation of hippocampal cell assemblies: effects of behavioral state, experience, and EEG dynamics. J Neurosci. 1999;19(10):4090–101.
55. McNaughton BL, Barnes CA, O'Keefe J. The contributions of position, direction, and velocity to single unit activity in the hippocampus of freely-moving rats. Exp Brain Res. 1983;52(1):41–9.
56. Sirota A, et al. Communication between neocortex and hippocampus during sleep in rodents. Proc Natl Acad Sci U S A. 2003;100(4):2065–9.
57. Skaggs WE, McNaughton BL. Replay of neuronal firing sequences in rat hippocampus during sleep following spatial experience. Science. 1996;271(5257):1870–3.
58. Wilson MA, McNaughton BL. Dynamics of the hippocampal ensemble code for space. Science. 1993;261(5124):1055–8.
59. Wilson MA, McNaughton BL. Reactivation of hippocampal ensemble memories during sleep. Science. 1994;265(5172):676–9.
60. Wirth S, et al. Single neurons in the monkey hippocampus and learning of new associations. Science. 2003;300(5625):1578–81.
61. Yokose J, et al. Overlapping memory trace indispensable for linking, but not recalling, individual memories. Science. 2017;355(6323):398–403.
62. Yu JY, et al. Specific hippocampal representations are linked to generalized cortical representations in memory. Nat Commun. 2018;9(1):2209.
63. Zhao J, Wang D, Wang J-H. Barrel cortical neurons and astrocytes coordinately respond to an increased whisker stimulus frequency. Mol Brain. 2012;5(12):1–10.
64. Nikolenko V, Poskanzer KE, Yuste R. Two-photon photostimulation and imaging of neural circuits. Nat Methods. 2007;4(11):943–50.
65. Stosiek C, et al. In vivo two-photon calcium imaging of neural networks. Proc Natl Acad Sci U S A. 2003;100(12):7319–24.
66. Wang KH, et al. In vivo two-photon imaging reveals a role of arc in enhancing orientation specificity in visual cortex. Cell. 2006;126(2):389–402.
67. Feng J, et al. Cell-specific plasticity associated with integrative memory of triple sensory signals in the barrel cortex. Oncotarget. 2018;9(57):30962–78.

68. Zhang G, et al. Upregulation of excitatory neurons and downregulation of inhibitory neurons in barrel cortex are associated with loss of whisker inputs. Mol Brain. 2013;6(1):2.
69. Gao Z, et al. Associations of unilateral whisker and olfactory signals induce synapse formation and memory cell recruitment in bilateral barrel cortices: cellular mechanism for unilateral training toward bilateral memory. Front Cell Neurosci. 2016;10(285):1–16.
70. Paxinos G, Watson C. The mouse brain: in stereotaxic coordinates. In: Paxinos G, Watson C, editors. The rat brain: in stereotaxic coordinates. London: Elsevier Academic Press; 2005.
71. Hama H, et al. Scale: a chemical approach for fluorescence imaging and reconstruction of transparent mouse brain. Nat Neurosci. 2011;14(11):1481–8.
72. Yan F, et al. Coordinated plasticity between barrel cortical glutamatergic and GABAergic neurons during associative memory. Neural Plast. 2016;2016(ID 5648390):1–15.
73. Bliss T, Lomo T. Long-lasting potentiation of synaptic transmission in the dentate area of the anaesthetized rabbit following stimulation of the perforant path. J Physiol Lond. 1973;232:331–56.
74. Debanne D, Poo MM. Spike-timing dependent plasticity beyond synapse – pre- and post-synaptic plasticity of intrinsic neuronal excitability. Front Synaptic Neurosci. 2010;2:21.
75. Stanton PK, Sejnowski TJ. Associative long-term depression in the hippocampus induced by hebbian covariance. Nature. 1989;339(6221):215–8.
76. Aizenmann C, Linden DJ. Rapid, synaptically driven increases in the intrinsic excitability of cerebellar nuclear neurons. Nat Neurosci. 2000;3:109–11.
77. Daoudal D, Debanne D. Long-term plasticity of intrinsic excitability: learning rules and mechanisms. Learn Mem. 2003;10:456–65.
78. Campanac E, Debanne D. Plasticity of neuronal excitability: Hebbian rules beyond the synapse. Arch Ital Biol. 2007;145(3–4):277–87.
79. Sourdet V, et al. Long-term enhancement of neuronal excitability and temporal fidelity mediated by metabotropic glutamate receptor subtype 5. J Neurosci. 2003;23(32):10238–48.
80. Zhang M, et al. Calcium signal-dependent plasticity of neuronal excitability developed postnatally. J Neurobiol. 2004;61:277–87.
81. Banerjee SB, et al. Perineuronal nets in the adult sensory cortex are necessary for fear learning. Neuron. 2017;95(1):169–179 e3.
82. Doucette W, et al. Associative cortex features in the first olfactory brain relay station. Neuron. 2011;69(6):1176–87.
83. Khodagholy D, Gelinas JN, Buzsaki G. Learning-enhanced coupling between ripple oscillations in association cortices and hippocampus. Science. 2017;358(6361):369–72.
84. Li H, et al. Experience-dependent modification of a central amygdala fear circuit. Nat Neurosci. 2013;16(3):332–9.
85. Liu X, et al. Optogenetic stimulation of a hippocampal engram activates fear memory recall. Nature. 2012;484(7394):381–5.
86. Raymond JL, Lisberger SG, Mauk MD. The cerebellum: a neuronal learning mechacine? Science. 1996;272:1126–30.
87. Otis JM, et al. Prefrontal cortex output circuits guide reward seeking through divergent cue encoding. Nature. 2017;543:103–7.
88. Pape HC, Pare D. Plastic synaptic networks of the amygdala for the acquisition, expression and extinction of conditioned fear. Physiol Rev. 2010;90(2):419–63.
89. Timmann D, et al. The human cerebellum contributes to motor, emotional and cognitive associative learning. A review. Cortex. 2010;46(7):845–57.
90. Weinberger NM. Specific long-term memory traces in primary auditory cortex. Nat Rev Neurosci. 2004;5:279–90.
91. Xu W, Sudhof TC. A neural circuit for memory specificity and generalization. Science. 2013;339(6125):1290–5.
92. Reder LM, Park H, Kieffaber PD. Memory systems do not divide on consciousness: reinterpreting memory in terms of activation and binding. Psychol Bull. 2009;135(1):23–49.
93. McGaugh JL. Time-dependent processes in memory storage. Science. 1966;153(3742):1351–8.

94. McLaughlin B. "Intentional" and "incidental" learning in human subjects: the role of instructions to learn and motivation. Psychol Bull. 1965;63:359–76.
95. Squire LR. Declarative and nondeclarative memory: multiple brain systems supporting learning and memory. J Cogn Neurosci. 1992;4(3):232–43.
96. Squire LR, Knowlton B, Musen G. The structure and organization of memory. Annu Rev Psychol. 1993;44:453–95.
97. Adler LL, Berkowitz PH. Influencing associative thinking and imagery in emotionally disturbed children. Psychol Rep. 1976;39(1):183–8.
98. Gordon R, Silverstein ML, Harrow M. Associative thinking in schizophrenia: a contextualist approach. J Clin Psychol. 1982;38(4):684–96.
99. Glassman RB. A working memory "theory of relativity": elasticity in temporal, spatial, and modality dimensions conserves item capacity in radial maze, verbal tasks, and other cognition. Brain Res Bull. 1999;48(5):475–89.
100. Martins J, Mendes RV. Neural networks and logical reasoning systems: a translation table. Int J Neural Syst. 2001;11(2):179–86.
101. Procyk E, Joseph JP. Problem solving and logical reasoning in the macaque monkey. Behav Brain Res. 1996;82(1):67–78.
102. Ramsey NF, et al. Excessive recruitment of neural systems subserving logical reasoning in schizophrenia. Brain. 2002;125(Pt 8):1793–807.
103. Wang J-H, et al. Neurons in barrel cortex turn into processing whisker and odor signals: a novel form of associative learning. Soc Neurosci. 2013;653(14):WW11.
104. Wang J-H, et al. Both glutamatergic and Gabaergic neurons are recruited to be associative memory cells. Biophys J. 2016;110(3):supplement 481a.
105. Zhang F, et al. mGluR1,5 activation improves network asynchrony and GABAergic synapse attenuation in the amygdala: implication for anxiety-like behavior in DBA/2 mice. Mol Brain. 2012;5(1):20.
106. Guo R, et al. Associative memory extinction is accompanied by decayed plasticity at motor cortical neurons and persistent plasticity at sensory cortical neurons. Front Cell Neurosci. 2017;11(168):1–12.
107. Liu Y, et al. Activity strengths of cortical glutamatergic and GABAergic neurons are correlated with transgenerational inheritance of learning ability. Oncotarget. 2017;8(68):112401–16.
108. Zhao X, et al. Coordinated plasticity among glutamatergic and GABAergic neurons and synapses in the barrel cortex is correlated to learning efficiency. Front Cell Neurosci. 2017;11(221):1–12.
109. Saneyoshi T, Fortin DA, Soderling TR. Regulation of spine and synapse formation by activity-dependent intracellular signaling pathways. Curr Opin Neurobiol. 2009;20(1):108–15.
110. Spitzer NC. Activity-dependent neuronal differentiation prior to synapse formation: the functions of calcium transients. J Physiol Paris. 2002;96(1–2):73–80.
111. Ueda H, et al. Distinct roles of cytoskeletal components in immunological synapse formation and directed secretion. J Immunol. 2015;195(9):4117–25.
112. Tagawa Y, Hirano T. Activity-dependent callosal axon projections in neonatal mouse cerebral cortex. Neural Plast. 2012;2012:797295.
113. Albright TD. On the perception of probable things: neural substrates of associative memory, imagery, and perception. Neuron. 2012;74(2):227–45.
114. Yang Z, et al. Functional compatibility between Purkinje cell axon branches and their target neurons in the cerebellum. Oncotarget. 2017;8(42):72424–37.
115. Yu J, et al. Quantal glutamate release is essential for reliable neuronal encodings in cerebral networks. PLoS One. 2011;6(9):e25219.
116. Yu J, Qian H, Wang JH. Upregulation of transmitter release probability improves a conversion of synaptic analogue signals into neuronal digital spikes. Mol Brain. 2012;5(1):26.
117. Cai DJ, et al. A shared neural ensemble links distinct contextual memories encoded close in time. Nature. 2016;534(7605):115–8.

118. Naya Y, Yoshida M, Miyashita Y. Forward processing of long-term associative memory in monkey inferotemporal cortex. J Neurosci. 2003;23(7):2861–71.
119. Takehara-Nishiuchi K, McNaughton BL. Spontaneous changes of neocortical code for associative memory during consolidation. Science. 2008;322(5903):960–3.
120. Viskontas IV. Advances in memory research: single-neuron recordings from the human medial temporal lobe aid our understanding of declarative memory. Curr Opin Neurol. 2008;21(6):662–8.
121. Wang J-H, Guo R, Wei Z. Associative memory extinction is accompanied by decays of associative memory cells and their plasticity at motor cortex but not sensory cortex. Soc Neurosci. 2017;81(09):10385.
122. Milo R. What is the total number of protein molecules per cell volume? A call to rethink some published values. BioEssays. 2013;35(12):1050–5.
123. Grewe BF, et al. Neural ensemble dynamics underlying a long-term associative memory. Nature. 2017;543(7647):670–5.
124. Xu C, et al. Distinct hippocampal pathways mediate dissociable roles of context in memory retrieval. Cell. 2016;167(4):961–972 e16.
125. Baldi E, Bucherelli C. Brain sites involved in fear memory reconsolidation and extinction of rodents. Neurosci Biobehav Rev. 2015;53:160–90.
126. Okuyama T, et al. Ventral CA1 neurons store social memory. Science. 2016;353(6307):1536–41.
127. Kitamura T, et al. Engrams and circuits crucial for systems consolidation of a memory. Science. 2017;356(6333):73–8.
128. Hubner C, et al. Ex vivo dissection of optogenetically activated mPFC and hippocampal inputs to neurons in the basolateral amygdala: implications for fear and emotional memory. Front Behav Neurosci. 2014;8:64.
129. Zhao J, Wang D, Wang JH. Barrel cortical neurons and astrocytes coordinately respond to an increased whisker stimulus frequency. Mol Brain. 2012;5:12.

Chapter 6
Plasticity of Associative Memory Cells

Abstract Neuronal plasticity has been observed in aspects of synapse transmission and neuron encoding capability, which has been thought to be relevant to learning and memory. The joint acquisitions, the integrative storages, and the reciprocal retrievals of associated signals are most commonly occurring in the life. Synaptic plasticity in a neural pathway may not reflect to associative processes in memory, though the plasticity of the neuronal encoding may be involved in the capability of signal integration in the cell body of neurons. Synaptic and neuronal plasticity is better to be studied in memory cells, which may be relevant to their recruitment and refinement. With the mouse model of associative learning by pairing cross-modal sensory signals, memory cells that encode the associated sensory signals have been functionally and morphologically identified in sensory cortices, prefrontal cortex, hippocampus, and motor cortex. In these associative memory cells, the excitatory synaptic transmission is upregulated, the inhibitory synaptic transmission is downregulated, the glutamatergic neuronal excitability is upregulated, and the GABAergic neuronal excitability is downregulated. These alternations may facilitate to shift the balance between excitation and inhibition toward more excitatory state in neural networks, in order to strengthen the recruitment of more associative memory cells through their mutual synapse innervations based on their coactivation. To prevent over-excitation at associative memory cells in the engram circuit, there are enhanced mutual innervations between glutamatergic and GABAergic associative memory cells. Moreover, there are functional conversions from silent synapse pathways to be functional or vice versa, as well as homosynaptic versus heterosynaptic plasticity in associative memory cells. These enriched forms of synaptic plasticity may be relevant to the complicated patterns of memory formation and cognitive events. In addition, the functional fluctuation of neuronal and synaptic plasticity in memory output cells may work for the memory formation, extinction, and reemergence.

Keywords Activity-dependent plasticity · Associative memory cells · Glutamate and γ-aminobutyric acid

J.-H. Wang, *Associative Memory Cells: Basic Units of Memory Trace*,
https://doi.org/10.1007/978-981-13-9501-7_6

6.1 Introduction

The plasticity of neurons and synapses have been observed in the central nervous system, which are presumably involved in learning and memory, such as long-term potentiation and long-term depression of synaptic transmission and/or neuronal excitability [1–4]. Numerous studies have provided evidences for the relationship between long-term plasticity and memory formation [5]. For instance, long-term potentiation of synaptic transmission is associated with memory formation. The downregulation of intracellular signaling pathways by pharmacological reagents and gene knockout attenuates both memory formation and long-term synaptic potentiation. The upregulation of intracellular molecules by pharmacological regents and gene overexpression facilitates both long-term synaptic potentiation and memory formation. These experiments suggest that synapse plasticity is featured as engrams for memory formation [6, 7], though some studies argue against this conclusion. It is noteworthy that the studies about neural plasticity are not conducted in memory cells, synaptic plasticity in a neural pathway may not reflect to the integrative storage of associated signals, as well as synaptic and neuronal plasticity can be observed under certain pathological conditions, such as epilepsy and anxiety [8–11]. In addition, there are no specific characters and quantitative changes of synaptic and neuronal plasticity in relevance to associative memory. Overly emphasizing the correlation between neural plasticity and memory formation should be cautioned. Instead, the physiological impacts of neural plasticity in memory cells on memories as well as memory-relevant cognition and emotion should be paid attention.

In a mouse model of associative learning by pairing whisker, olfactory, and tail signals, associative memory cells are recruited for the integrative storage and reciprocal retrieval of these associated signals. Their functional and morphological alternations are investigated by electrophysiological recording, two-photon cell imaging, and neural tracing in vivo and in vitro. In slice preparations of sensory cortices including barrel and piriform cortices, excitatory synaptic transmission on glutamatergic associative memory cells is upregulated, inhibitory synaptic transmission is downregulated, spike-encoding capability at glutamatergic neurons is upregulated, and spike-encoding capability at GABAergic neurons is downregulated [12–18]. Their functional plasticity may shift the balance between excitation and inhibition toward an excitatory state in the networks of associative memory cells and in turn facilitate the recruitment of more associative memory cells as well as strengthen their capability to encode spikes for memory retrieval. In the combined analysis of neural plasticity and the onset efficiency of associative memory, there are the positive correlations between memory efficiency and plasticity strength at excitatory neurons or synapses as well as the negative correlations between memory efficiency and plasticity strength at inhibitory neurons and synapses in mice and their offspring [15, 18]. Therefore, plasticity at associative memory cells assists the further recruitment of associative memory cells as well as the efficiency of associative memory.

By using two-photon cell imaging to study associative memory cells in vivo, their plasticity in the different pathways to carry signals appears complicated. After intensive stimulations to the different inputs, such as whisker, tail, and olfactory signals, some associative memory cells in response to these inputs are upregulated, and others in response to these inputs are downregulated. Moreover, the upregulated or downregulated responses to these inputs may be induced by either homosynaptic or heterosynaptic in nature, especially in secondary associative memory cells in the prefrontal cortex [19]. This result indicates that those complicated functions of the prefrontal cortex may be caused by the regulation of associative memory cells in a cell-by-cell manner.

In terms of plasticity at memory output cells, their functional regulation is closely related to pair-trainings. Their functional upregulation may be decayed quickly without the pair-training for a week as well as reemergence shortly after the pair-training, in comparison with persistent upregulation in primary associative memory cells [14, 20]. This dynamic change based on the pair-training may be related to the temporal extinction of memory and the quick recovery by learning cues [21].

Taken these data together, there are two types of activity-dependent plasticity at associative memory cells, neural plasticity coupled with the recruitment of associative memory cells and plasticity at associative memory cells induced by repetitively administering the learning cues. Plasticity at the primary and secondary associative memory cells appears to be maintained relatively longer, in comparison with plasticity at memory output cells that is decayed without pair-training [21]. In order to investigate neural plasticity coupled with the recruitment of associative memory cells, there are two approaches, functional and morphological comparisons between associative memory cells in memory formation in mice and cortical neurons in mice without associative learning as well as these comparisons between neurons in mice before associative learning and associative memory cells recruited from these neurons after memory formation. In the study of neural plasticity at associative memory cells evoked by the learning cues, the strategy will be the identification of associative memory cells after memory formation and then examination of their plastic changes induced by repetitively administering the learning cues.

6.2 Plasticity at Primary Associative Memory Cells Linked with Their Recruitment

Cell assemblies formed by their connection strengthening during the coincidence activity of presynaptic and postsynaptic cells presumably work for learning and memory [22]. This hypothesis is matched with synaptic and neuronal plasticity based on neuronal activity [23–26], e.g., long-term potentiation and depression in synaptic transmission [1, 2] or neuronal excitability [4, 27]. The studies of synapse and neuron plasticity are not conducted in memory cells despite brain regions in these studies presumably relevant to memory. Tetanus-induced LTP that expresses in a bundle of axons and their innervated synapses may be due to the decrease of their spike threshold, instead of its expression at individual synapses. If the strengthening of synaptic transmission expresses in a given neural pathway, this potentiation cannot explain the integrative storage of associated signals carried through different synaptic inputs in associative memory cells. Moreover, tetanus-induced LTP usually cannot express at synapses in unmatured animals, such that it is not relevant to learning and memory in early life when the learning is efficient and memory is impressive. Long-term depression at synapses is induced by low-frequency stimulation in that 1 Hz stimulations do not mimic any physiological spike patterns in neurons as well as results from the extensive depletion of presynaptic vesicles, especially the synapses in young ages [28–30]. The associations of potentiation in some neural pathways and depression in other neural pathways [31] may be relevant to the contrast in memories [21]. These uncertainties question how these data about neural plasticity are written in the profile of cellular mechanisms underlying associative memory.

The potentiation of synaptic transmission may be caused by the conversion of developmentally unmatured synapses into natured ones, the conversion of silent into functional synapses, the conversion of inactive into active synapses, and the upregulated compatibility between presynaptic and postsynaptic partners [21]. During postnatal development, the formation of glutamatergic synapses is characterized as the emergence of NMDA receptors in the central area of postsynaptic density and the subsequent insertion of AMPA receptors around [32–34]. The complete form of glutamatergic synapses includes the embedding of NMDA and AMPA receptors in the postsynaptic density, i.e., matured glutamatergic synapses that are critical for the fulfilment of synapse plasticity. Even though both NMDA and AMPA receptors are localized in the postsynaptic density, their functions emerge sequentially to be different, in which AMPA receptors are activated by glutamates for fast synaptic transmission and the membrane depolarization facilitates the activation of NMDA receptors to mediate slow synaptic transmission [35]. With this complete form of the glutamatergic synapses, they have a nature in the conversion of silent synapses into functional synapses during synaptic plasticity, in which membrane depolarization activates NMDA receptors and subsequently AMPA receptors to have synapses be converted from silence to function. In fact, even though synapses include both kinds of receptors, they remain to be inactive. Once calmodulin-dependent protein kinase II is activated in postsynaptic neurons, the inactive synapses are converted into active synapses, especially this process is matured during the postnatal development [36]. Because this kinase plays an important role in LTP induction and maintenance [37, 38], the conversion of inactive synapses into active synapses is likely the mechanism underlying synaptic potentiation (Fig. 4.2). Finally, there is a functional compatibility between presynaptic and postsynaptic partners [39] (Fig. 4.1). The upregulated compatibility between presynaptic and postsynaptic partners may be critical for the refinement of synapses.

Based on the analyses above, the studies of synaptic plasticity may not need to apply either high-frequency tetanus stimulation or low-frequency stimulation, especially for plasticity coupled with the recruitment of associative memory cells. In the study of neuronal plasticity linked with the recruitment of associative memory cells, functional and morphological comparisons are conducted between neurons in mice without associative learning and associative memory cells in memory formation in mice that are trained by pairing whisker and odor stimulations. As mouse brains used in studies are genetically labeled by fluorescent proteins, such as glutamatergic neurons by yellow fluorescent protein and GABAergic neurons by green fluorescent protein, neural plasticity is studied in the cell-specific manner [12–18].

In mice showing associative memory, i.e., odorant-induced whisker motion and whisker-induced olfactory response, the neural plasticity is studied at glutamatergic and GABAergic neurons in barrel and piriform cortices. By whole-cell recording at these neurons in cortical slices, their response to various synaptic inputs and spike-encoding capability can be monitored and analyzed [40]. The function of GABAergic synapses is inhibited to study the activity of glutamatergic excitatory synapses. On the other hand, the function of glutamatergic synapses is inhibited to study the activity of GABAergic synapses.

To excitatory synapses, the amplitude and frequency of excitatory postsynaptic currents are upregulated on glutamatergic associative memory cells in the barrel cortex, in comparison with those in neurons from control mice [14, 15, 17, 18]. The amplitude of synaptic responses is mainly influenced by the quantum size of released transmitters and the responsiveness of postsynaptic receptors. The frequency of synaptic responses is mainly controlled by the release probability of presynaptic transmitters and the number of active synapses [36, 41–43]. Therefore, excitatory synapses on glutamatergic associative memory cells are upregulated in their number, release probability, and/or receptor responsiveness. This indication is strengthened by morphological studies, in which the number of innervated synapses and spherical spines is increased in glutamatergic associative memory cells [16, 17]. The upregulation of excitatory synapses on glutamatergic associative memory cells will increase the excitatory driving force to these memory cells.

The spike-encoding capability of glutamatergic associative memory cells in the barrel cortex is upregulated, in which the input-output curve shifts toward up-left and inter-spike intervals are shortened, in comparison to those in neurons from control mice. These alternations are based on the attenuation of threshold potential to fire spikes and the shortening of refractory periods to fire sequential spikes [16, 17]. The increase of spiking capability in glutamatergic associative memory cells indicates the importance of this upregulation in the recruitment of associative memory cells.

To inhibitory synapses, the amplitude and frequency of inhibitory postsynaptic currents are downregulated on glutamatergic associative memory cells, compared to those in neurons from control mice [16, 17]. Therefore, inhibitory synapses on glutamatergic associative memory cells are downregulated in their number, release probability, or receptor responsiveness. The downregulation of inhibitory synapses on glutamatergic associative memory cells will facilitate the excitatory driving force to activate associative memory cells.

In terms of the plasticity at inhibitory associative memory cells, the amplitude and frequency of excitatory postsynaptic currents on GABAergic associative memory cells are upregulated. Therefore, excitatory synapses on GABAergic associative memory cells are upregulated in their number, release probability, or receptor responsiveness. This indication is strengthened by morphological study, in which dendritic branches to receive synapse innervations and synapse innervations from glutamatergic associative memory cells are increased in the glutamatergic associative memory cells [16, 17]. Moreover, the spike-encoding capability at these GABAergic associative memory cells is downregulated. That is, the upregulated driving force from excitatory synapses and the downregulated spike-encoding capability may maintain the functions of GABAergic associative memory cells to be homeostatic in the network of associative memory cells.

It is noteworthy that results from glutamatergic and GABAergic associative memory cells in the piriform cortex are similar to these results in the barrel cortex. Excitatory synapses on glutamatergic associative memory cells are upregulated, spike-encoding capability is upregulated, and inhibitory synapses are downregulated in the piriform cortex. These results are granted by morphological study. In addition, excitatory synapses of GABAergic associative memory cells are upregulated, and their spike-encoding capability is downregulated [15]. Therefore, neural plasticity in relevance to the recruitment of associative memory cells is confirmed by the studies in barrel and piriform cortices, which may be extended into other sensory cortices. Moreover, these alternations have been seen in glutamatergic and GABAergic associative memory cells in barrel cortices from mice that have expressed associative memory trained by pairing triple signals, i.e., whisker, olfactory, and tail stimulations [12].

Taken all of these studies together, the functions of glutamatergic associative memory cells are upregulated based on the increased driving force from excitatory synapses, the increased capability of spike-encodings, and the decreased function of inhibitory synapses. However, these glutamatergic associative memory cells will not be overexcited since inhibitory synapse innervations on these memory cells are raised [17]. To GABAergic associative memory cells, the increased driving force from excitatory synapses, the increased number of excitatory synapses, and the decreased capability of spike-encoding will maintain them to be homeostatic [15, 17]. As illustrated in Fig. 5.5, the plasticity in neuronal network consisting of glutamatergic and GABAergic associative memory cells is coordinated in relevance to the recruitment of associative memory cells. In general, glutamatergic associative memory cells are functionally upregulated for the recruitment of more associative memory cells during associative learning and memory. These glutamatergic associative memory cells are not overexcited due to the increase of synapse innervations from GABAergic associative memory cells. The mutual effect by an increased excitatory driving force and a decreased spike-encoding capability in GABAergic associative memory cells makes them to be homeostatic in memory cells in engram circuits [21].

In terms of physiological impacts of plasticity at glutamatergic and GABAergic associative memory cells on the memory-relevant behaviors, the following points should be emphasized, such as the efficiency of associative learning and memory, the persistent maintenance of associative learning and memory, the unilateral learning for bilateral memory, as well as the inheritance of associative learning and memory [13–15, 44, 45]. These impacts, on the other hand, prove that the plasticity at those glutamatergic and GABAergic associative memory cells is relevant to associative memory, in comparison with previous studies solely for neural plasticity, by which memory is presumably caused.

The full onset of associative memory is variable in the efficiency among mice that have received the pair-training of odor and whisker signals. For instance, the full onset of associative memory varies from 1 up to 2 weeks. In the meantime, the levels of neuronal plasticity in glutamatergic and GABAergic associative memory cells are variable. Interestingly, the plasticity levels for excitatory synapses and inhibitory synapses on glutamatergic and GABAergic associative memory cells are linearly correlated with the efficiency of associative memory [18]. This result strengthens the conclusion about the relevance of the plasticity at associative memory cells to memory formation [18, 21]. In addition, the filials cross-mated from mice that express high learning efficiency or low learning efficiency demonstrate various full onset of associative memory. In general, the ranks of learning efficiency in these filials are the offspring from male and female with high learning efficiency, the offspring from female with high learning efficiency and male with low learning efficiency, as well as the offspring from male with high learning efficiency and female with low learning efficiency. Moreover, plasticity levels for excitatory synapses and inhibitory synapses on glutamatergic and GABAergic associative memory cells are linearly correlated with the efficiency of associative memory in three groups of filials [15]. This result further strengthens the conclusion about the relevance of the plasticity at associative memory cells to memory formation [15, 21].

In a mouse model of associative learning by pairing whisker and olfactory signals unilaterally, the bilateral effects of odorant-induced whisker motion and whisker-induced olfactory responses, i.e., the reciprocal retrievals of associated signals, are observed. In this unilateral learning for bilateral memory formation, associative memory cells are recruited in both sides of barrel cortices. In the meantime, the function of excitatory synapses on glutamatergic associative memory cells is upregulated, the capability of spike-encoding on glutamatergic associative memory cells is elevated, as well as the function of inhibitory synapses on glutamatergic associative memory cells is attenuated [13]. Taken this result with the data that plasticity levels of associative memory cells are correlated with learning and memory efficiency [18], it will be concluded that neural plasticity at associative memory cells is correlated to memory formation and associative memory cell recruitment [12–18].

6.3 Plasticity at Secondary Associative Memory Cells

In addition to neural plasticity linked to associative memory and associative memory cell recruitment, the functions of associative memory cells may undergo further plastic changes [19], similarly to the plasticity of neuronal excitability that can be multiple grades [4]. This functional plasticity at the associative memory cells induced by learning cue is studied in the prefrontal cortex by two-photon cell imaging in vivo. After mice express associative memory induced by simultaneously pairing whisker, odor, and tail stimulations, in which the reciprocal retrieval of associated signals includes odorant-induced whisker motion and tail swing, tail-induced whisker motion and olfactory response, as well as whisker-induced olfactory response and tail swing, two-photon cell imaging is conducted at the prefrontal cortex by loading calcium-sensitive dye, Oregon Green-BAPTA. Initially, associative memory cells are identified by examining their responses to learning cues, such as whisker, tail, and odor signals. Prefrontal cortical neurons in response to two of these learning cues are defined to be associative memory cells including full and incomplete associative memory cells [19, 46]. As their responses to learning cues are blocked by inhibiting the functions of barrel, piriform, and S1-tail cortices, these associative memory cells are called as secondary associative memory cells [44, 45].

In response to learning cues including whisker, odor, and tail stimulations, secondary associative memory cells express functional plasticity variously in three pathways. After intensive activation is given to an input pathway, some input pathways of these associative memory cells express the upregulation in response to learning cues, and others show the downregulation in their responses. Moreover, some pathways are converted from silence to function or from inactive to active state in their responses or vice versa. Furthermore, functional plasticity in input pathways on these associative memory cells can be homosynaptic and heterosynaptic patterns [19, 21]. The complicated patterns of neural plasticity at the secondary associative memory cells indicate the coordination of the networks that consist of secondary associative memory cells in the prefrontal cortex to execute the high-order cognitions. Neuronal plasticity in various patterns has been reported previously, such as some neurons with upregulation and others with depression [31]. It is noteworthy that this plasticity is regulated by monoamine transmitters [19].

6.4 Plasticity at Memory Output Cells

Compared with synaptic and neuronal plasticity at primary associative memory cells, which is maintained to be relatively long, neuronal and synaptic plasticity at memory output cells appears to be decayed without further pair-training [14]. For instance, a study has compared neuronal plasticity at memory output cells in the motor cortex and primary associative memory cells in the barrel cortex in mice that express associative memory, i.e., odorant-induced whisker motion and whisker-induced olfactory response, trained by pairing whisker and odor stimulations. In slice preparations from these mice with associative memory, excitatory synapses on glutamatergic and GABAergic associative memory cells are upregulated, inhibitory synapses on glutamatergic associative memory cells are downregulated, and the capability of spike-encoding is upregulated in the barrel cortex. In the meantime, these types of neural plasticity are also observed in memory output cells in the motor cortex. After this pair-training stops for a week, the reciprocal retrieval of associative memory disappears, e.g., odorant-induced whisker motion is back to the baseline before training. The types of neural plasticity above still express in primary associative memory cells in the barrel cortex, but not in the motor cortex [14]. This result indicates the persistence of neural plasticity at primary associative memory cells in the sensory cortices and the fluctuation of neural plasticity at memory output cells in the motor cortex, which is based on the pair-training for the consolidation [21].

This fluctuation of neural plasticity at memory output cells in the motor cortex is similar to the functional fluctuation of memory output cells in response to learning cues (please see Chap. 5) [20]. As these fluctuated changes at memory output cells in response to learning cues as well as in functional plasticity are correlated with the levels of associative memory efficiency, the data above strengthen the role of fluctuated neural plasticity at memory output cells in memory retrieval and extinction [21].

6.5 Summary About Plasticity at Associative Memory Cells

Based on discussions above, the plasticity in associative memory cells is classified into two forms. In other words, two forms of plasticity among associative memory cells are the upregulation during their recruitment and the refinement induced by learning cues after their recruitment. Their recruitment-related refinement confers them to be coordinated with each other for their compatibility during the recruitment. Their cue-induced refinement confers them to be dominantly retrieving specific signals. These two forms of structural and functional plasticity are induced by the coactivation among associative memory cells and ready-recruited neurons [13, 16, 17, 19, 47–49]. The high level of recruitment-related refinement may facilitate to recruit more associative memory cells. The high level of learning cue-induced refinement may elevate the functional state of associative memory cells. Both outcomes will assist to strengthen memory depth and maintenance as well as memory retrievals [21].

In the recruitments of associative memory cells from cortical ready-recruited neurons by their coactivation and mutual synapse innervations, the number and function state of excitatory synapses on glutamatergic associative memory cells and GABAergic associative memory cells are upregulated, the encoding capability of glutamatergic associative memory cells is strengthened, and the spike output of GABAergic associative memory cells is weakened [12, 15–18]. Active intrinsic properties, e.g., refractory periods and threshold potential, at glutamatergic associative memory cells are upregulated. These active intrinsic properties at GABAergic associative memory cells are downregulated [15–18]. Synapse interconnections are raised between glutamatergic and GABAergic associative memory cells [16, 17]. With the increased driving force from excitatory synapses, the upregulated excitability and the decreased driving force from inhibitory synapses on glutamatergic associative memory cells will shift the functional state of their networks toward more excitation for memory strengthening and maintenance for future retrieval. The high activities of glutamatergic associative memory cells will activate more neighboring GABAergic associative memory cells. Their coordinated activities attract more synapse innervations from coactivated brain areas, recruit more glutamatergic and GABAergic associative memory cells, promote their function state to the optimal level for the information storage, as well as facilitate the activation of these associative memory cells for the reciprocal retrievals of associated signals [13, 14, 17, 49]. These upregulations in the number and function state of excitatory synapses among associative memory cells can strengthen their spike-encoding capability and precision [42, 50, 51] for efficient memory formation and precise retrievals. If glutamatergic associative memory cells are overly active, they activate neighboring inhibitory associative memory cells to prevent the hyperactivity themselves by recurrent negative feedback [31, 42, 52] (Fig. 5.5).

The refinement of excitation at associative memory cells includes the downregulation of threshold potential to produce spikes and the upregulation of spike-encoding capability to generate sequential spikes. These changes in the excitatory state of associative memory cells can be well used to interpret how their refinement is involved in the integrative storage and the reciprocal retrieval of associated signals. Intensive activity at cortical neurons by the high-frequency stimulus, similarly to neuronal coactivation during associative learning, shifts spiking threshold potential toward the resting membrane potential. The shortened distance between membrane potentials and resting membrane potential facilitates the production of neuronal spikes [4]. High-frequency neuron spikes can also enhance the capability to fire sequential spikes [25, 27]. Both mechanisms upregulate the ability of ready-recruited neurons as well as of associative memory cells that are ready to be recruited as advanced associative memory cells to encode sequential spikes. The upregulated spikes may strengthen the chain reaction from intensive sequential spikes to epigenetic processes. The altered expression of certain microRNAs can regulate the expression of genes and proteins relevant to axon prolongation, synapse formation, and channel-receptor activities. Subsequently, cellular changes facilitated by these molecules include the recruitments of mutual synapse innervations among ready-recruited neurons and associative memory cells as well as their refinement. Such recruited and refined associative memory cells confer the integrative storage and reciprocal retrieval of the associated signals [16, 17, 47–49, 53]. Therefore, the functional upregulation of neuronal excitability may play one of central roles in associative learning and memory [13–18, 49]. This point of view has been reiterated by a current review [54].

Synaptic potentiation has been observed at engram cells in slices of the prefrontal cortex, hippocampus, and amygdala [55], and excitation enhancement at B51 neurons is isolated from *Aplysia* [56]. In the study of cue-induced plasticity at associative memory cells by those learning cues to stimulate sensory inputs in vivo, activity-dependent potentiation of associative memory cells in response to these learning cues (i.e., associated signals) is induced at synapse input pathways in active groups of primary and secondary associative memory cells. In addition, activity-dependent conversion from silent to functionally active neural pathways or the other way around, in response to these associated signals, is initiated in inactive groups of associative memory cells; these potentiation and conversion can be homosynaptic or heterosynaptic manner [19, 21, 45, 46]. In terms of physiological impacts, this potentiated response to associated signals enables some associative memory cells to be more excitable. Compared to neighboring associative memory cells, these highly active associative memory cells are more sensitive to the excitatory driving force from learning cues. This alternation increases a chance that more associative memory cells respond to all of the associated signals, i.e., the recruitment of full associative memory cells from incomplete associative memory cells through their increased mutual synapse innervations [16, 17]. This activity-dependent potentiation at homosynaptic and heterosynaptic pathways of associative memory cells may also facilitate the reciprocal retrieval and logical reasoning of associated signals. In terms of mechanisms underlying this activity-dependent potentiation at associative memory cells in response to the associated signals in vivo, these new mutual synapse innervations may result from the strengthening of single synapses, the conversion from inactive to active synapses, and the conversion from silent to functional synapses [35, 36]. This hypothesis is based on observations that new mutual synapse innervations are morphologically formed among the associative memory cells [16, 17, 19, 45–49]. In primary associative memory cells, activity-dependent upregulation facilitates the retrieval of exogenous-associated signals. In secondary associative memory cells, activity-dependent potentiation assists the memory retrieval of endogenous signals produced during cognitive events and emotional reactions. Thus, the impulsive or cue-induced retrievals of these associated signals are emerged for the rehearsal of cognitive events and emotional reactions. Recruitment-relevant refinement and activity-dependent refinement at associative memory cells are supported by a fact that the enhancement of neuronal excitability is multigrade in nature [4].

The recruitments of primary associative memory cells in sensory cortices by learning cues and of secondary associative memory cells in cognition-/emotion-relevant brain areas during thinking mainly endorse the specificity of the stored associative signals [13, 16, 17, 44, 47–49]. The number and functional state of associative memory cells affect memory strength, maintenance, and retrieval [13, 14, 47, 48]. The coordinated upregulation among the subcellular compartments of these associative memory cells influences whether associative memory cells sensitively detect input signals, precisely integrate the storage of these associated signals, as well as efficiently activate memory output cells to retrieve these memorized signals [16, 17]. The maintenance of activity-dependent upregulation at associative memory cells grants how long these associative memory cells are sensitive to the learning cues for information retrieval. It is noteworthy that the recruitment and upregulation of associative memory cells are based on their coactivity [13, 16, 17, 47, 49]. The activities of associative memory cells as a central point constitute coactivity-dependent positive cycle in the recruitment and refinement of associative memory cells, i.e., activity together, synapse interconnection together, and strengthening together. The highly active neurons while receiving the associated signals are recruited as associative memory cells and are upregulated in their functions. The upregulated population and functions of associative memory cells including those full and incomplete associative memory cells during repetitive learnings and rehearsals will further upregulate their active state to be full associative memory cells as well as will recruit more associative memory cells from ready-recruited neurons [21, 45]. Based on the function compatibility between the presynaptic axon branches and the postsynaptic neuronal partners [39], an activity-dependent positive cycle in the recruitment and refinement of associative memory cells can interpret realistic practices. There is the proportional correlation among the learning times, associative memory cells, and memory strength. In other words, the more learning times are, the more associative memory cells are, and the more impressive memory is under normal consciousness and attentions. Although associative memory cells are upregulated in their functions, i.e., active neurons in the brain, active neurons labeled by nonspecific immediate early genes may not be memory cells [21].

Cellular changes in relevance to the activity-dependent positive cycle in the recruitment and refinement of associative memory cells may include the following steps. In early life, the learning of associated signals induces the ready-recruited cortical neurons to be mutually innervated through axon prolongations and synapse formations, such that associative memory cells are recruited for the integrative storages and reciprocal retrievals of associated signals. After the learning of numerous unitary-associated signals, many groups of associative memory cells are recruited to encode the storages of unitary signals. In later life, although the ability of axon prolongation in long distance appears decayed, the coactivations of associative memory cells during learnings can make these axon terminals to be grown and extended locally. With the local growth and extension of axon terminals of associative memory cells from other regions, the formation of new synapse innervations on associative memory cells via en passant synapses strengthens the driving force onto these associative memory cells to upregulate their functional states, i.e., the activity-dependent refinement of associative memory cells. The formation of new synapse innervations on the ready-recruited neurons in the neighboring of these associative memory cells recruits more associative memory cells, the activity-dependent recruitment of associative memory cells. The repeated coactivation of associative memory cells leads to the positive cycle of their recruitment and refinement, which is activity-dependent [21] (Fig. 5.4).

To the functional states at primary and secondary associative memory cells influenced by synapse inputs, the number and strength of inputted synapses are proportional to the excitation level of the associative memory cells [14, 15, 18, 47, 48]. When excitatory synapse innervations are increased and upregulated by repeatedly giving the learning cues, the two elevated driving forces enable associative memory cells to become more excitable. Their more excitation driven by the learning-input pathway will facilitate the retrieval of specific memory contents and the recruitment of full associative memory cells. The elevated activities of synapse innervations from the arousal system may also boost associative memory cells to become more excitable. Their more excitation through this way assists memory retrieval in nonspecific manner. The increased excitability or the decreased spiking threshold in associative memory cells enables them to be activated easily. Their more excitation through changing their excitability facilitates memory retrieval nonspecifically and the recruitment of more associative memory cells [16, 17]. The diagram in Fig. 6.1 illustrates the correlations among the efficiency of associative memory, the number of associative memory cells, the encoding ability of associative memory cells, and the number of synapse innervations on associative memory cells.

In the brain, what kinds of neurons are the ready-recruited neurons for associative memory cells remains to be examined. In the principle, the coactivations among ready-recruited neurons are essential for their mutual synapse innervations and their recruitment to be associative memory cells. Moreover, activity-dependent positive cycle in the recruitment and refinement of associative memory cells is another primary driving force for ready-recruited neurons to be recruited as associative memory cells as well as for associative memory cells to be upgraded to some advanced associative memory cells [21, 45]. Therefore, the neurons with high strengths of excitation and synapse innervations are preferentially recruited as associative memory cells. That is, cortical neurons, which have the lower spiking threshold caused by their activities as well as the stronger synapse inputs driven by the attention cues that are previously learned signals with the synapse formation or driven by the arousal system for full consciousness plus memories, are favorably recruited to be associative memory cells. The active neurons are always easily recruited to be associative memory cells at the first grade. The activation and recruitment of the first grade of associative memory cells can trigger the neighboring neurons through their synapse connections to be more active and become associative memory cells in the second grade. This mechanism may induce the recruitment of the third or even fourth grade of associative memory cells This preferential cascade in the recruitment of associative memory cells may lead to a time sequence for the groups of cortical ready-recruited neurons to become associative memory cells, when learners expose to multiple associated signals sequentially, e.g., words by words in sentences or articles and pictures by pictures in vision or video [21, 45].

In terms of development regulation, the establishment of associative memory has shown initial increase and then decrease with postnatal ages. The recruitment and refinement of associative memory cells may also follow this regulation [21, 49]. It has been found that synapse and neuron plasticity matures during postnatal development [4, 36]. The data indicate that the recruitment and refinement in associative memory cells execute their function differently in the postnatal development. The activity-dependent recruitment of associative memory cells may play the dominant role in associative memory during early and young age, whereas the activity-dependent refinement of associative memory cells works dominantly after these stages. The knowledge learned in young age is relatively simple, i.e., unitary signals, whereas the knowledge learned in matured age is complicated signals that are reorganized from unitary signals. In this regard, associative memory cells recruited in young age memorize unitary signals, and associative memory cells refined in matured age work for learning complicated signals reorganized from unitary signals [45].

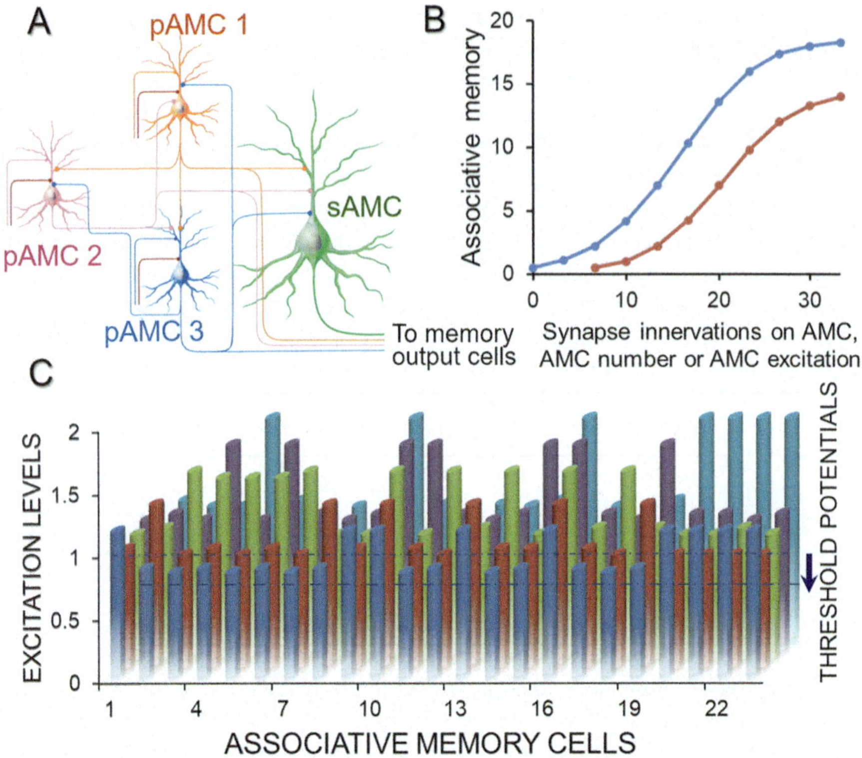

Fig. 6.1 The relationship among the efficiency of associative memory, the number of associative memory cells, the encoding capability of associative memory cells, and the number of synapse innervations on associative memory cells. These factors play the important role in the integrative storages and reciprocal retrievals of specific associated signals. (**a**) Illustrates associative memory cells and their interconnections. Three primary associative memory cells (pAMC) are recruited based on their mutual synapse innervation, pAMC1 (orange), pAMC 2 (pink), and pAMC 3 (blue). Each associative memory cell receives synapse innervations from innate inputs (their colors are identical to those for cell bodies), inputs from the arousal system (red), as well as their mutual axon projection and synapse innervations. The axons of these pAMCs convergently project toward secondary associative memory cells (sAMC, green) and make synapse innervations. All of these associative memory cells synaptically innervate onto memory output neurons. Mutual synapse innervations among pAMCs, convergent innervations onto sAMC, and their synapse innervations onto memory output cells confer the integrative storage and reciprocal retrieval of associated signals as well as logical reasoning and associative thinking. (**b**) Illustrates the sigmoid correlations among the efficiency of associative memory, the number of associative memory cells, the excitation of associative memory cells, and the number of synapse innervations on associative memory cells. The excitation of associative memory cells is affected by the number and functional state of their synapses and by their own excitability. If the excitability of associative memory cells increases, this sigmoid curve (red) shifts to left (blue) as well as the efficiency of associative learning and memory increases. (**c**) Illustrates the relationship between different associative memory cells and their excitatory levels. If spiking threshold potential lowers, the relative excitatory level of associative memory cells rises, and more ready-recruited neurons are coactivated for the recruitment and upregulation of associative memory cells in memory formation and retrieval. (This illustration is modified from a reference [21])

6.6 Associative Memory Cells Are Modulated by the Arousal System

As discussed in Chapters 5 and 6, primary associative memory cells in sensory cortices receive specific new synapse innervations and their innate inputs that take in the associated signals learned from exogenously natural and social environments for their integrative storage and reciprocal retrieval. Secondary associative memory cells in the brain regions relevant to cognition and emotion receive the convergent and divergent synapse innervations from primary associative memory cells for logical reasoning, associative thinking, computation, and so on (Fig. 5.3). These associative memory cells may receive other synaptic inputs and neurotransmitters for the modulation of their functional state [45]. In the brain and spinal cord, the neurons receive synapse innervations from the arousal system in the brain stem for maintaining their basal activities. The ascending reticular-activating pathway in the arousal system passes through the relay in the thalamus and then widely innervates cerebral neurons [57, 58]. The ascending activating pathways from the neuronal axons of the locus coeruleus, midbrain raphe nuclei, and cholinergic nuclei also innervate the forebrain in spreading manner [59–62]. By their released neurotransmitters, e.g., acetylcholine, serotonin, and norepinephrine, this system is able to elevate excitatory state of almost all of neurons in the central nervous system to maintain wakefulness and confer consciousness. As the neurons, associative memory cells should receive synapse innervations from the arousal system, such that memory-relevant behaviors can be well executed under the wakefulness. In other words, these neurotransmitters modulate the excitations of ready-recruited neurons and associative memory cells. The activity of this arousal system facilitates the coactivated ready-recruited neurons to be recruited as associative memory cells as well as maintains the basal activity of primary and secondary associative memory cells under the conditions of wakefulness and consciousness (Fig. 5.3). In other words, the arousal system facilitates the recruitment and refinement of associative memory cells [45].

It has been demonstrated that learning and memory are modulated by neurotransmitters released from the arousal system, such as norepinephrine, serotonin, and acetylcholine [63–67], in the locus coeruleus, midbrain raphe nuclei, and cholinergic nuclei [59–62]. For example, learning process and memory formation are facilitated by a direct activation of acetylcholine M1-type receptors on hippocampal interneurons [68]. Memory formation is strengthened by infusing norepinephrine or adrenoceptor agonists into the amygdala or prefrontal cortex. This memory enhancement matches up an action of stress hormone [69]. The increased activity of serotonin neurons enhances memory and cognition. The attenuated activity of serotonin neurons weakens the memory and cognition [70, 71]. Stimulations type-I and type-V dopamine receptors in the hippocampus and forebrain facilitate spatial learning and memory [72, 73]. There is a coordinated strengthening effect of serotonin and norepinephrine on associative memory cells to facilitate associative learning and memory [19]. Therefore, neurotransmitters including acetylcholine, serotonin, norepinephrine, and dopamine as well as stress hormones regulate learning and memory.

Mechanisms underlying the influences neurotransmitters and hormones on associative memory cells remain to be examined, though the presumable targets include the presynaptic inputs of associative memory cells to regulate their neurotransmitter release or the encoding capability of associative memory cells to regulate their coactivations. In fact, the studies have indicated that neuron excitability and neuron responses to synaptic inputs are upregulated by serotonin [74, 75]. In addition, synaptic bouton formation and postsynaptic neuron activity are upregulated by stimulating dopaminergic neurons [76]. Based on these data, monoaminergic neurons presumably regulate the recruitment and refinement of primary and secondary associative memory cells and in turn influence memory formation and memory-relevant cognitions. This regulation grants the facts that the efficiencies of associative learning and memory retrieval are upregulated by high levels of wakefulness, consciousness, attention and motivation, and that both of them are based the active monoaminergic and cholinergic neurons [45].

References

1. Bliss T, Lomo T. Long-lasting potentiation of synaptic transmission in the dentate area of the anaesthetized rabbit following stimulation of the perforant path. J Physiol Lond. 1973;232:331–56.
2. Stanton PK, Sejnowski TJ. Associative long-term depression in the hippocampus induced by hebbian covariance. Nature. 1989;339(6221):215–8.
3. Wang JH, Ko G, Kelly PT. Cellular and molecular bases of memory: synaptic and neuronal plasticity. J Clin Neurophysiol. 1997;14:264–93.
4. Zhang M, et al. Calcium signal-dependent plasticity of neuronal excitability developed postnatally. J Neurobiol. 2004;61:277–87.
5. Bliss TVP, Lynch MA. Long-term potentiation of synaptic transmission in the hippocampus: properties and mechanisms. In: Landfield PW, Deadwyler SA, editors. Long-term potentiation: from biophysics to behavior. New York: Alan R. Liss, Inc; 1988. p. 3–72.
6. Nicoll RA, Malenka RC. Contrasting properties of two forms of long-term potentiation in the hippocampus. Nature. 1995;377:115–8.
7. Poo MM, et al. What is memory? The present state of the engram. BMC Biol. 2016;14:40.
8. Bannerman DM, et al. Hippocampal synaptic plasticity, spatial memory and anxiety. Nat Rev Neurosci. 2014;15:181–92.
9. Sgobio C, et al. Hippocampal synaptic plasticity, memory, and epilepsy: effects of long-term valproic acid treatment. Biol Psychiatry. 2010;67(6):567–74.
10. Wen B, et al. A portion of inhibitory neurons in human temporal lobe epilepsy are functionally upregulated: an endogenous mechanism for seizure termination. CNS Neurosci Ther. 2015;21(2):204–14.
11. Zhang F, et al. mGluR1,5 activation improves network asynchrony and GABAergic synapse attenuation in the amygdala: implication for anxiety-like behavior in DBA/2 mice. Mol Brain. 2012;5(1):20.
12. Feng J, et al. Cell-specific plasticity associated with integrative memory of triple sensory signals in the barrel cortex. Oncotarget. 2018;9(57):30962–78.

13. Gao Z, et al. Associations of unilateral whisker and olfactory signals induce synapse formation and memory cell recruitment in bilateral barrel cortices: cellular mechanism for unilateral training toward bilateral memory. Front Cell Neurosci. 2016;10(285):1–16.
14. Guo R, et al. Associative memory extinction is accompanied by decayed plasticity at motor cortical neurons and persistent plasticity at sensory cortical neurons. Front Cell Neurosci. 2017;11(168):1–12.
15. Liu Y, et al. Activity strengths of cortical glutamatergic and GABAergic neurons are correlated with transgenerational inheritance of learning ability. Oncotarget. 2017;8(68):112401–16.
16. Liu Y, et al. Piriform cortical glutamatergic and GABAergic neurons express coordinated plasticity for whisker-induced odor recall. Oncotarget. 2017;8(56):95719–40.
17. Yan F, et al. Coordinated plasticity between barrel cortical glutamatergic and GABAergic neurons during associative memory. Neural Plast. 2016;2016(ID5648390):1–20.
18. Zhao X, et al. Coordinated plasticity among glutamatergic and GABAergic neurons and synapses in the barrel cortex is correlated to learning efficiency. Front Cell Neurosci. 2017;11(221):1–12.
19. Wang JH, et al. Secondary associative memory cells and their plasticity in the prefrontal cortex. Biophys J. 2019;116(3):427a.
20. Wang J-H, Guo R, Wei Z. Associative memory extinction is accompanied by decays of associative memory cells and their plasticity at motor cortex but not sensory cortex. Soc Neurosci. 2017;81(09):10385.
21. Wang J-H. Searching basic units of memory traces: associative memory cells. F1000Res. 2019;8(457):1–28.
22. Hebb DO. The organization of behavior, a neuropsychological theory. New York: Wiley; 1949.
23. Bailey CH, Kandel ER, Harris KM. Structural components of synaptic plasticity and memory consolidation. Cold Spring Harb Perspect Biol. 2015;7(7):a021758.
24. Bliss TVP, Collingridge GL. A synaptic model of memory: LTP in the hippocampus. Nature. 1993;361(Jan. 7):31–9.
25. Dubnau J, Chiang AS, Tully T. Neural substrates of memory: from synapse to system. J Neurobiol. 2003;54(1):238–53.
26. Malenka RC, Nicoll RA. Long-term potentiation – a decade of progress? Nat Neurosci. 1999;285:1870–4.
27. Viskontas IV. Advances in memory research: single-neuron recordings from the human medial temporal lobe aid our understanding of declarative memory. Curr Opin Neurol. 2008;21(6):662–8.
28. Bear M. A synaptic basis for memory storage in the cerebral cortex. Proc Natl Acad Sci U S A. 1996;93:13453–9.
29. Linden DJ. The return of the spike: postsynaptic action potentials and the induction of LTP and LTD. Neuron. 1999;22(4):661–6.
30. Trussell LO, Fischbach GD. Glutamate receptor desensitization and its role in synaptic transmission. Neuron. 1989;3(2):209–18.
31. Chen N, Chen X, Wang J-H. Homeostasis established by coordination of subcellular compartment plasticity improves spike encoding. J Cell Sci. 2008;121(17):2961–71.
32. Liao D, et al. Regulation of morphological postsynaptic silent synapses in developing hippocampal neurons. Nat Neurosci. 1999;2:37–43.
33. Petralia RS, et al. Selective acquisition of AMPA receptors over postnatal development suggests a molecular basis for silent synapses. Nat Neurosci. 1999;2:31–6.
34. Takumi Y, et al. Different modes of expression of AMPA and NMDA receptors in hippocampal synapses. Nat Neurosci. 1999;2:618–24.
35. Liao D-Z, Hessler NA, Malinow R. Activation of postsynaptic silent synapses during pairing-induced LTP in CA1 region of hippocampal slice. Nature. 1995;375:400–4.
36. Wang J-H, Kelly PT. Ca2+/CaM signalling pathway up-regulates glutamatergic synaptic function in non-pyramidal fast-spiking neurons of hippocampal CA1. J Physiol Lond. 2001;533(2):407–22.

37. Malinow R, Schulman H, Tsien RW. Inhibition of postsynaptic PKC or CaMKII blocks induction but not expression of LTP. Science. 1989;245:862–6.
38. Wang J-H, Kelly PT. Balance between postsynaptic Ca2+–dependent protein kinase and phosphatase activities controlling synaptic strength. Learn Mem. 1996;3:170–81.
39. Yang Z, et al. Functional compatibility between Purkinje cell axon branches and their target neurons in the cerebellum. Oncotarget. 2017;8(42):72424–37.
40. Zhang G, et al. Upregulation of excitatory neurons and downregulation of inhibitory neurons in barrel cortex are associated with loss of whisker inputs. Mol Brain. 2013;6(1):2.
41. Stevens CF. Presynaptic function. Curr Opin Neurobiol. 2004;14(3):341–5.
42. Wang JH, et al. The gain and fidelity of transmission patterns at cortical excitatory unitary synapses improve spike encoding. J Cell Sci. 2008;121(17):2951–60.
43. Zucker RS. The calcium hypothesis and modulation of transmitter release by hyperpolarizing pulses. Biophys J. 1987;52:347–50.
44. Wang JH, Cui S. Associative memory cells: formation, function and perspective. F1000Res. 2017;6:283.
45. Wang JH, Cui S. Associative memory cells and their working principle in the brain. F1000Res. 2018;7:108.
46. Wang J-H, et al. Prefrontal cortical neurons are recruited as secondary associative memory cells for associative memory and cognition. Biophys J. 2018;114(3):155a.
47. Feng J, et al. Barrel cortical neuron integrates triple associated signals for their memory through receiving epigenetic-mediated new synapse innervations. Cereb Cortex. 2017;27(12):5858–71.
48. Lei Z, et al. Synapse innervation and associative memory cell are recruited for integrative storage of whisker and odor signals in the barrel cortex through miRNA-mediated processes. Front Cell Neurosci. 2017;11(316):1–11.
49. Wang D, et al. Neurons in the barrel cortex turn into processing whisker and odor signals: a cellular mechanism for the storage and retrieval of associative signals. Front Cell Neurosci. 2015;9:320.
50. Yu J, et al. Quantal glutamate release is essential for reliable neuronal encodings in cerebral networks. PLoS One. 2011;6(9):e25219.
51. Yu J, Qian H, Wang JH. Upregulation of transmitter release probability improves a conversion of synaptic analogue signals into neuronal digital spikes. Mol Brain. 2012;5(1):26.
52. Grienberger C, et al. Inhibitory suppression of heterogeneously tuned excitation enhances spatial coding in CA1 place cells. Nat Neurosci. 2017;20:417–26.
53. Pignatelli M, et al. Engram cell excitability state determines the efficacy of memory retrieval. Neuron. 2019;101(2):274–284.e5.
54. Titley HK, Brunel N, Hansel C. Toward a neurocentric view of learning. Neuron. 2017;95(1):19–32.
55. Ryan TJ, et al. Memory. Engram cells retain memory under retrograde amnesia. Science. 2015;348(6238):1007–13.
56. Mozzachiodi R, et al. Changes in neuronal excitability serve as a mechanism of long-term memory for operant conditioning. Nat Neurosci. 2008;11(10):1146–8.
57. Kayama Y, et al. Tonic and phasic components of the ascending reticular activating system. Fukushima J Med Sci. 1991;37(2):59–74.
58. Neylan TC. Physiology of arousal: Moruzzi and Magoun's ascending reticular activating system. J Neuropsychiatry Clin Neurosci. 1995;7(2):250.
59. Adell A, et al. Origin and functional role of the extracellular serotonin in the midbrain raphe nuclei. Brain Res Brain Res Rev. 2002;39(2–3):154–80.
60. Aston-Jones G, Cohen JD. An integrative theory of locus coeruleus-norepinephrine function: adaptive gain and optimal performance. Annu Rev Neurosci. 2005;28:403–50.
61. Chandler DJ, Gao WJ, Waterhouse BD. Heterogeneous organization of the locus coeruleus projections to prefrontal and motor cortices. Proc Natl Acad Sci U S A. 2014;111(18):6816–21.
62. Steriade M, et al. Neuronal activities in brain-stem cholinergic nuclei related to tonic activation processes in thalamocortical systems. J Neurosci. 1990;10(8):2541–59.

63. Hasselmo ME. The role of acetylcholine in learning and memory. Curr Opin Neurobiol. 2006;16(6):710–5.
64. Levey AI. Muscarinic acetylcholine receptor expression in memory circuits: implications for treatment of Alzheimer disease. Proc Natl Acad Sci U S A. 1996;93(24):13541–6.
65. Meneses A, Liy-Salmeron G. Serotonin and emotion, learning and memory. Rev Neurosci. 2012;23(5–6):543–53.
66. Roberts RB, Flexner JB, Flexner LB. Some evidence for the involvement of adrenergic sites in the memory trace. Proc Natl Acad Sci U S A. 1970;66(2):310–3.
67. Tully K, Bolshakov VY. Emotional enhancement of memory: how norepinephrine enables synaptic plasticity. Mol Brain. 2010;3:15.
68. Yi F, et al. Direct excitation of parvalbumin-positive interneurons by M1 muscarinic acetylcholine receptors: roles in cellular excitability, inhibitory transmission and cognition. J Physiol. 2014;592(16):3463–94.
69. Ferry B, Roozendaal B, McGaugh JL. Role of norepinephrine in mediating stress hormone regulation of long-term memory storage: a critical involvement of the amygdala. Biol Psychiatry. 1999;46(9):1140–52.
70. Wagatsuma A, et al. Locus coeruleus input to hippocampal CA3 drives single-trial learning of a novel context. Proc Natl Acad Sci U S A. 2017;115:E310–6.
71. Xu S, et al. Dorsal raphe serotonergic neurons control intertemporal choice under trade-off. Curr Biol. 2017;27(20):3111–3119 e3.
72. Kempadoo KA, et al. Dopamine release from the locus coeruleus to the dorsal hippocampus promotes spatial learning and memory. Proc Natl Acad Sci U S A. 2016;113(51):14835–40.
73. Sarinana J, Tonegawa S. Differentiation of forebrain and hippocampal dopamine 1-class receptors, D1R and D5R, in spatial learning and memory. Hippocampus. 2016;26(1):76–86.
74. Achee N, Zoran M. Serotonin-induced modulation of excitability in an identified *Helisoma trivolvis* neuron. J Exp Biol. 1997;200(Pt 10):1537–48.
75. Malyshev A, et al. Effects of serotonin levels on postsynaptically induced potentiation of snail neuron responses. Neurosci Behav Physiol. 1998;28(5):556–63.
76. Mastwal S, et al. Phasic dopamine neuron activity elicits unique mesofrontal plasticity in adolescence. J Neurosci. 2014;34(29):9484–96.

Chapter 7
Working Principles of Associative Memory Cells

Abstract Compared with neurons, associative memory cells receive synapse innervations that bring the associated signals during learning activities and are recruited from neurons during their coactivation within intramodal cortices and among cross-modal cortices. Synapse innervations among associative memory cells as well as their characters constitute working principles of associative memory cells. The connection patterns among associative memory cells determine the direction of their information flow and the specificity of the memory contents. The weight of synapse innervations for individual associative memory cells influences their functional states to be activated during further recruitments and memory retrieval. The interactions among glutamatergic and GABAergic associative memory cells influence homeostasis in their local network for memory contrast and over-excitation prevention. As the family of neurons, associative memory cells possess neuronal excitability that is set by threshold potentials and influences the spiking capability. The excitability of associative memory cells mainly influences whether they are activated driven by excitatory synapses from the innate inputs, the coactivated neurons, and the arousal system, as well as how spike patterns have been produced for encoding memory contents and triggering memory retrieval strength. Recruitment-relevant and activity-induced refinements of associative memory cells in their synapse innervations and excitability influence their functional states to encode the integrative storage and reciprocal retrieval of associated signals. With these characters and working principles of associative memory cells, memory traces that include primary associative memory cells in sensory cortices, secondary associative memory cells in association cortices, and cognition-relevant brain areas as well as memory output cells in the motor cortex work for the brain functions in relevance to memories in the aspects of physiology and psychology.

Keywords Memory · Associative memory cell · Cross-modal · Intramodal · Neuron and synapse

J.-H. Wang, *Associative Memory Cells: Basic Units of Memory Trace*,
https://doi.org/10.1007/978-981-13-9501-7_7

7.1 Introduction

Associative memory is featured by the integrative storage of associated signals in the brain and their reciprocal retrievals by recall, representation, and recollection [1–4]. Associative memory is essential for cognitive processes and emotional reactions. Psychologically, these activities based on associative memory are complicated, such as various memory patterns including explicit memory versus implicit memory, episodic memory versus semantic memory, perceptual memory versus working memory, spatial memory versus temporal memory, and eidetic memory versus false memory. Memory-relevant cognitive processes include consciousness, attention, associative thinking, imagination, logical reasoning, computation, trust, pleasure, dissatisfaction, and so on. Since these events are memory-relevant, how the characters of associative memory cells closely work for the cognitive processes and emotional reactions needs to be figured out [5].

In addition to general characters for the neuron, associative memory cells receive synapse innervations that bring the associated signals in intramodal or cross-modal manners. Synapse innervations among associative memory cells and the characters of neurons constitute working principles of associative memory cells [5]. Connection patterns among associative memory cells determine the direction of their signals flow and the specificity of memory contents including explicit or implicit memory, episodic or semantic memory, and perceptual or working memory. The strength of synapse innervations at individual associative memory cells influences their function state for memory retrieval, consciousness, attention, and other cognitive events. The interactions among glutamatergic and GABAergic associative memory cells regulate their coordination in the local network for memory contrast and over-excitation prevention. Moreover, the excitability of associative memory cells influences whether they are activated by excitatory synapses from different synapse innervations, for instance, the driving force from coactivated neurons for specific memory contents and the driving force from the arousal system for the consciousness and attention. The upregulation of associative memory cells strengthens memory, cognition, and emotions [5].

In perceptual memory, associated signals come from a single modality, such as visual, auditory, olfactory, taste, or tactile sensory system, leading to intramodal associative memory. The associated signals also come from the multiple sensory modalities, leading to cross-modal associative memory. In the intramodal associative memory, the integrative storage of associated signals is fulfilled by mutual synapse innervations among associative memory cells in the single modality of the sensory cortex, such as visual cortex, auditory cortex, olfactory cortex, gustatory cortex, and somatosensory cortex. In the cross-modal associative memory, the integrative storages of associated signals are processed by mutual synapse innervations among associative memory cells across multiple sensory cortices [6]. Certain types of intramodal associative memory are complicated in the integration of unitary signals and the storage of these integrated complex signals, such as photon beams in visual images and letters in words and sentences, which are never known in our knowledge [5].

Another complication appears the conversion of association memory into high-order brain functions, such as consciousness, attention, and other integrative cognitions. Consciousness can be thought as the functional state of the brain that individuals are wakeful and recognize their environments and themselves, i.e., memory retrievals and cognitions under wakefulness condition. As the level of memory retrievals varies, the levels of consciousness state are different. If individuals are wakeful and recognize unitary signals, their direct relatives or themselves, in which their memory retrievals are mainly episodic in nature, they fall into moderate consciousness state. If individuals are able to recall knowledge and theories in semantic memory, they are in high consciousness state as both primary and secondary associative memory cells are activated. With this full consciousness, individuals are able to pay attention to specific events by having more ready-recruited neurons or incomplete associative memory cells for their further recruitment to be full associative memory cells. In other words, attention is the combination of consciousness plus ready-learning or ready-cognition in the presence of topic cues for the alert [5]. In this chapter, author intends to discuss the working principles of associative memory cells in these memory-relevant events.

7.2 Associative Memory Cells in Their Circuits and Functions

As discussed in Chap. 5, associative memory cells are defined as those nerve cells that receive synapse innervations from the coactivated brain areas as well as encode the integrative storage and the reciprocal retrieval of associated signals. The characters of associative memory cells include mutual synapse innervations among coactive neurons and general features, such as intrinsic active property and plasticity [5]. Specific mutual synapse innervations can be used as criteria to identify whether the neurons detected in experiments are associative memory cells. General features in neurons can be used to predict the potential activities of associative memory cells, such as activity-dependent plasticity in morphology and functions. As the coactivation by synchronous activities of cortical neurons that are readily recruited to be memory units triggers their mutual synapse innervations and recruits them as associative memory cells, specific characters of associative memory cells are presented below [5–7].

Associative memory cells receive new synapse innervations from the coactivated cortical neurons for their mutual connections alongside innate sensory input. Associative memory cells are able to encode the newly learned signal and innate or previous learnt signals for their integrative storage and reciprocal retrievals. Their axons project to and synapse innervations onto neurons in brain areas relevant to cognitive events, emotion reaction, and behaviors in convergent and divergent manner. The mutual synapse innervations among cross-modal associative memory cells confer the reciprocal retrieval of associated signals by one of learning cues as well as the automatic conversion of signal retrievals among different modalities, such as visual signals in images are represented by verbal language, and verbal signals in stories are recalled by diagrams. Synapse convergences onto downstream associative memory cells and the activation of associative memory cells confer associative thinking and logical reasoning, computation, and so on. Therefore, synapse innervations determine the specificity of memory contents [5, 8–10] (Figs. 5.2 and 5.3).

To general features, the number and function state of associative memory cells influence memory strength and maintenance. The recruited number of associative memory cells is proportional to the number of mutual synapse innervations produced by neuron coactivation strength and repetitive coactivations during learning, which is affected by the development [3, 8]. The functional state of associative memory cells is regulated by the strengths of innate and new synapse inputs as well as synapse inputs from the arousal system, their somatic capability to convert synaptic analog signals into digital spikes, and their axonal capability to propagate spikes out [11–14]. Glutamatergic associative memory cells suppress the activity of other memory cells through GABAergic associative memory cells and lateral inhibition to have themselves to be dominantly active for memory contrast [15, 16].

Synapse innervations to associative memory cells determine memory specificity. The number and function state of associative memory cells as well as the connection and activity strengths in their synapse inputs and axonal output partners influence memory power and persistence [8, 10, 17, 18]. For instance, associative memory cells in the piriform cortex receive new synapse innervations from associative memory cells in the barrel cortex after associative learning alongside the innate innervations from the olfactory bulb. Synapse activity in the pathway of whisker signal will drive piriform cortical associative memory cells toward spiking threshold under the basal activity driven by olfactory bulb inputs. Once spike threshold reaches, the spikes trigger memory output cells for whisker-induced olfactory responses. With these associative memory cells in sensory cortices [3, 8, 10], their axon-innervated downstream neurons are able to encode the associated signals [19–24]. Stimulations to any of these areas in neural circuits from sensory cortices to behavior- and emotion-relevant brain nuclei induce memory retrievals [25–30]. There are about 10,000 types of proteins in living cells [31], which is much less than millions of unitary signals remembered in life. On the other hand, more than ten billion of neurons reside in the brain. Those neurons with their synapse interconnections, i.e., associative memory cells as well as their networks, should provide the sufficient number of basic units for memories to specific contents.

Each of sensory cortical neurons receives one type of innate sensory inputs via sensory afferent axons and their synapse innervations as well as encodes this particular innate sensory signal. Once this group of sensory cortical neurons is coactivated with sensory cortical neurons that encode other sensory signals in cross-modal or intramodal manner during learning processes, their mutual synapse innervations recruit them to be primary associative memory cells that encode the integrative storage and the reciprocal retrieval of innate signal and associatively learned signals [5]. The types of associatively memorized signals in each sensory cortical neuron are dependent on how many kinds of sensory cortical neurons have been coactivated during various associative learning processes. In other words, more associative learnings the individuals experience, and more types of associated signals each of the associative memory cells encode, such that each of associative memory cells in the individual's brain can more efficiently perform the associations in relevance to further associative learning, new associative memory, reciprocal retrieval, logical reasoning, associative thinking, and so on [6, 7].

Secondary associative memory cells are recruited in cognition-related brain areas and association cortices. They receive new axon projection and synapse innervations from primary associative memory cells in coactivated sensory cortices in convergent and divergent manners during cognitive events and emotional reactions. Secondary associative memory cells encode endogenous associated signals propagated from the axons of primary associative memory cells for their integrative storages. The associations of cognitive events and emotional reactions can induce mutual synapse innervations among secondary associative memory cells. These secondary associative memory cells in turn innervate onto memory-output cells in the motor cortex for memory retrieval through language and writing. The number of secondary associative memory cells recruited during the cognitions is proportional to mutual synapse innervations evoked by coactivation strength and repetition. Their number is also influenced by developmental stage [5]. Similar to neurons, the activity strength of secondary associative memory cells is affected by the following factors, including synaptic innervation, the ability to convert synaptic signals into neuronal spikes, and the ability to propagate spikes out toward memory output cells [6]. In terms of physiological impact, the specificity of memory contents during cognitive events and emotional reaction is assigned by the sources of synapse innervation on secondary associative memory cells. Memory strength and maintenance are influenced by the quantity, excitability, and innervations of secondary associative memory cells. Their subsequent activations result in the rehearsal of associative thinking, logical reasoning, and mood reaction. Moreover, secondary associative memory cells innervate the hypothalamus and extrapyramidal system. These brain areas influence the balance of sympathetic and parasympathetic nerves, food ingestion, body temperature, and hormones, which can regulate mood reaction and behavioral performance [5, 7] (Figs. 5.2 and 5.3).

Based on these characters of primary and secondary associative memory cells including glutamatergic and GABAergic neurons and astrocytes, a diagram about their connections and interconnections is given in Figs. 5.2 and 5.3. How these different grades of associative memory cells work together to execute the brain functions for physiology and psychology remains to be figured out.

A linear connection of primary associative memory cells, secondary associative memory cells, and memory output cells constitutes the signal transfers through station-by-station during learning, memory, and cognition [23, 24, 32]. The connection from sensory organs to sensory cortices works for the acquisition of exogenous associated signals. The connection from primary associative memory cells in sensory cortices to secondary associative memory cells in the brain areas related to cognition and emotion converts these exogenous signals into endogenous signals for cognitive processes and emotional reactions. Those connections from both primary and secondary associative memory cells to memory output cells in the motor cortex execute the memory retrievals by language, gesture, and countenance. These linear pathways from sensory systems are critical to fulfill the acquisition and storage of exogenous information, the conversion of exogenous signals into endogenous signals to be memorized and provide the resources of information for cognition and emotion, as well as the encoding of memory orders for information retrievals [5–7].

There are convergent and divergent connection patterns emerged in the linear connections from one station to another [24, 32]. Convergent connections are existed from primary associative memory cells in sensory cortices to secondary associative memory cells in cognition-/emotion-relevant brain areas. These convergences make the association of exogenous sensory signals that have memorized into sensory cortices to be used for cognitive processes and emotional reactions in an integrative manner. For instance, those signals used by secondary associative memory cells for logical reasoning and associative thinking may come from the auditory and visual cortices for their integration. The signals used by secondary associative memory cells in the amygdala and the ventral tegmental area for either fear or pleasure events may be integrated signals from visual and auditory cortices. The signals used by secondary associative memory cells in the hippocampus for spatial locations may be integrated from visual and auditory cortices. Furthermore, the signals used by motor cortical memory output cells for memory retrieval by various responses to cues are integrated from sensory signals, cognitive signals, and emotion signals. In the meantime, the convergent association of those exogenous sensory signals in secondary associative memory cells makes them to be converted into endogenous signals and be memorized for the future presentations, in which these signals have been told to be thought, reasoning, and emotional reaction by individuals [5]. It is noteworthy that these processes need the mutual synapse innervations among secondary associative memory cells (please see below).

On the other hand, divergent connections are also present from primary and secondary associative memory cells, for instance, the axons of primary associative memory cells project toward the hippocampus and the prefrontal cortex [24, 32]. This divergent connection allows signals from a sensory source to be used to multiple cognitive events and emotional reactions. Visual signals can be used by secondary associative memory cells in the prefrontal cortex for cognitions and by secondary associative memory cells in the hippocampus for spatial localization. Olfactory signals can be used by secondary associative memory cells in the prefrontal cortex for cognitions and by secondary associative memory cells in the amygdala for emotional reactions. In addition to the physiological impacts in real life, the divergent connections make signals to be memorized in different brain regions, such that memory loss under certain pathological conditions can be prevented moderately [5].

The featured connections among associative memory cells are their mutual synapse innervations after their coactivation [3, 8, 9, 16, 33]. In sensory cortices, primary associative memory cells are recruited from coactivated neurons. For instance, the pairing of whisker and odor stimulations leads to mutual synapses innervations between barrel and piriform cortices [3, 15, 16, 33, 34]. In the meantime, mutual synapse innervations are also seen between left and right hemispheres [9]. These mutual innervations make the primary associative memory cells among different sensory cortices to encode the integrative storage and reciprocal retrieval of associated signals. Plus their convergence to secondary associative memory cells [24, 32], any learning cue is able to activate these groups of primary associative memory cells in sensory cortices for associative perceptual memory, and in turn to activate secondary associative memory cells for integrative cognitions, such as associative thinking and logical reasoning [5]. For instance, after the associations of visual and auditory stimulations, visual signals directly activate associative memory cells in the visual cortex. With a basal activation of auditory cortical associative memory cells by auditory afferents, visual signals can activate primary associative memory cells in the auditory cortex based on their mutual synapse innervations. As the axons of these primary associative memory cells convergently project and innervate to secondary associative memory cells, these two groups of primary associative memory cells in visual and auditory cortices can activate secondary associative memory cells for cognition and emotion. Through the mutual synapse innervations of primary associative memory cells and their convergence onto secondary associative memory cells, a learning cue can activate all of the primary associative memory cells being interconnected and secondary associative memory cells being convergently innervated, as long as synapse innervations are actively enough to drive membrane potentials to the spiking threshold potential (Fig. 6.1). Any cue occurred in previously associative learning may trigger various feelings, perceptions, cognitions, and emotional reactions in relevance to this specific associative memory [5].

In addition, there are mutual synapse innervations among secondary associative memory cells in brain areas in relevance to the cognition and emotion [5] (Figs. 5.1 and 5.2). This type of mutual synapse innervations is formed in associative thinking, logical reasoning, imagination, and emotional reactions. Signals and contents generated during these events are memorized in secondary associative memory cells based on their mutual synapse innervations in the intramodal and cross-modal manner. Intramodal secondary associative memory cells in cognition-related brain areas are used to encode cognitive processes and their memories. Cross-modal secondary associative memory cells based on their interconnection between cognition- and emotion-relevant brain areas work for the cognition associated with emotional reactions, such as associative thinking with enjoyable perception or with upset feeling.

It should be noteworthy that the convergence of primary associative memory cells or secondary associative memory cells onto memory output cells in the motor cortex makes different sources of memories to use a common pathway for their retrievals through language, gesture, and countenance. Because of the limit space for each of motor neurons, there are competitions for the axons of these primary and secondary associative memory cells to innervate these memory output cells. Based on a rule of activity together, connection together, and strengthening together, the repeated rehearsals of associative memory are needed for these associative memory cells to have more and stronger innervations onto memory output cells. As the memory output cells are commonly shared by numerous associative memory cells, these memory output cells may have intrinsic mechanisms to prevent themselves from the overload through switching down those infrequently used pathways and synapses to be inactive. So, memory output cells are not overloaded, and the function of their synapse innervations is decayed without the rehearsal trainings [5, 17, 23].

Mutual synapse innervations among associative memory cells also have a role of feedforward and feedback connections [5–7, 35], e.g., mutual synapse innervations between glutamatergic and GABAergic associative memory cells [15, 16, 33, 34]. Glutamatergic associative memory cells can make synapse innervations to GABAergic associative memory cells. These GABAergic associative memory cells may innervate back onto the same group of glutamatergic associative memory cells, i.e., indirect feedback connection constitutes feedback inhibition to prevent the over-excitation in this group of glutamatergic associative memory cells. These GABAergic associative memory cells may innervate forward to other groups of glutamatergic associative memory cells, i.e., indirect feedforward connection constitutes feedforward inhibition (lateral inhibition) to weaken other glutamatergic associative memory cells, such that the memory by this group of glutamatergic associative memory cells will be dominant, i.e., memory contrast. In addition to the recurrent and feedforward inhibitions, the excitation may be fulfilled through mutual synapse innervations among glutamatergic associative memory cells, i.e., positive feedback and feedforward interactions [5] (Fig. 5.5).

It is noteworthy that mutual synapse innervations between primary and secondary associative memory cells as well as between associative memory cells and memory output cells are critically important [5]. If these mutual synapse innervations between primary and secondary associative memory cells work for the positive interaction, memories to specific contents with the sources of sensations and perceptions will be strengthened. These positive interactions for enjoyable memories need to be consolidated, or vice versa. If mutual synapse innervations between primary and associative memory cells execute a negative interaction through lateral inhibition, memories to specific contents with the sources of sensations and perceptions will be weakened. These negative interactions in the memory of important contents need to be debilitated, or vice versa. In this regard, GABAergic associative memory cells appear to be important to set up the appropriate interactions among associative memory cells. It may be major reasons that many neurological diseases and psychiatric disorders in relevance to memory deficit are associated with the dysfunctions of GABAergic neurons [5, 36–46].

Mutual synapse innervations among intramodal primary associative memory cells and their convergences onto secondary associative memory cells are important for the integration of unitary signals and their integrative memory [5]. Although it is hard to study the morphological interconnections among associative memory cells in a single modality, the pair-recording in somatosensory cortical neurons has been well done [11, 13, 14]. With their mutual synapse innervations, unitary signals from the sensory afferents can be integrated in intramodal sensory cortices [47, 48]. As discussed below, visual signals and images are dissected as photon beams by cone cells in the retina, and auditory signals in the words and sentences are dissected as sound frequencies by hair cells on base membrane of the cochlea. Through the sensory afferents, these unitary signals can be integrated in their correspondent sensory cortices, where the integration of such unitary signals to complicated images or sentences is fulfilled by mutual synapse innervations among these intramodal primary associative memory cells. Subsequently, these primary associative memory cells in an intramodality convergently innervate onto the association cortices for these complicated signals to be condensed and memorized. The mutual synapse innervations among primary associative memory cells in cross-modal sensory cortices or among secondary associative memory cells in cross-modal association cortices, e.g., visual and auditory cortices, confer them work together for the reciprocal retrieval of associated signals by automatic conversion among modalities as well as for the understanding of associated signals from different modalities [5].

One associative memory cell or one group of associative memory cells can become the common connection with all of other associative memory cells, which can fall in intramodal or cross-modal manner. The mechanism for this core of associative memory cells is proposed below. In addition to the innate synapse innervations, the new and mutual synapse innervations are generated during the coactivation of ready-recruited neurons with others. Once the new and mutual synapse innervations are formed among two or more groups of ready-recruited neurons, associative memory cells are recruited from these neurons. The specificity of associated signals stored in associative memory cells or their group is based on their synapse innervations that carry in these signals. In future learnings, if one group of associative memory cells is associatively activated with other groups of ready-recruited neurons or of associative memory cells that memorize different types of associated signals, this group of associative memory cells may form the mutual synapse innervations with other groups of associative memory cells. With convergent innervations from other groups of associative memory cells, this group of associative memory cells may take a position as the core of other groups of associative memory cells (please refer to Chap. 5). This type of interconnections among groups of associative memory cells may work for the integrative storage of those associated signals that share one of common characters. Furthermore, all these groups of associative memory cells can be mutually innervated if they have been coactivated, which may work for the integrative storage of associated signals that are grouped in similar concepts or knowledge. Therefore, the associated episodic signals taken by other groups can be summarized in one group or all groups of associative memory cells as well as semantic memory (please refer to Chap. 8).

Taken all of these messages together, we summarize the following points. The linear connection of primary associative memory cells, secondary associative memory cells, and memory output cells executes information transfer. Convergent connections from primary associative memory cells onto secondary associative memory cells work for the integration and association of multiple signals at high levels for cognitions and emotions. Divergent connections from primary associative memory cells to secondary associative memory cells work for signal extension to the wide brain areas, such that memory traces are widely distributed. Mutual synapse innervations among associative memory cells are critically important for the interaction and integration of multiple signals as well as the integrative storage and reciprocal retrieval of associated signals at primary and secondary levels. Feedforward innervations plus lateral inhibition facilitate memory contrast. Feedback connections in positive and negative manner are critical for positive and negative regulations among associative memory cells. Therefore, the feedback connection from the secondary associative memory cells to the primary associative memory cells confers the regulations of cognition and emotion to associative memory. Most importantly, these synapse interconnections among associative memory cells or ready-recruited neurons are based on the rule, activity together, connection together, and strengthening together [5–7].

The patterns of synapse innervations among associative memory cells are critically important for them to execute various physiological activities in relevance of associative memory. Another important issue is whether the synapse innervations are sufficient to activate these associative memory cells. In order to activate these associative memory cells, the number of synapses, the functional states of synapses, the spike threshold potential of associative memory cells, and the capability of associative memory cells to convert synapse analog signals to sequential digital spikes are critical on associative memory cells [5–7]. The number of synapse innervations is influenced by the learning efficiency and times [3, 8, 10, 18], i.e., active together and wiring together. The functional state of synapses on associative memory cells is affected by their maturation and potential upregulation [11, 49], i.e., active together and strengthening together. The spike threshold potentials of associative memory cells are influenced by voltage-gated sodium channels [50–56] as well as their plasticity at multiple grades [57]. The capability of associative memory cells to convert synapse analog signals to sequential digital spikes is mainly influenced by voltage-gated sodium channels [51, 52, 58–62]. It is noteworthy that molecules to regulate these neuronal active intrinsic properties are commonly done by the intracellular calcium signaling pathway [11, 49, 57, 63]. Therefore, the coactivation and coactivity of associative memory cells determine the patterns, number, and function state of synapse innervations, as well as influence their spike threshold and spike-encoding capability. The coactivity of associative memory cells elevates calcium signaling pathways for the upregulation of their synapses and encoding capability [11, 49, 57, 63] as well as activates the chain reaction from cell spikes to microRNA-regulated expressions of genes and proteins for their axon prolongation and synapse formation [8, 10, 16].

In summary, associative memory cells with their innervation patterns and encoding capability are essential for memory formation, memory retrieval, cognitive events, and emotional reactions [3, 8, 10, 18, 23, 24, 33, 64]. Their characters and working principle can be used to figure out working maps related to associative memory in cross-modal or intramodal manner, such as the efficiency of associative learning, the integrative storage of multiple signals, the strength and preservation of associative memory, the efficiency of memory retrievals, the transformation of the unitary signals to complicated signals in their storage, the temporal sequence of learning and memory to multiple signals, and the correlation of associating memory to cognitive events and emotional reaction. The working map of associative memory cells also assists to reveal the high-order cognitions, such as consciousness, attention, as well as various memory patterns, including explicit versus implicit memory, episodic versus semantic memory, and transformation between these patterns.

The coactivation among the neurons from distinct brain regions is critical to induce new synapse formation and recruit associative memory cells. For instance, the coactivation of the neurons among cross-modal or intramodal sensory cortices induces their mutual synapse innervation. In the meantime, these interconnected neurons are able to encode multiple associated signals. Thus, such sensory cortical neurons are recruited as primary associative memory cells [3, 8, 16, 33]. The coactivations of primary associative memory cells also drive their axon prolongation and synaptically innervate onto ready-recruited neurons in cognition- and emotion-relevant brain areas, recruiting them as secondary associative memory cells during logical reasoning and associative thinking [5, 7, 23, 24, 65]. Associative memory cells with their synapse inputs and mutual synapse innervations constitute basic units in memory traces specifically to those associated signals. The activity-dependent positive cycle in the recruitment and refinement of associative memory cells will recruit more associative memory cells to strengthen the memory depth and maintenance [7]. The interconnections among primary associative memory cells concede the reciprocal retrieval of associated signals. The interconnections among secondary associative memory cells and their synapses convergently from primary associative memory cells endorse associative thinking and logical reasoning. These characteristics about the union together through coactivity-induced interconnections and connection strengthening among associative memory cells upgrade Hebb's hypothesis. He has proposed that the repetitive coactivity of interconnected neurons strengthens their connection for memory formation [66].

Mechanisms underlying activity-dependent positive cycle in the recruitment and refinement of associative memory cells may include the following step. In early life, the learning of associated signals induces ready-recruited neurons to be mutually innervated through axon prolongation and synapse formation, such that associative memory cells are recruited for the integrative storage and reciprocal retrieval of these associated signals. By the learning of numerous unitary associated signals, many groups of associative memory cells are recruited to encode the storages of unitary signals. In later life, although the capability of axon prolongation in long distances appears decayed, the coactivations of associative memory cells during learning can make these axon terminals to be grown and extended locally. The local growth and extension of axon terminals of associative memory cells from other regions will make new synapse formations onto ready-recruited neurons and associative memory cells by *en passant* synapses. More synapses will strengthen the driving force to these associative memory cells to upregulate their functional states, i.e., the activity-dependent refinement of associative memory cells. More new synapse innervations onto ready-recruited neurons in the neighboring of these associative memory cells recruit more associative memory cells or the activity-dependent recruitment of associative memory cells (Fig. 5.4). The repeated coactivations of associative memory cells will lead to the activity-dependent positive cycle of their recruitment and refinement [5].

7.3 Associative Memory Cells for Intramodal Associative Memory

As discussed above, associative memory formed by the association of multiple signals from cross-modal sensory modalities is commonly seen in life, such as the associations between visual and auditory signals and smell and taste signals. Memory retrievals can be achieved by the automatic conversion of these associated signals, e.g., visual signals into verbal signals or the other way around. Memory can also be retrieved reciprocally by either cue of associated signals [7]. For example, pictures in movies or videos can be recollected by the verbal language. Contents in verbal stories can be represented by imageries and diagram. These reciprocal retrievals induced by cues or automatic conversions among different sensory modalities are based on synapse interconnections among primary associative memory cells across sensory cortices. In addition to this cross-modal form of associative memory cells across distinct brain areas that have been coactivated, intramodal associative memory cells are present in a single modality of sensory cortices. Pair-coding neurons that program two visual pictures have been recorded in animal visual cortices in vivo [47, 48]. Their activity property is similar to the encoding property of cross-modal associative memory cells. Such pair-coding neurons within an intramodal sensory cortices program the integrative memory of associated signals inputted from a single sensory modality. For example, photon beams included in images are associatively to stimulate the visual system. Various odor signals associatively activate the olfactory system. Letters and words are associatively to trigger the auditory system. These pair-encoding neurons can be thought of as intramodal associative memory cells. Mutual synapse innervations among intramodal primary associative memory cells in sensory cortices and their convergences to secondary associative memory cells in association cortices are important for the integration of unitary signals and their integrative memory [5]. Although it is difficult to study morphological interconnections among primary associative memory cells in a single modality, the pair-recording in somatosensory cortical neurons to profile their characters has been well performed [11, 13, 14, 67–70]. With their mutual synapse innervations, unitary signals from sensory afferents can be integrated in intramodal sensory cortices.

The association of intramodal multiple unitary signals commonly occurs in the information acquisition. For instance, the associations of different objects in a single view or different parts in an object are detected by the visual system. The associations of different letters in a word or of words in a sentence are detected by the auditory system. Intramodal associative learning refers to the associations of unitary signals in a sensory system. Primary associative memory cells by their mutual synapse innervations [7] or pair-encoding neurons [47, 48] in single sensory cortex endorse memory retrieval in a picture-by-picture or word-by-word manner. The signals detected by the visual system and auditory system are usually complicated. For instance, a picture consists of numerous photon beams that possess different light strengths and colors. A sentence is composed of distinct words and letters. These photon beams with spatial location and light strength in a picture are detected by different cone cells in the retina. After the conversion of photon signals into neural signals at these cone cells, visual nerves transmit these neural signals that carry photon signals to visual cortical neurons in the point-by-point manner. Sound wavelength signals, especially frequencies from words and letters, are converted into electrical signals by hair cells. These hair cells on different segments of the cochlea base membrane correspond to different sound frequencies. Electrical signals from hair cells are propagated on auditory nerves. Passing through a few relays, electrical signals from each of auditory nerves reach to neurons within the auditory cortex in a point-by-point manner [71]. It is largely unknown how these unitary signals included in an image or a sentence are reintegrated into the image or the sentence in the form of neuron-encoded signals by a group of the associative memory cells in sensory cortices and/or association cortices, as well as specifically memorized in these cortices and other brain areas [7].

There is a rule about coactivity-induced union in the connection and strengthening among neurons [5–7]. The coactivations of auditory cortical neurons that encode words or letters induce mutual synapse innervations among these neurons. Their interconnections recruit intramodal primary associative memory cells. These associative memory cells in the auditory cortex become able to encode the integrative storage of these unitary sound signals in early life. How do these primary associative memory cells in the auditory cortex are recruited? It has been found that cortical neurons have a few folds of their excitatory activities as well as show discharge and afterdischarges in response to stimulations [11, 45, 72, 73]. These cortical neurons with the highest excitability are preferentially activated by synapse inputs, or vice versa. In a group of auditory cortical neurons that receive innervations divergently from hair cells on cochlea base membrane with the specific sound frequency, the most active neurons are preferentially activated by this sound frequency. This situation is also present in other groups of auditory cortical neurons. In the learning of a word or a sentence, certain groups of auditory cortical neurons are activated by sounds sequentially. These groups of auditory cortical neurons include those in response to the first letter or word up to the last ones. The afterdischarge of the neurons triggered by the first letter or word coincides with the discharge of those neurons triggered by the second letter or word. Similarly, the afterdischarge of those neurons by the second letter or word coincides with the discharge of those neurons by the third one. This coincidence occurs in all pairs of sequential letters or word in the learning. The coincident activities of the auditory cortical neurons induce their synapse interconnection, such that they become a new group of neurons for this word or sentence. These auditory cortical neurons based on their mutual synapse innervations become able to integrate letters in a word or words in a sentence into the word or sentence in the form of neuronal encoding. In the meantime, these interconnected neurons are recruited as primary associative memory cells in the auditory cortex for the integrative storage of unitary sound signals. In the repeated learnings or rehearsals of this sentence or word, these auditory cortical associative memory cells in this new group are strengthened by the refinement of their mutual synapse innervations, excitability, and encoding capability. Furthermore, the portion of auditory cortical associative memory cells in this new group is upregulated by the recruitment of more associative memory cells through activity-dependent positive cycle in the recruitment and refinement of associative memory cells [7]. The coactivity-dependent recruitment and refinement in this group of primary associative memory cells in the auditory cortex confer the consolidated memory of this word or sentence for the later retrieval. In the early life, more groups of primary associative memory cells in relevance to various sound spectra of words or sentences are recruited based on their mutual synapse innervations if more sound signals from words or sentences are listened. With more primary associative memory cells in auditory cortices, young individuals will quickly learn more new sound signals as well as more complicated signals in their later life or called as smart young individuals [5].

In the later time of life span, sound signals transmitted to the auditory system for learning and memory are complicated, which are reorganized from unitary sound signals including letters and words learned previously. The learning of a complicated sound signal will coactivate some primary associative memory cells in the auditory cortex. These primary associative memory cells have previously memorized unitary sound signals that are included in this complicated signal. The coactivation of these auditory cortical associative memory cells will strengthen their mutual synapse innervations and excitability. The strengthening of mutual synapse innervations that have been onset in the early life may be based on the conversion of silent into functional synapses [74] or of inactive into active synapses [49]. The elevation of their excitability can be the multiple grades in nature [57]. These upregulated associative memory cells in the auditory cortex are able to integrate unitary sound signals in these newly listened and complicated words or sentences, as well as to encode their integrative storages. These upregulated associative memory cells will be preferentially activated during memory retrievals by cues due to high excitatory state [5]. As the excitability of cortical neurons is upregulated in multiple grades [57], each of associative memory cells with this property is able to integrate with other associative memory cells for learning different sound spectra, provided their mutual synapse innervations have been established in early life. Furthermore, the glutamatergic associative memory cells that memorize these newly learned words or sentences can suppress the activity of other glutamatergic associative memory cells through GABAergic associative memory cells and lateral inhibition in order to have themselves to be activated preferentially for the retrieval of memorized words or sentences [15, 16]. The glutamatergic associative memory cells can limit themselves not to be overexcited through recurrent feedback inhibition from GABAergic associative memory cells [15, 16].

The coactivity of visual cortical neurons that receive point-by-point synapse innervations from retina cone cells in early life induces mutual synapse innervations among these neurons. Visual cortical neurons with the mutual innervation are recruited to be intramodal associative memory cells in the visual cortex or visual cortical associative memory cells. These primary associative memory cells in the visual cortex can integrate unitary signals, i.e., photon beams with different intensity, color, and spatial distribution, in visual images for their storage. It has been found a proportional relationship between activity strength at primary associative memory cells and stimulus intensity [16, 33]. Primary associative memory cells in the visual cortex that receive the light beams with stronger intensity and more fresh colors in a visual image are more active. In line with a rule of neurons that are active together, connecting together, and strengthening together [7], the number and strength of mutual synapse innervations are formed dominantly among associative memory cells with high activity. These more active primary associative memory cells in the visual cortex are recruited and organized to be a group of intramodal primary associative memory cells. Their interconnections and encoding capability fulfill the integration and storage of light beams with more strong intensity and fresh color in this visual image. Thus, this visual image in the front of eyes is converted into the images in the form of neuronal encoding in the visual cortex on the basis of a local neural network that consists of associative memory cells in an engram particular for this visual image. Although this visual image has been dissected by retina cone cells into different photon beams and transmitted by visual nerves to visual cortical neurons in a point-by-point manner in medium steps, associative memory cells in the visual cortex integrate them as an image in neural encoding format [5].

In the meantime, the primary associative memory cells in the visual cortex may send their axons to project toward visual association cortices [75, 76] and make convergent synapse innervations onto association-cortical neurons to recruit them as secondary associative memory cells. The coactivity of these secondary associative memory cells makes mutual synapse innervation among them [5, 32]. This linear connection with convergence from primary to secondary associative memory cells will transfer neural form image signals, especially unitary signals with stronger photon beams in this visual image, into the visual association cortex for their integrative storage by secondary associative memory cells based on their mutual synapse innervation. This transferring allows primary associative memory cells in the visual cortex to be able to take subsequent new signals without loss of previous image signals. As visual cortical neurons receive synapse innervations from the retina cone cells in a point-by-point manner, whether intramodal primary associative memory cells in the visual cortex receive strong or weak (major or minor) synapse innervations from cone cells is influenced by the activity strength of their connected cone cells. The activity strength in primary associative memory cells also influences their mutual synapse innervations among them. In other words, the primary associative memory cells that receive strong inputs from retina cone cells and more mutual synapse innervations will be more active. Secondary associative memory cells in the visual association cortices are activated by primary associative memory cells with strong synapse innervations and active synapses converted from inactive synapses, so that major features with strong light intensity and fresh color in this visual image are transferred into the visual association cortices for integrative storages and subsequent retrievals [5]. Author's point of view is granted by current report that visual association regions are recruited during memory formation [77]. In subsequent associative learning based on the reorganization of unitary signals in various new images, the proportion of primary associative memory cells reactivated by new images based on the reorganization of unitary signals will be reintegrated together through the conversion of inactive synapses into active synapses among them to fulfill the integrative storage of new associated signals [5, 49].

Memory to olfactory signals or taste signals can be intramodal. The simulations of odorants or tastes will recruit primary associative memory cells in the olfactory cortex or the gustatory cortex [5]. The odorants or tastes can stimulate the olfactory system or the gustatory system in unitary or mixture manner, such that primary associative memory cells in single sensory modality are recruited in relevance to encode unitary or mixture signals. For instance, each group of primary associative memory cells in the gustatory cortex based on their mutual synapse innervations can encode sweet, sour, or salty signal. Three groups of these primary associative memory cells in the gustatory cortex will establish their mutual synapse innervations, if a food including these three tastes is given, to coordinately encode their integrative storage and reciprocal retrieval. Similarly, each group of primary associative memory cells in the olfactory cortex based on their mutual synapse innervations can encode perfume or stinky signal. Two groups of these primary associative memory cells in the olfactory cortex will establish their mutual synapse innervations, if these two smells are simultaneously given, to encode their integrative storage and reciprocal retrieval.

How are the intramodal signals from primary associative memory cells in a sensory cortex retrieved by the recall, representation, and recollection? In order to retrieve these intramodal signals, they require to be specifically associated with words, phrases, and sentences generated in the auditory cortex. The coactivations of these intramodal associative memory cells within sensory cortices with auditory cortical associative memory cells will induce their mutual synapse innervations, such that cross-modal primary associative memory cells between the auditory cortex and visual, olfactory, gustatory, or somatosensory cortex are recruited [5]. By the association of verbal language with gestures and countenance, so-called secondary order conditioning [78], gestures, and countenances can replace the role of verbal languages under certain conditions. Associative memories to these intramodal signals with auditory signals will result in the clear definition of those unitary signals in visual, olfactory, gustatory, or somatosensory signals. It is noteworthy that unitary signals in the visual system at this time are unitary objects, but not photon beams, since associative memory cells in visual cortex and association cortex are able to integrate associated photon beams into objects for their storages. The further learning of complicated signals reorganized from unitary signals can lead to their advanced associations. Otherwise, memories to these intramodal signals can only be represented by saying sight, taste, smell, or touch before. In order to fulfill these types of cross-modal associative memory, mutual synapse innervations between primary associative memory cells in the auditory cortex and primary associative memory cells in the visual, olfactory, gustatory, or somatosensory cortices should be established to recruit these intramodal associative memory cells into cross-modal associative memory cells, which remains to be examined experimentally [6, 8]. Similar to working principles of associative memory cells in other sensory cortices, the integrative storages of the visual, olfactory, somatosensory, or gustatory signals with auditory signals can be completed in the association cortices correspondent to these functions through their interconnections for their future reciprocal retrievals and the understanding. Furthermore, memories to olfactory and gustatory signals are often associated with emotion-relevant perceptions and reactions, so that various feelings and emotional reactions are expressed during the retrieval of odor and gustatory signals.

7.4 Associative Memory Cells for Cross-Modal Associative Memory

In fact, intramodal and cross-modal associative learning occurs in the same time. Memories to these cross-modal and intramodal associated signals transpire concurrently. In particular, visual and auditory signals are jointly acquired and associatively memorized. Unitary objects organized by photon beams in visual picture are associated to verbal signals during social communications. In these family events, personal interactions, and classroom learnings, apiece feature of visual pictures is defined clearly by words or sentences. In social activities, numerous associations are formed between unitary signals in visual picture and words or phrases. The associations of unitary signals among cross-sensory modalities confer the learning of complicated signals, such as video images, with the reorganization of verbal definitions for unitary signals. Memories to these complicated signals can be fulfilled by the reorganization of the unitary signals and the reorganized integration of associative memory cells that have memorized unitary signals across sensory cortices. After cross-modal associative memory is formed, people are able to achieve the reciprocal retrieval of these associated signals. A signal induces the recall of its associated signals or the other way around. In addition, the automatic conversion of retrieval signals can be fulfilled, in which signals learned from one modality are retrieved through another modality. For instance, signals in pictures are converted into verbal recall or the other way around [5, 7].

Mechanisms underlying cross-modal associative memory include the recruitment of associative memory cells across sensory cortices and their refinement. In early life, the acquisition of those associated signals from a few modalities can coactivate their sensory cortical neurons and induce their synaptic interconnections. In the learning of visual signals and auditory signals, visual cortical associative memory cells that encode unitary signals in a picture interconnect auditory cortical associative memory cells that encode words or phrases. For instance, the associative memory cells that integrate and encode unitary signals in an image lemon interconnect the associative memory cells that encode words lemon, yellow, and oval, when they are coactivated during initial learning [7]. Numerous associations between unitary signals from visual modality and words in social interactions evoke mutual synapse innervations between visual and auditory cortical neurons. Dozens and dozens of the pairs of cross-modal associative memory cells are recruited. These pairs of cross-modal primary associative memory cells can integrate the cross-modal associated signals and encode their integrative storage. Their active states influenced by the number of mutually innervated synapses and the capability of spike encoding will determine whether memories to these cross-modal associated signals can be retrieved. The accumulation in distinct groups of associative memory cells that encode various pairs of unitary signals will facilitate the learning of complicated signals reorganized from these unitary signals as well as the recollection of these category-relevant signals. During postnatal development, the capabilities of axon growth and synapse formation are gradually attenuated [3]. The learning of complicated signals during the later stage may utilize another mechanism for their memory, the coactivity-dependent upregulations of associative memory cells in their excitability as well as mutually innervated synapses [7, 16, 32] as well as activity-dependent conversions of inactive synapses into active synapses [49]. However, their functional upregulation depends on the previous formation of mutual synapse innervations. With the repeated coactivations of these associative memory cells, activity-dependent positive cycle in their recruitment and refinement upregulates the number and function state of associative memory cells. Provided the upregulation of their number and functional state reaches a sufficiently high level with long time, the retrievals of these complicated signals can be done automatically or by cues [5].

With the coactivity-dependent recruitments and upregulations of primary associative memory cells, people can gradually memorize associated signals in the level of unitary signals up to the level of complicated signals, that is, the transformation of simple to complicated information storage in topic-relevant manner [6]. In an initial step, the associations of simple images in different intramodal features with words based on letters can coactivate visual and auditory cortical neurons, respectively. Their synapse interconnections recruit intramodal and cross-modal primary associative memory cells. They encode the integrative storages of pictures, letters, as well as picture and words. The repeated coactivations among these associative memory cells by practices and rehearsals can induce their activity-dependent upregulation and recruit more associative memory cells. Their coactivity leads to the positive cycle between the recruitment and refinement of associative memory cells. They are recruited from sensory cortical neurons at the first grade [7]. When the different groups of associative memory cells to memorize various unitary signals are accumulated, they are ready to be recruited as advanced associative memory cells that encode the integrative storages of complicated associated signals. In the learning of complicated signals from visual and auditory modalities, the first grade of associative memory cells in visual and auditory cortices can be coactivated. Their synapse interconnections and coactivity-dependent upregulation recruit the second grade of associative memory cells. With these advanced associative memory cells, those complicated pictures and sentences reorganized from unitary signals can be encoded and associatively memorized. The continuous learning will recruit numerous sets of the first and second grades of associative memory cells accumulatively. The advance learning will recruit multiple grades of associative memory cells that encode more complicated signals in the life span. With the accumulation of different groups and grades of associative memory cells, further learning in late life may be based on their activity-dependent functional upregulation. The high excitability of associative memory cells enables them to be easily activated for quick memory formation and quick conversional retrievals among distinct modalities. For instance, visual pictures and verbal language signals are associatively memorized. These mechanisms facilitate the memory of complicated signals in images and sentences quickly. Reading book or looking images induces intensive activities in certain groups of associative memory cells that encode the storage of these sentences and images, which leads to their coactivity-dependent functional upregulation. The low-threshold potential of these associative memory cells to fire spikes and active synapse inputs to drive these associative memory cells will permit the cues dominantly to reactivate them for the reciprocal retrievals of pictures and sentences and will confer their spontaneous activation to drive secondary associative memory cells for free associative thinking. The activities of these associative memory cells lead to memory retrieval by behaviors if they successfully activate memory output cells in the motor cortex [5, 7].

In addition to mutual synapse innervations among primary associative memory cells as well as advanced innervations among these multigrades of associative memory cells, there is another mechanism for the cross-modal associative memory of unitary and complicated signals. Association cortices are thought to be able to integrate the signals from the different sensory modalities [77, 79–81]. The coactivation of primary associative memory cells and multigrade associative memory cells during repeated learning and memory retrievals will make their convergent and divergent innervations onto neurons in these association cortices [5]. The integrative storage of associated signals from different modalities in association cortices will confer the translation among those associated signals from different modalities and the understanding of these signals through mutual translations among sensory modalities, such as the conversion between visual and auditory signals, as well as mixed perception. The lesion of these association cortices will lead to the memory deficits [82, 83], such as the correlation of Wernicke's area lesion with understanding inability and cross-modal memory impairment [84–86].

In addition to the integration of unitary signals into complicated signals for their storage, the complicated signals can be dissected and memorized by recruiting different groups of associative memory cell and reorganizing them. In the learning of complicated signals, the coactivation of ready-recruited neurons in different activity strength may lead to mutual synapse innervations in different extents. The associative memory cells may receive one, two, or more new synapse innervations and encode multiple signals with different capabilities [8]. The complicated signals consist of numerous unitary signals. When they are learned by different sensory systems, the intramodal and cross-modal sensory cortical neurons detect and sense them through a dissection approach, i.e., cell-by-cell corresponds to unitary-by-unitary signals. In the learning of complicated signals, associative memory cells are recruited to encode these unitary signals. Their recruitment is based on the random association of these unitary signals to coactivate their correspondent ready-recruited neurons and to induce their synapse interconnections in different extents. Associative memory cells with different encoding and integrative capability to associated signals are recruited, such as encoding one, two, or more associated signals [8]. In memory retrieval, the activation of different portions of these associative memory cells leads to the selective recall of these complicated signals. The reorganization of these associative memory cells with different encoding capability may be the reason why memory retrievals may not be exactly identical to original signals [7].

Associative memory cells can be activated by three resources of synapse innervation. These synapse inputs include innate synapse innervations formed during development, synapse innervations from the arousal system, and new synapse innervations from coactivated brain regions. The synapse-driving forces from innate and arousal inputs activate the neurons and make them ready-recruited to be new associative memory cells. The arousal system consists of the ascending activating pathways and the reticular formation in the brain stem. This reticular formation receives synapse inputs from lateral branches of different sensory inputs in the brain stem. Their axons through the thalamus widely innervate the entire cerebral brain to confer the wakefulness and consciousness [87–89]. The ascending activating pathways are composed of those neuronal axons in the cholinergic nuclei; midbrain raphe nuclei and locus coeruleus innervate the forebrain to maintain the alertness and consciousness by releasing acetylcholine, serotonin, and norepinephrine [90–92]. This arousal system can maintain the basal activity of associative memory cells as well as confer them to integrate innate and new synaptic inputs specifically and to memorize these associated signals. This arousal system may also activate ready-recruited neurons to influence the efficiencies of learning and memory, activate associative memory cells to assist memory retrievals, as well as strengthen the interactions between primary and secondary associative memory cells to permit the associations of memory with cognitive process and emotional reactions.

7.5 Associative Memory Cells for Memory Efficiency and Cognition

Learning efficiency is affected by neuronal excitability, synapse innervations, synapse responsiveness, and ready-recruited neurons [7]. During the learning in early life, ready-recruited neurons for storing new associated signals may be neurons that have received synapse innervations from the arousal system and innate input pathways. The coactivation of these ready-recruited neurons will induce their mutual synapse innervations and recruit primary and secondary associative memory cells. The number of associative memory cells in this recruitment is largely dependent upon the excitatory state of these ready-recruited neurons, in which the activity strength of the arousal system and the activation of innate synapse inputs play important role. With the accumulation of associative memory cells in the different groups, ready-recruited neurons may be those associative memory cells that have been able to encode the storage of previously learned signals carried by specific synapse innervations. Their memorized signals may be closely related to those associated signals that will be learnt. These associative memory cells ready to be recruited as the full or advanced associative memory cells can be activated by giving topic-relevant cues to attract learner's attention. The number of the ready-recruited neurons influences how the information is acquired and stored easily as well as how complicated signals can be efficiently learnt. That is a reason why the efficiency of associative learning is influenced by learner's knowledge in the topic to be learned [5]. In addition, cortical neurons are diversified in their synapse inputs and intrinsic property [11]. The associative memory cells with more synapse innervations and lower threshold potential are easily activated to produce sequential spikes for high learning efficiency [3, 18]. These sequential spikes can trigger a chain reaction started from intensive spikes. Intensive spikes will elevate intracellular calcium, which upregulates calcium signaling pathway, such as calmodulin-dependent protein kinase and phospholipid-dependent protein kinase, as well as microRNA expression changes. The former pathway phosphorylates receptor-channels to upregulate neuron and synapse functions [93–98]. The changed microRNA expression alternates the expressions of genes and proteins in relevance to axon prolongation and synapse formation [8, 10]. In addition, intensive activities at glutamatergic synapses will activate their metabotropic receptors, which upregulate cAMP-dependent protein kinases through G-protein [99–102]. This kinase can phosphorylate receptor-channels to upregulate neuron and synapse functions [103, 104] (Fig. 7.1). Therefore, the activity-dependent upregulation in neuron excitability and synapse innervations will assist to recruit more associative memory cells and to upregulate the efficiencies of associative learning and memory.

The memory retrieval is affected by associative memory cells and the arousal system. As discussed above, the arousal system activates and maintains cortical neurons including associative memory cells to be functional, i.e., wakefulness and consciousness. Therefore, the efficiency of memory retrievals will be elevated by fully wakefulness and consciousness. The number of associative memory cells is affected by the weight of mutual synapse innervation on ready-recruited neurons during their recruitment. The encoding capability of associative memory cells is affected by their excitability and the active state driven by various synapse inputs. In addition, there is a coactivity-dependent positive cycle in the recruitment and refinement of associative memory cells [7]. With ordinary wakefulness and consciousness, there is the positive relationship between the number of associative memory cells recruited associative learning and the number of associative memory cells activated during memory retrieval. In this regard, the efficiency of memory retrievals is consistent with the efficiency of associative learning [3, 18]. The encoding capability of associative memory cells influences how these cells are easily activated by the cues during memory retrieval or spontaneously during free thinking [16, 33]. The coactivity-dependent positive cycle in the recruitment and upregulation of associative memory cells will add more associative memory cells into memory traces. In addition, whether the memorized signals can be effectively retrieved is also influenced by the encoding capability of memory output cells in the motor cortex. It has been found that the decay of encoding capability in memory output cells due to lack of continuous pair-training may lead to the inability of

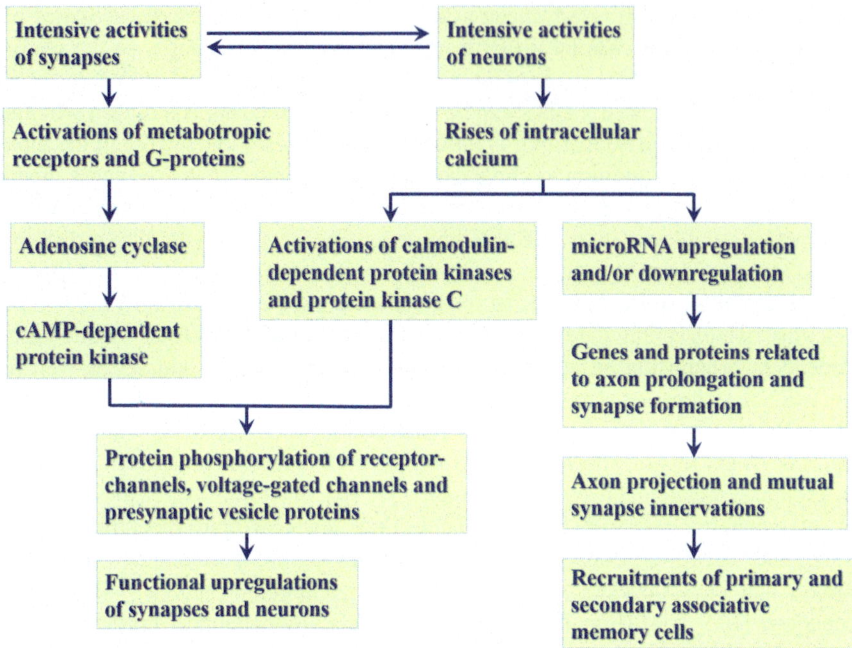

Fig. 7.1 The molecular mechanisms for activity-dependent positive cycle in the recruitment and upregulation of associative memory cells. Intensive activities at glutamatergic synapses activate their metabotropic receptors and in turn activate cAMP-dependent protein kinase via G-proteins. Intensive spike activities on neuron membrane activate voltage-gated calcium channels and elevate the intracellular Ca^{2+} level, which in turn activates intracellular calcium signaling pathway and changes microRNA expression. Calcium signaling pathway includes calmodulin-dependent protein kinase, phospholipid-dependent protein kinase, and so on. These three kinds of protein kinases will upregulate neuron and synapse functions. The changed microRNA expression upregulates the expressions of genes and proteins in relevance to axon prolongation and synapse formation, in order to facilitate the recruitment of associative memory cells. These recruitment and upregulation of associative memory cells further raise their intensive activities to form a positive cycle

memory retrievals (i.e., memory extinction), though those primary associative memory cells in sensory cortices are well-maintained under normal function [17, 23]. Therefore, the high quantity and the active encoding capability of associative memory cells in memory trace are able to initiate automatic memory retrieval. This cue-free memory retrieval may be based on a coactivity-dependent positive cycle in the recruitment and refinement of associative memory cells, after repeated learnings and thinking [5].

The precision of memories to objects, events, and locations includes their spatial precision and time precision. The spatial precision of associative memory is largely dependent on the function of the sensory system. For instance, the sensory precision is regulated by the density of sensory receptors, the size of receptive field, the lateral inhibition, and the number of sensory cortical neurons [71]. Moreover, the divergent connection of sensory afferent nerves onto sensory cortical neurons makes a sensory nerve to innervate many sensory cortical neurons. As these cortical neurons show different encoding capabilities [11, 45], the diversity of a sensory signal by these sensory cortical neurons enables the sensation to be precise spatially [5]. This encoding precision of sensations in sensory cortices confers primary associative memory cells that encode the integrative storage and reciprocal retrieval of associated signals to be spatial precision. In terms of time precision of associative memory, the enriched experience and cognitive capability in the time estimation for activities, operations, and processes are critically important. Even though there is no time signal to be associated in memory formation, people can often estimate rough time for these activities, operations, and processes.

In the conversion of exogenous signals into endogenous signals and their integrative memory [7, 23, 24, 32], the efficiency of making the correlation of associative memory with cognitive events and emotional reactions is a critical issue. In this process, the interactions between primary and secondary associative memory cells via their mutual synapse innervations (Figs. 5.1 and 5.2) as well as the number and encoding capability of these associative memory cells should be taken into account during logical reasoning, associative thinking, and other integrative cognitions [7]. In general, cellular architecture and working principle involved in the efficacy of associative learning memory as well as the efficacy of retrieving exogenous associated signals may be similar to the conversion of exogenous into endogenous signals.

How are associative memory cells correlated to different memory patterns? Memory patterns are classified to declarative memory versus nondeclarative memory, episodic memory versus semantic memory, and others. Their correlation will be discussed in the next chapter in detail. Despite these psychological classifications, there is no the clear borderline to separate them. Explicit memory refers to memory retrievals with consciousness and intention, whereas implicit memory stands to memory retrievals without consciousness awareness [4, 105, 106]. In fact, explicit memory and implicit memory share similar processes during the learning of associated signals. People who represent implicit memory have initially paid attention to learning the operations and procedures for the specific skill, such as riding bicycles. Through the repetitive practices to be skilled, the execution of these operations and procedures does not need conscious effort. Before being skilled, people execute these operations and procedures with consciousness and intention. On the other hand, people who represent explicit memory have learned knowledge and concepts with great attention. After their frequent uses, the representations of these knowledge and concepts do not need consciousness and intention, likely the representation of implicit memory. In other words, the formations of explicit and implicit memories may share similar mechanisms. Whether memory contents are retrieved with conscious effort (explicit memory) or without consciousness (implicit memory) depends on the familiarity or impression to memory contents. The repeated learning, rehearsal, and practices will concurrently coactivate primary and secondary associative memory cells, which can recruit more associative memory cells and upregulate their excitability [3, 7, 16, 32, 33, 65], as well as intensify synapse connection from associative memory cells to memory output cells in the motor cortex [17, 23]. So, the retrieval of explicit memory can be converted into the retrieval of implicit memory. In other words, there may be an inverse relationship between the requirement of consciousness and the number or encoding capability of associative memory cells. This may be the reason why impressive memory in mind is usually associated with its replay during the sleep [5]. Implicit memory based on more associative memory cells that are easily activated is supported by phenomena that implicit memory can usually be expressed spontaneously. In explicit memory, episodic memory in individual events can be converted into semantic memory after the repeated associative thinking and logical reasoning. These repeated cognitions strengthen the groups of secondary associative memory cells that have stored the common signal of these events by central synapse innervations or gather secondary associative memory cells that have stored those events in similar topics (please refer to Figs. 8.1 and 8.2), such that the new group of secondary associative memory cells is reorganized for general concept, knowledge, and principle through convergent innervations to the advanced grade of associative memory cells in an abstraction manner [5, 7].

Consciousness is the combinational state of wakefulness and memory for individuals to be aware and identify themselves and objects in the environment [7, 107]. Consciousness is not all-or-none state. If subjects are wakeful and able to identify familiar objects, such as closely relatives and themselves, they fall into a moderate consciousness. If subjects are able to represent and recollect the complicated information and knowledge as well as to play logical reasoning and associative thinking, they fall into full consciousness. The normal consciousness may be based on the basal activation of associative memory cells by the arousal system for wakefulness and the specific activation of associative memory cells from their associated inputs for memory and cognition triggered by sensory cues [7]. Thus, the number and encoding capability of associative memory cells are proportional to the state of consciousness. The activation of primary associative memory cells by the arousal system and specific mutual synapse innervations, which can activate memory output cells in the motor cortex, may work for a moderate consciousness. If all of primary associative memory cells in sensory cortices, secondary associative memory cells in cognition-/emotion-related brain areas, as well as memory output cells in the motor cortex are activated, the subjects will be engaged in full consciousness [5].

The combination of consciousness and the specific alert signals constitutes the attention, in which subjects are ready to learn something and/or play cognitive activities and emotional reactions. In other words, the attention is a state of full consciousness plus the specific alertness. In terms of cellular architecture for the attention, all of those associative memory cells can be activated for memories and cognitions as well as a specific group of the associative memory cells is activated to the even higher level for memory retrieval in relevance to the alertness, in which these associative memory cells in relevance to the alertness are ready to be activated and recruited as advanced associative memory cells for learning the alertness-relevant signals [5]. When individuals are conscious, they have two forms of logical reasoning and associative thinking, either critical or creative style. In critical thinking, secondary associative memory cells are activated for the evaluation. In creative thinking, new secondary associative memory cells are recruited for inspiration [5, 7].

Awareness refers to the quality to perceive, know, and understand something and can also be equivalent to consciousness that includes weak, moderate, and full states. The sleeping appears unawareness with a slow wave and weak consciousness with fast wave [107]. How do the different sets of associative memory cells work for the dream during fast wave sleep? Dreaming is characterized to be active in electronic encephalograph and behaviors. The characters include rapid eye movement, breath, heat beat, and muscle twitch, which indicate high activities in the forebrain. In addition, associative memories to specific events that have been frequently thought in daytime are retrieved. In the daytime, primary and secondary associative memory cells have been intensively active for memory retrieval, associative thinking, logical reasoning, and emotional reactions. By the activity-dependent upregulation in their excitation, these associative memory cells are easily activated when the brain activity is high during dreams. Their coactivity leads to the coactivity-dependent positive cycle in the recruitment and upregulation of associative memory cells during dreams. So, the events stored in these associative memory cells are playbacks. The dream contents can even be remembered by newly recruited associative memory cells. Because of the inverse correlation between the upregulation of associative memory cells and the requirement of consciousness, associative memory cells with large population and upregulated function due to repeated learning and thinking can be activated under a weak consciousness. However, these associative memory cells are not activated by all of the synapse innervations relevant to real events, since they are partially activated by the nonspecific arousal system under a weak consciousness. The replay of events is incompletely identical to those realistic events. In addition, the coactivation of different groups of associative memory cells, which are more active currently, leads their mutual synapse innervations or interconnection upregulation during dreaming. New associative memory cells are recruited for encoding dream contents and their memory [7]. As the replays of events during dreams can be recalled and recollected, associative thinking and logical reasoning in the integration of endogenous signals based on primary and secondary associative memory cells are achieved under low consciousness [6]. This viewpoint is supported by an observation that temporal sequences of place cell activity in the novel spatial experience are detected during the resting or sleeping period preceding experiences. This replay occurs in the disjunction to sequences of replay in a familiar experience. These data suggest that internal neuronal dynamics during resting or sleep organize cellular assemblies into temporal sequences that contribute to encode a relevant novel experience in the future [108]. In this regard, the dreaming can be thought of as the reassociation of associative memory cells that have stored learned signals and have been active currently under weak consciousness. The reassociation is based on the population of associative memory cells with high excitation [5].

In associative thinking and logical reasoning, signals imbedded in video pictures, odorants, tastes, and other events are presented by verbal language. In initial learning, the sensations, perception, and events are associated with words for their definitions. Cross-modal associative memory cells to program the integration of these processes and words are recruited. When these signals are retrieved in sequential playbacks, their correspondent words memorized in these associative memory cells are initiated to substitute complicated pictures and events. The words to describe the complicated signals can speed up memory retrieval and cognition. The substitution of the complicated signals by words is achieved based on the recruitment of more associative memory cells and their upregulations in coactivity-dependent positive cycle manner by repeated practices. However, if words and signals are associated improperly, the correction of these associations is difficult because the recruited synapse innervations, associative memory cells, and their circuits are hard to be ruined [7].

In addition to the convergent synapse innervations onto secondary associative memory cells from primary associative memory cells, the coactivation of secondary associative memory cells can induce their mutual synapse innervation [5] (Figs. 5.1 and 5.2). These convergent and mutual synapse innervations onto secondary associative memory cells confer them to associate and integrate these endogenous signals at high level of cognitions, such as associative thinking, logical reasoning, imagination, and other cognitive processes. Associative thinking refers to the association of endogenous signals that are stored from previous learning without the necessary requirement of spatial and temporal sequences. Logical reasoning stands for the association of endogenous signals from previous memorized knowledge in certainty as well as references with the requirement of spatial and temporal sequences acquired during learning. Imagination refers to the integration and reorganization of previous memorized experiences into those expected sequential or three-dimensional images in the mind. It is noteworthy that the signals formed during associative thinking, logical reasoning, imagination, and other cognitive activities can be memorized at secondary associative memory cells in brain areas relevant to cognitions for future representation and recollection [5–7].

References

1. Kandel ER, Pittenger C. The past, the future and the biology of memory storage. Philos Trans R Soc Lond Ser B Biol Sci. 1999;354(1392):2027–52.
2. Suzuki WA. Associative learning signals in the brain. Prog Brain Res. 2008;169:305–20.
3. Wang D, et al. Neurons in the barrel cortex turn into processing whisker and odor signals: a cellular mechanism for the storage and retrieval of associative signals. Front Cell Neurosci. 2015;9:320.
4. Wasserman EA, Miller RR. What's elementary about associative learning? Annu Rev Psychol. 1997;48:573–607.
5. Wang J-H. Searching basic units of memory traces: associative memory cells. F1000Res. 2019;8(457):1–28.
6. Wang JH, Cui S. Associative memory cells: formation, function and perspective. F1000Res. 2017;6:283.

7. Wang JH, Cui S. Associative memory cells and their working principle in the brain. F1000Res. 2018;7:108.
8. Feng J, et al. Barrel cortical neuron integrates triple associated signals for their memory through receiving epigenetic-mediated new synapse innervations. Cereb Cortex. 2017;27(12):5858–71.
9. Gao Z, et al. Associations of unilateral whisker and olfactory signals induce synapse formation and memory cell recruitment in bilateral barrel cortices: cellular mechanism for unilateral training toward bilateral memory. Front Cell Neurosci. 2016;10(285):1–16.
10. Lei Z, et al. Synapse innervation and associative memory cell are recruited for integrative storage of whisker and odor signals in the barrel cortex through miRNA-mediated processes. Front Cell Neurosci. 2017;11(316):1–11.
11. Wang JH, et al. The gain and fidelity of transmission patterns at cortical excitatory unitary synapses improve spike encoding. J Cell Sci. 2008;121(17):2951–60.
12. Yang Z, et al. Functional compatibility between Purkinje cell axon branches and their target neurons in the cerebellum. Oncotarget. 2017;8(42):72424–37.
13. Yu J, et al. Quantal glutamate release is essential for reliable neuronal encodings in cerebral networks. PLoS One. 2011;6(9):e25219.
14. Yu J, Qian H, Wang JH. Upregulation of transmitter release probability improves a conversion of synaptic analogue signals into neuronal digital spikes. Mol Brain. 2012;5(1):26.
15. Wang J-H, et al. Both glutamatergic and Gabaergic neurons are recruited to be associative memory cells. Biophys J. 2016;110(3):supplement 481a.
16. Yan F, et al. Coordinated plasticity between barrel cortical glutamatergic and GABAergic neurons during associative memory. Neural Plast. 2016;2016(ID5648390):1–20.
17. Guo R, et al. Associative memory extinction is accompanied by decayed plasticity at motor cortical neurons and persistent plasticity at sensory cortical neurons. Front Cell Neurosci. 2017;11(168):1–12.
18. Zhao X, et al. Coordinated plasticity among glutamatergic and GABAergic neurons and synapses in the barrel cortex is correlated to learning efficiency. Front Cell Neurosci. 2017;11(221):1–12.
19. Cai DJ, et al. A shared neural ensemble links distinct contextual memories encoded close in time. Nature. 2016;534(7605):115–8.
20. Naya Y, Yoshida M, Miyashita Y. Forward processing of long-term associative memory in monkey inferotemporal cortex. J Neurosci. 2003;23(7):2861–71.
21. Takehara-Nishiuchi K, McNaughton BL. Spontaneous changes of neocortical code for associative memory during consolidation. Science. 2008;322(5903):960–3.
22. Viskontas IV. Advances in memory research: single-neuron recordings from the human medial temporal lobe aid our understanding of declarative memory. Curr Opin Neurol. 2008;21(6):662–8.
23. Wang J-H, Guo R, Wei Z. Associative memory extinction is accompanied by decays of associative memory cells and their plasticity at motor cortex but not sensory cortex. Soc Neurosci. 2017;81(09):10385.
24. Wang J-H, et al. Prefrontal cortical neurons are recruited as secondary associative memory cells for associative memory and cognition. Biophys J. 2018;114(3):155a.
25. Li H, et al. Experience-dependent modification of a central amygdala fear circuit. Nat Neurosci. 2013;16(3):332–9.
26. Liu X, et al. Optogenetic stimulation of a hippocampal engram activates fear memory recall. Nature. 2012;484(7394):381–5.
27. Otis JM, et al. Prefrontal cortex output circuits guide reward seeking through divergent cue encoding. Nature. 2017;543:103.
28. Pape HC, Pare D. Plastic synaptic networks of the amygdala for the acquisition, expression and extinction of conditioned fear. Physiol Rev. 2010;90(2):419–63.
29. Xu W, Sudhof TC. A neural circuit for memory specificity and generalization. Science. 2013;339(6125):1290–5.

30. Yokose J, et al. Overlapping memory trace indispensable for linking, but not recalling, individual memories. Science. 2017;355(6323):398–403.
31. Milo R. What is the total number of protein molecules per cell volume? A call to rethink some published values. BioEssays. 2013;35(12):1050–5.
32. Wang JH, et al. Secondary associative memory cells and their plasticity in the prefrontal cortex. Biophys J. 2019;116(3):427a.
33. Liu Y, et al. Piriform cortical glutamatergic and GABAergic neurons express coordinated plasticity for whisker-induced odor recall. Oncotarget. 2017;8(56):95719–40.
34. Wang JH, et al. Upregulation of glutamatergic receptor-channels is associated with cross-modal reflexes encoded in barrel cortex and piriform cortex. Biophys J. 2014;106(2):supplement 191a.
35. Klyachko VA, Stevens CF. Excitatory and feed-forward inhibitory hippocampal synapses work synergistically as adaptive filter of neural spike trains. PLoS Biol. 2006;4(7):e207.
36. Benes FM, Berretta S. GABAergic interneurons: implications for understanding schizophrenia and bipolar disorder. Neuropsychopharmacology. 2001;25(1):1–27.
37. Cotter D, et al. The density and spatial distribution of GABAergic neurons, labelled using calcium binding proteins, in the anterior cingulate cortex in major depressive disorder, bipolar disorder, and schizophrenia. Biol Psychiatry. 2002;51(5):377–86.
38. Liu B, Feng J, Wang J-H. Protein kinase C is essential for kainate-induced anxiety-related behavior and glutamatergic synapse upregulation in prelimbic cortex. CNS Neurosci Ther. 2014;20:982–90.
39. Lloyd KG, et al. The gabaergic hypothesis of depression. Prog Neuro-Psychopharmacol Biol Psychiatry. 1989;13(3–4):341–51.
40. Ma K, et al. Impaired GABA synthesis, uptake and release are associated with depression-like behaviors induced by chronic mild stress. Transl Psychiatry. 2016;6(e910):1–10.
41. Maciag D, et al. Reduced density of calbindin immunoreactive GABAergic neurons in the occipital cortex in major depression: relevance to neuroimaging studies. Biol Psychiatry. 2010;67(5):465–70.
42. Rajkowska G, et al. GABAergic neurons immunoreactive for calcium binding proteins are reduced in the prefrontal cortex in major depression. Neuropsychopharmacology. 2007;32(2):471–82.
43. Wen B, et al. A portion of inhibitory neurons in human temporal lobe epilepsy are functionally upregulated: an endogenous mechanism for seizure termination. CNS Neurosci Ther. 2015;21(2):204–14.
44. Xu A, Cui S, Wang J. Incoordination among subcellular compartments is associated to depression-like behavior induced by chronic mild stress. Int J Neuropsychopharmacol. 2015;19(5):pyv122.
45. Zhang F, et al. mGluR1,5 activation improves network asynchrony and GABAergic synapse attenuation in the amygdala: implication for anxiety-like behavior in DBA/2 mice. Mol Brain. 2012;5(1):20.
46. Zhu Z, et al. GABAergic neurons in nucleus accumbens are correlated to resilience and vulnerability to chronic stress for major depression. Oncotarget. 2017;8(22):35933–45.
47. Albright TD. On the perception of probable things: neural substrates of associative memory, imagery, and perception. Neuron. 2012;74(2):227–45.
48. Wirth S, et al. Single neurons in the monkey hippocampus and learning of new associations. Science. 2003;300(5625):1578–81.
49. Wang J-H, Kelly PT. Ca2+/CaM signalling pathway up-regulates glutamatergic synaptic function in non-pyramidal fast-spiking neurons of hippocampal CA1. J Physiol Lond. 2001;533(2):407–22.
50. Chen N, et al. The refractory periods and threshold potentials of sequential spikes measured by whole-cell recordings. Biochem Biophys Res Commun. 2006;340:151–7.
51. Chen N, et al. Sodium channel-mediated intrinsic mechanisms underlying the differences of spike programming among GABAergic neurons. Biochem Biophys Res Commun. 2006;346:281–7.

52. Chen N, et al. After-hyperpolarization improves spike programming through lowering threshold potentials and refractory periods mediated by voltage-gated sodium channels. Biochem Biophys Res Commun. 2006;346:938–45.
53. Goldman L. Stationarity of sodium channel gating kinetics in excised patches from neuroblastoma N1E 115. Biophys J. 1995;69:2364–8.
54. Hanson JE, Smith Y, Jaeger D. Sodium channels and dendritic spike initiation at excitatory synapses in globus pallidus neurons. J Neurosci. 2004;24(2):329–40.
55. Hodgkin AL, Huxley AF. The dual effect of membrane potential on sodium conductance in the giant axon of Loligo. J Physiol Lond. 1952;116:497–506.
56. Hodgkin A. The optimum density of sodium channels in an unmyelinated nerve. Philos Trans R Soc Lond Ser B Biol Sci. 1975;270(908):297–300.
57. Zhang M, et al. Calcium signal-dependent plasticity of neuronal excitability developed postnatally. J Neurobiol. 2004;61:277–87.
58. Bennett V, Lambert S. Physiological roles of axonal ankyrins in survival of premyelinated axons and localization of voltage-gated sodium channels. J Neurocytol. 1999;28(4–5):303–18.
59. Catterall WA. From ionic currents to molecular mechanisms: the structure and function of voltage-gated sodium channels. Neuron. 2000;26:13–25.
60. Lu W, et al. Voltage-independent sodium channels emerge for an expression of activity-induced spontaneous spikes in GABAergic neurons. Mol Brain. 2014;7(1):38.
61. Stevens CF. Sodium channel structure-function relations. Soc Gen Physiol Ser. 1987;41:99–108.
62. Tian LM, Otoom S, Alkadhi KA. Endogenous bursting due to altered sodium channel function in rat hippocampal CA1 neurons. Brain Res. 1995;680:164–72.
63. Chen N, Chen X, Wang J-H. Homeostasis established by coordination of subcellular compartment plasticity improves spike encoding. J Cell Sci. 2008;121(17):2961–71.
64. Liu Y, et al. Activity strengths of cortical glutamatergic and GABAergic neurons are correlated with transgenerational inheritance of learning ability. Oncotarget. 2017;8(68):112401–16.
65. Feng J, et al. Cell-specific plasticity associated with integrative memory of triple sensory signals in the barrel cortex. Oncotarget. 2018;9(57):30962–78.
66. Hebb DO. The organization of behavior, a neuropsychological theory. New York: Wiley; 1949.
67. Markram H, et al. Regulation of synaptic efficacy by coincidence of postsynaptic APs and EPSPs. Science. 1997;275:213–5.
68. Regehr WG, Stevens CF. Physiology of synaptic transmission and short-term plasticity. In: Cowan WM, et al., editors. Synapses. Baltimore/London: The Johns Hopkins University Press; 2001.
69. Reyes A, et al. Target-cell-specific facilitation and depression in neocortical circuits. Nat Neurosci. 1998;1:279–85.
70. Silver RA, et al. High-probability uniquantal transmission at excitatory synapses in barrel cortex. Science. 2003;302(5652):1981–4.
71. Guyton AC, Hall JE. The nervous system: B. The special senses. In: Guyton AC, Hall AC, editors. Textbook of medical physiology. 11th ed. Printed in China: Elsevier; 2006.
72. Ge R, Qian H, Wang JH. Physiological synaptic signals initiate sequential spikes at soma of cortical pyramidal neurons. Mol Brain. 2011;4(1):19.
73. Ge R, et al. Input-dependent subcellular localization of spike initiation between soma and axon at cortical pyramidal neurons. Mol Brain. 2014;7(1):26.
74. Liao D-Z, Hessler NA, Malinow R. Activation of postsynaptic silent synapses during pairing-induced LTP in CA1 region of hippocampal slice. Nature. 1995;375:400–4.
75. Van Hoesen GW. The modern concept of association cortex. Curr Opin Neurobiol. 1993;3(2):150–4.
76. Zeki S. The visual association cortex. Curr Opin Neurobiol. 1993;3(2):155–9.
77. Rosen ML, et al. The role of visual association cortex in associative memory formation across development. J Cogn Neurosci. 2018;30(3):365–80.

78. Pavlov I. Conditioned reflexes: an investigation of the physiological activity of the cerebral cortex. Translated by Anrep GV. Nature. 1927;121(3052):662–4.
79. Goldman-Rakic PS. Topography of cognition: parallel distributed networks in primate association cortex. Annu Rev Neurosci. 1988;11:137–56.
80. Masterton RB, Berkley MA. Brain function: changing ideas on the role of sensory, motor, and association cortex in behavior. Annu Rev Psychol. 1974;25:277–312.
81. Torrealba F, Valdes JL. The parietal association cortex of the rat. Biol Res. 2008;41(4):369–77.
82. Colombo M, et al. Auditory association cortex lesions impair auditory short-term memory in monkeys. Science. 1990;247(4940):336–8.
83. Goldman-Rakic PS. Circuitry of the frontal association cortex and its relevance to dementia. Arch Gerontol Geriatr. 1987;6(3):299–309.
84. Ardila A, Bernal B, Rosselli M. How extended is Wernicke's area? Meta-analytic connectivity study of BA20 and integrative proposal. Neurosci J. 2016;2016:4962562.
85. Binder JR. Current controversies on Wernicke's area and its role in language. Curr Neurol Neurosci Rep. 2017;17(8):58.
86. DeWitt I, Rauschecker JP. Wernicke's area revisited: parallel streams and word processing. Brain Lang. 2013;127(2):181–91.
87. Kayama Y, et al. Tonic and phasic components of the ascending reticular activating system. Fukushima J Med Sci. 1991;37(2):59–74.
88. Neylan TC. Physiology of arousal: Moruzzi and Magoun's ascending reticular activating system. J Neuropsychiatry Clin Neurosci. 1995;7(2):250.
89. Yeo SS, Chang PH, Jang SH. The ascending reticular activating system from pontine reticular formation to the thalamus in the human brain. Front Hum Neurosci. 2013;7:416.
90. Adell A, et al. Origin and functional role of the extracellular serotonin in the midbrain raphe nuclei. Brain Res Brain Res Rev. 2002;39(2–3):154–80.
91. Aston-Jones G, Cohen JD. An integrative theory of locus coeruleus-norepinephrine function: adaptive gain and optimal performance. Annu Rev Neurosci. 2005;28:403–50.
92. Steriade M, et al. Neuronal activities in brain-stem cholinergic nuclei related to tonic activation processes in thalamocortical systems. J Neurosci. 1990;10(8):2541–59.
93. Barria A, et al. Regulatory phosphorylation of AMPA-type glutamate receptors by Ca2+/calmodulin-dependent protein kinase II during long-term potentiation. Science. 1997;276:2042–5.
94. Huang KP. The mechanism of protein kinase C activation. TINS. 1989;12(11):425–32.
95. Lisman J, Schulman H, Cline H. The molecular basis of CaMKII function in synaptic and behavioural memory. Nat Rev Neurosci. 2002;3(3):175–90.
96. McGlade-McCulloh E, et al. Phosphorylation and regulation of glutamate receptors by calcium/calmodulin-dependent protein kinase II. Nature. 1993;362:640–2.
97. Schulman H. Activity-dependent regulation of calcium/calmodulin-dependent protein kinase II location. J Neurosci. 2004;24(39):8399–403.
98. Yakel JL, et al. Identification of a Ca2+/calmodulin-dependent protein kinase II regulatory phosphorylation site in non-N-methyl-D-aspartate glutamate receptors. Proc Natl Acade Sci USA. 1995;92:1376–80.
99. Baskys A. Metabotropic receptors and 'slow' excitatory actions of glutamate agonists in the hippocampus. Trends Neurosci. 1992;15(3):92–6.
100. Hammond C. The metabotropic glutamate receptor. In: Hammond C, editor. Cellular and molecular neurophysiology. 3rd ed. Amsterdam: Academic; 2008.
101. Miller RJ. Metabotropic excitatory amino acid receptors reveal their true colors. Trends Pharmacol Sci. 1991;12(10):365–7.
102. Schoepp DD, Conn PJ. Metabotropic glutamate receptors in brain function and pathology. Trends Pharmacol Sci. 1993;14:13–20.
103. Raymond LA, Blackstone CD, Huganir RL. Phosphorylation and modulation of recombinant GluR6 glutamate receptors by cAMP-dependent protein kinase. Nature. 1993;361:637–41.
104. Wang LY, Salter MW, MacDonald JF. Regulation of kainate receptors by cAMP-dependent protein kinase and phosphatases. Science. 1991;253:1132–5.

105. Reder LM, Park H, Kieffaber PD. Memory systems do not divide on consciousness: reinterpreting memory in terms of activation and binding. Psychol Bull. 2009;135(1):23–49.
106. Squire LR. Declarative and nondeclarative memory: multiple brain systems supporting learning and memory. J Cogn Neurosci. 1992;4(3):232–43.
107. Hobson JA, Pace-Schott EF. The cognitive neuroscience of sleep: neuronal system, consciousness and learning. Nat Rev Neurosci. 2002;3:679–93.
108. Dragoi G, Tonegawa S. Preplay of future place cell sequences by hippocampal cellular assemblies. Nature. 2011;469(7330):397–401.

Chapter 8
Associative Memory Cells in Physiological Psychology

Abstract Associative memory and memory-related cognition have been extensively studied in psychology. Various memory patterns and theoretical models have been classified based on memory contents, content sources, memory accuracy, and consciousness states. Cellular mechanisms underlying these patterns and models are largely unknown since the experiments to reveal cellular processes cannot be studied in human beings up to now. The comprehensive cellular architectures of associative memory and memory-related behaviors may be figured out based on the profiles studied in rodents in order to understand memory regulation and to develop therapeutic strategies for memory deficit. Based on the working principles of associative memory cells and their networks in rodents, the author intends to provide some cellular architectures for explaining these patterns and theoretical models of associative memory. As discussed in Chap. 2, associative learning is a major approach in information acquisition. Associative memory is classified into distinct patterns and terms, such as explicit versus implicit memory, episodic versus semantic memory, perceptual versus working memory, intramodal versus cross-modal memory, and eidetic versus false memory. Furthermore, memory-relevant cognitions and emotions are complicated. Although associative thinking, logical reasoning, computation, imagination, pleasure, and fear are well defined, the expression of these processes is mixed in nature. In fact, these memory patterns and processes share certain common features, such as information-integrative memory, reciprocal memory retrieval, activity-dependent strengthening, and memory-unit linked cognitions. The synaptic and neuronal plasticity in a single pathway is not matching these features of memory and cognitions. Instead, associative memory cells and their featured interconnections may constitute the foundation of the memory patterns and memory-relevant behaviors. In this chapter, the potential links between associative memory cells in memory traces and these types of memories will be discussed.

Keywords Psychology · Physiology · Associative memory cells · Memory · Thinking and reasoning

J.-H. Wang, *Associative Memory Cells: Basic Units of Memory Trace*,
https://doi.org/10.1007/978-981-13-9501-7_8

8.1 Introduction

Associative memory is a major brain function, essentially for cognitive activities, emotional reactions, and other integrative behaviors in life [1–8]. Topics about learning efficiency, memory capacity, memory maintenance, and memory enhancement are commonly interested by normal individuals and patients suffering from memory deficits. Scientists in fields of physiology, psychology, pathology, pharmacology, and neuroscience have paid greater attention to reveal potential mechanisms underlying memory formation and its relevant behaviors [9]. There are different patterns about learning and memory that have been noticed and classified by the society for memoriology, especially psychologists. However, mechanisms underlying these memory patterns have not been clearly uncovered, though their pathological locations and origins have been studied in normal population and memory deficit patients by psychological analyses and neural imaging. Core and comprehensive cellular architecture and molecular substrates for learning and memory remain to be figured out. Efforts to abstract core and comprehensive diagrams about memory cells and their circuits for memory formation and memory-relevant behaviors are worthy to benefit to people's life.

The distinct patterns of associative memory have been defined based on memory content, location, retrieval, persistence, and efficiency. Compared to implicit memory for the execution of procedures and operations without the need of consciousness, explicit memory refers to the retrieval of memory contents under the consciousness condition. The consciousness is thought as wakefulness plus memory. Memory retrieval requires the storage of certain contents. The execution of procedures and operations without requiring consciousness indicates the deep storage of signals for implicit memory. The retrieval of memory contents in explicit form needs help from other memories. In other words, implicit memory is featured to be strengthened memory [10]. The storage of spatial and temporal events, specific knowledge, as well as their associated emotions is called as episodic memory in explicit form. Once episodic memories are generalized and summarized as views, knowledge, and theories, semantic memory is onset in the form of explicit memory [11–13]. Thus, a conversion of episodic memory into semantic memory is fulfilled by the assembly of memory contents based on their summary, generalization, and abstraction.

If memory involves a conversion of sensory guidance into behavior manipulations, this short-term memory from signals to motion is called as working memory. Long-term perceptual memory includes the storages of visual, auditory, and other perceptual signals [14]. Working memory appears to be based on the efficiency in the transferring of sensory signals to motion operations, which is influenced how many sources of sensory signals are given in this short period. Perceptual memory is dependent on the maintenance of the active state of associative memory cells and their circuits. As shown in our studies, these are convergent pathways, or synapse innervations, from associative memory cells in sensory cortices to memory output cells in the motor cortex, which may work for working memory [15].

In terms of the capacity and efficiency of retrievals about perceptual memory, eidetic memory is termed as that signals learned in short time can be retrieved vividly, especially in childhood stage, to check child's capacity and ability to remember something in short period [16]. On the other hand, the contents in people recall, presentation, or recollection are never experienced by these individuals. This form of memory has been called as false memory, which is usually produced with a need of long time [17]. The information generated during cognitive events, such as associative thinking, logical reasoning, and imagination, can be stored in secondary associative memory cells. As long as these false beliefs generated in the cognitive events are circulated many times to generate more and more secondary associative memory cells, false memory can be made [15].

As discussed in Chaps. 5, 6, and 7, basic units in memory traces, or engrams, are primary associative memory cells in sensory cortices, secondary associative memory cells in association cortices, and cognition-/emotion-relevant brain areas and memory output cells in the motor cortex. These memory-relevant cells constitute various connection circuits for different memory patterns, cognitions, and emotional reactions. If basic units in memory traces are these cellular components, the assumption is that various patterns of memory are influenced by cellular architectures and their circuitry styles. In this chapter, the author intends to sketch the simplified and common cores shared by various memory patterns and to figure out the roles of these associative memory cells and their circuits for these memory patterns.

8.2 Associative Memory Cells in Various Memory Patterns

Although learning may be non-associative, the most common way for information acquisition is associative learning and memory [18–20]. In other words, multiple signals, objects, activities, or experiences are associatively acquired by cross-modal sensory modalities or intramodal sensory modality. These multiple signals are reinforced to each other and gathered together for their integrative storage and reciprocal retrieval, i.e., associative memory [8, 15, 21, 22]. Associative learning is classified as operant conditioning, classical conditioning, observation learning, imprinting, play, multimedia learning, episodic learning, meaningful learning, rote learning, enculturation, drug addiction, and so on [20, 21, 23–41]. After the associated signals are acquired, whether they have been memorized is tested by their retrievals including recalls, representations, and recollections [10, 42]. There are a few patterns of associative memory that are classified based on memory content, location, retrieval, and efficiency, e.g., declarative versus nondeclarative memory, episodic versus semantic memory, short-term versus long-term memory, working versus perceptual memory, spatial versus temporal memory, and eidetic versus false memory [15].

In order to memorize these associatively learned signals for their integrative storage and reciprocal retrieval, memory traces, engrams, or cell assemblies are formed to encode these signals and receive new synapse innervations carrying in these signals. This hypothesis is based on the fact that associative learning and memory are influenced by internal and external factors, which are presumably based on neurons, their interconnected synapses, and the arousal system. Primary associative memory cells have been identified in sensory cortices for the integrative storage and the reciprocal retrieval of associated signals. These associative memory cells encode multiple associated signals on the basis of the synapses from innate inputs and new synapse innervations from coactivated brain regions [43–50]. In addition, secondary associative memory cells that receive synapse innervations from primary associative memory cells have been detected in the prefrontal cortex, hippocampus, and motor cortex [51–53]. These associative memory cells are expected to well explain these types of associative memory.

8.2.1 Associative Memory Cells for Declarative and Nondeclarative Memory

Declarative memory stands for memory retrievals with conscious and intention components to represent and recollect previously learned knowledge, experiences, and information [10]. Declarative memory depends on conceptually driven and top-down sequential process logically. In order to have declarative memory, individuals have to do something with their experiences and knowledge previously learned by thinking, talking, writing, and studying them. The more individual practices, the better individual remember. As individuals actively associate explicit signals with endogenous signals in their mind, these endogenous cues can be used to initiate the spontaneous retrieval of these signals [9]. In summary, declarative memory can be thought to be the memory of the information including relevant subjects and contents by conscious and intension organization. The well retrieval of entire information is helped by more rehearsals and practices. Several neural structures for the explicit memory have been proposed, especially the prefrontal cortex, the amygdala, the hippocampus, and the rhinal cortex. Nuclei in the thalamus are included since it connects between the prefrontal cortex and temporal cortex [54]. These structures to constitute neural circuits for declarative memory receive the axon and synapse inputs from the neocortex as well as the brain stem system including acetylcholine, serotonin, and noradrenaline systems [55]. It remains unclear about the cellular architectures for declarative memory. As declarative memory is associative in nature, associative memory cells in engram circuit may be candidates of neural substrates for explicit memory [15]. In explicit memory, episodic memory refers to memory and retrieval to events and experiences in individual's life. Episodic memory is necessary for time traveling by remembering individual's past and imagining his or her future or a unique human quality that depends on the maturation. Semantic memory consists of all explicit memories by summary and generalization, such as the knowledge organized from historical events, figures, and experiences, the capability to recognize friends and associates, as well as the information learned about the specialized vocabularies, reading, writing, and mathematics in schools. As knowledge and information in semantic memory are reorganized from single events and experiences in episodic memory, the cellular units in engrams for episodic memory may be integrated with convergent inputs and outputs in their architectures for semantic memory [15] (Figs. 8.1 and 8.2).

Nondeclarative memory stands for the retrieval of previous experiences to perform a specific task without consciously knowing these experiences [10]. People use implicit memory every day in the form of procedures that allow them to remember how to tie their shoes or ride bicycle without consciously thinking about these activities. There are a few subcategories in implicit memory, such as illusion of truth and procedure memory. Illusion of truth effect refers to a phenomenon that an individual is more likely to believe a familiar statement than an unfamiliar one. As

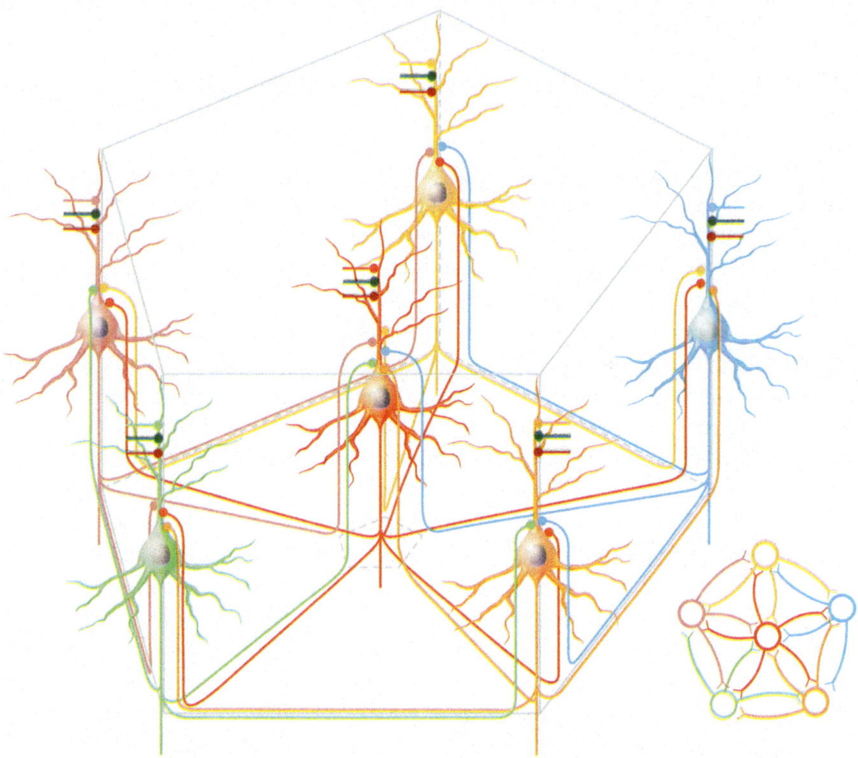

Fig. 8.1 Associative memory cells and their connection pattern in a common center. An associative memory cell in the center area (red) interconnects with five associative memory cells in surround by mutual synapse innervations. Their interconnection pattern is showed as a horizontal section in bottom view. This connection pattern of secondary associative memory cells is presumably relevant to logical reasoning with a core concept. These associative memory cells receive their innate synapse input (identical color with soma), the arousal system input (green), and the new synapse input (red)

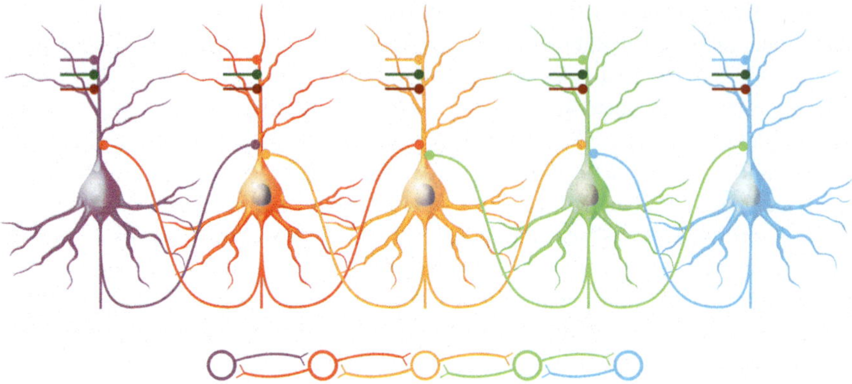

Fig. 8.2 Associative memory cells and their connection pattern in a sequential manner. An associative memory cell in the center area (red) interconnects with its neighboring associative memory cell through mutual synapse innervations in a sequential manner. Their interconnection pattern is showed as horizontal section in bottom view. This connection pattern is presumably relevant to logical reasoning step-by-step. These associative memory cells receive their innate synapse input (identical color with soma), the arousal system input (green), and the new synapse input (red)

illusion of truth effect occurs without explicit knowledge, it is a direct result of implicit memory [56]. Procedural memory allows to perform some actions, such as writing or riding a bike, without consciousness thinking about it [57, 58].

Main difference between explicit and implicit memory appears whether retrievals to memorized signals are associated with consciousness and intension. This classification ignores signals' acquisition processes [15]. In the learning of these associated signals regardless of their future retrievals in explicit or implicit manner, the consciousness and attentions are needed. The formation of nondeclarative memory in operations and procedures, such as riding bicycles and playing piano, has been practiced by numerous times, in comparison with the formation of declarative memory in knowledge and experiences. There appear to be more practicing, more familiar, and easier to be recalled in memory contents. Similar to the situation for people to ride bicycles without the requirement of consciousness and intension, teachers experienced in certain knowledge that has been taught uncounted times do not need to pay particular attentions to their speeches about such knowledge. This speech is more likely an operation or procedures. In addition, the contents in implicit memory can be stated consciously if the teaching of these operations and procedures is required. The operations and procedures in implicit memory need to be done with cautions under the consciousness if these activities are potentially intervened by environments. It seems to be that there are mutual conversions between explicit memory and implicit memory in the individuals based on their purposes, environments, and practices [15].

In terms of cellular architecture for explicit and implicit memories, the repeated activation of memory cells that store explicit knowledge or implicit procedures may lead to their recruitment and function upregulation further. As long as these learned signals activate cortical neurons by specific afferent pathways to recruit them to be memory cells, these memory cells can be activated by cues in memory retrievals. Those memory cells historically recruited may be reactivated before this learning session by attention cues and are reactivated many times during rehearsal. As showed in recent studies, the coactivations of cortical neurons induce their mutual synapse innervations and recruit them to be associative memory cells and functional upregulation. Based on a principle of active together, connecting together, and strengthening together, activity-dependent positive cycle in recruitments and refinements of associative memory cells makes those memory cells historically recruited to be more active, so that their storage of statements is dominantly retrieved. This activity-dependent positive cycle may work for this conversion from explicit memory to implicit memory [8, 15, 22]. In addition, each of associative memory cells receives three resources of synapse innervations, i.e., innate synapse input, synapse input from the arousal system, and synapse innervations from coactivated brain regions [8, 22]. All these synapse innervations work together to drive and activate associative memory cells. Through the coactivity-dependent positive cycle in the recruitment and upregulation of associative memory cells, the encoding capability and *en passant* synapses on these associative memory cells are highly increased. The activation of these associative memory cells may not need the help of synapse inputs from the arousal system, such that memory retrievals can be fulfilled without consciousness and particular intensions to know these procedures, experiences, and even knowledge. Associative memory cells that carry the repeatedly learned and much familiar signals can also be activated during dreams with subconsciousness for their retrievals. On the other hand, the activation of associative memory cells in relevance to implicit form without the enriched number during the initial learning needs the help from the arousal system for consciousness, regardless of knowledge or procedures [15].

8.2.2 Associative Memory Cells for Episodic and Semantic Memory

Based on the natures of memory contents, declarative memory is classified into episodic memory and semantic memory [11–13, 59–63]. Episodic memory refers to memory and retrieval to single event and experience in individual's life. Episodic memory is necessary for remembering individual's past and imagining his or her future. Semantic memory consists of all explicit memories by summary and generalization, such as knowledge organized from historical events and figures, capability to recognize friends and associates, as well as information learned about specialized vocabularies, reading, writing, and mathematics in schools [11, 12]. Despite this classification of episodic and semantic memory, these two types of declarative memory are related to each other under certain conditions [11, 13, 63]. Their differences are obvious [56, 60]. Episodic memory is the one-shot learning mechanism, i.e., remember an affair through one exposure to an episode. On the other hand, semantic memory is thought of multiple exposures to each of references and is updated on each exposure. Furthermore, episodic memory can be understood as storing each of items in a map that links items together in semantic memory. For example, semantic memory tells what a bicycle looks and its functions are, while episodic memories in all new experiences about this bicycle modify the single semantic retrieval of bicycle. It is presumed that episodic memory can be converted into semantic memory over time [64]. In this process, episodic memory about a particular event is generalized, and the context of specific events is missing. If someone tells and retells one story repeatedly, he or she may feel that the event is no longer remembered, but the pre-written story is emerged in his or her recall, such as personal semantics [63]. Further hypothesis would be a chain of conversion of episodic memory, semantic memory, and implicit memory [15].

The features of episodic memory include that memory contents are individual affairs and facts in specific location and time, memory contents are related to properties of single object and event, and its formation can be fulfilled through an exposure as well as episodic memories to an affair or fact in multiple exposures can be generalized and converted into semantic memory [11, 13, 63, 64]. Episodic memories are usually associated with emotional reactions. Although relationships between emotion and memory are complicated, the emotion tends to increase a likelihood that an event will be remembered easily and vividly, in which flashbulb is an example [65]. In terms of neuronal substrates, the formation of new episodic memories requires the hippocampus, prefrontal cortex, and limbic system [66–72]. The prefrontal cortex works for organizing signals for more efficient storage or for semantic strategies, such as thinking about the meaning of learned materials or rehearsing in working memory [73–75]. In addition, episodic memory is modulated by neurotransmitters, e.g., serotonin, norepinephrine, dopamine, and acetylcholine [76–82]. These studies only indicate the involvement of neuronal activities in episodic memory.

Semantic memory is memories to meanings, understandings, principles, and generalized knowledge as well as the conscious retrieval of factual information and general knowledge [11, 13]. Contents in semantic memory include generalized knowledge, principle, and formula that do not involve the memory of specific events [10, 60, 64, 83]. In terms of mechanisms underlying semantic memory, various models have been proposed, such as a model of link-connected nodes with spreading activations for semantic memory [84–86] and associative mode in that each cell in the matrix corresponds to the strength of associations of row item and column item [87–90]. To locations of semantic memory in the brain, semantic memory may be stored in the same brain systems for episodic memory. The hippocampus encodes memories, while the cortex stores memories after the initial encoding process is completed [11, 91, 92]. Neural substrates for semantic memory are largely unknown [11, 91, 92]. According to examining the correlation between the impairment of semantic memory and the damage of specific brain areas, there are four types of specific semantic impairments, such as perceptual/functional features, topographic organization, informativeness, and intercorrelations [93, 94]. The experiment data are still inconclusive [92, 95–100]. Neural imaging studies have suggested that a large distributed network of semantic retrievals include the extensive areas of ventral (form and color knowledge) and lateral (motion knowledge) temporal cortex, parietal cortex (size knowledge), and premotor cortex (manipulation knowledge). Other areas, such as more anterior regions of temporal cortex, may be involved in the representation of non-perceptual (verbal) conceptual knowledge in some categorically organized fashion [101, 102].

Although certain brain areas are correlated to episodic and semantic memory, cellular architectures at levels of individual neurons and their circuits remain unknown. Semantic memory can be read out in humans and primates, but not rodents. However, the rodents can be used to study cellular mechanisms underlying associative memory. The proposed approaches are the uses of knowledge about memory cells and their circuits studied in rodents to interpret semantic versus episodic memory in human beings [15]. Because semantic memory is memories to meaning, understanding, and concept-based knowledge generalized to multi-facts, semantic memory may include cognitive processes, such as associative thinking and logical reasoning, to organize episodic memory [11, 12]. In comparison with the memory of spatial and temporal features for specific objects in episodic memory, memories to generalized knowledge from multiple objects in semantic memory may come from the abstraction and generalization through repeated practices, i.e., a feature about more practices, more summary, and more potential conversion of episodic memory into semantic memory. Are there memory cells meeting these criteria for episodic and semantic memory?

The coactivations in a population of neurons induce their encoding of intensive spikes and a chain reaction from spikes to microRNA-regulated expression of genes and proteins in relevance to axon prolongation and synapse formation. These processes facilitate the recruitments of new synapse innervations among these neurons and in turn recruit them to be associative memory cells [43, 46, 47, 50, 103] as well as the retrieval of associated signals [104]. In addition, the coactivation of these associative memory cells elevates intracellular calcium, which evokes coordinated alternations in these memory cells and their synapse innervations [105, 106]. The recruitment and refinement of associative memory cells work together for the associative memory [22, 51–53, 107]. In other words, associative memory cells with their synapse inputs and mutual synapse innervations constitute memories specific to those associated signals in episodic memory. Activity-dependent positive cycle in the recruitment and refinement of associative memory cells recruits more associative memory cells to upregulate the memory strength and maintenance [22]. Mutual synapse innervations among primary associative memory cells endorse the reciprocal retrieval of associated signals in episodic memory, whereas secondary associative memory cells based on their mutual synapse innervation and synapse convergences from primary associative memory cells execute associative thinking and logical reasoning related to semantic memory. To make episodic memory being generalized and semantic memory being summarized from episodic memory contents, linear connections from primary associative memory cells to secondary associative memory cells with the convergence and the divergence of synapse innervations are needed across such memory cells [15]. Figures 8.1 and 8.2 illustrates diagrams for the connections of different associative memory cells. One associative memory cell as a common connection with all other associative memory cells will facilitate the summary of multiple episodic signals. The interconnections of different associative memory cells in a chain manner work for logical sequence to upgrade episodic memory to semantic memory [8, 22, 43, 44, 47] (Figs. 8.1 and 8.2).

In addition to coactivity-dependent mutual synapse innervations among neurons for them to be recruited as associative memory cells, activity-dependent positive cycle in the recruitment of associative memory cells and their upregulations play critical roles in their interconnection patterns. In early life, the learning of associated signals may induce ready-recruited cortical neurons to be mutually innervated through axon prolongation and synapse formations, such that associative memory cells are recruited for the integrative storage and reciprocal retrieval of episodic events. By the learning of numerous episodic-associated signals, many groups of associative memory cells are recruited to encode the storage of these episodic signals. Once their connections are formed in a central group with the surround of other groups of associative memory cells and/or sequential styles among these groups of associative memory cells, the associations among episodic events are potentially granted. In later life, the coactivations of associative memory cells during learning make these axon terminals to be grown and extended locally. With this local expansion of axon terminals of associative memory cells from other regions, new synapse innervations on associative memory cells are formed by *en passant* synapses, which strengthens the driving force to these associative memory cells to upregulate their functional states, i.e., the activity-dependent refinement of associative memory cells. New synapse innervations on ready-recruited neurons around these associative memory cells recruit more associative memory cells or activity-dependent recruitment of associative memory cells. The repeated coactivation of associative memory cells leads to the positive cycle of their recruitment and refinement, which reconstructs neural circuits of associative memory cells for conversions of episodic memory into semantic memory, and even nondeclarative memory [15].

8.2.3 Associative Memory Cells for Eidetic Memory and False Memory

According to the fidelity of memory contents and the efficiency of memory formation and retrieval, declarative memory is classified into eidetic memory and false memory in psychology. Eidetic memory refers to the situation that subjects are able to recall images, sound, or objects with extreme accuracy and abundant volume [108]. False memory stands for memories to events or knowledge that have never happened or that are distorted memories [109, 110]. In addition, eidetic memory is examined immediately after images, sound, or objects are given, and false memory is usually examined with the delay. In this regard, false memory is influenced by many facts, such as time and cognitive state, beyond memory mechanisms.

Eidetic memory is featured by the vivid recall of visual images and word language with extraordinary detail, especially in young children [108, 111]. Famous artists and composers are thought to have eidetic memory. However, their memories are highly trained in their respective fields, and they have devoted much time to improve their capability. The focus and attention to their individual arts may improve the related parts of their memory. Like other memories, the intensity of eidetic memory recall may be influenced by the duration and frequency of exposure to stimuli, the conscious observation, the relevance to persons, and so on. Therefore, the time of learning and the attention to learning under the consciousness are critical for eidetic memory. If memory formation is based on memory cells that memorize unitary signals in images, verbal, and other sensory dimensions, the vivid recall of these integrated signals may be caused by the efficiently mutual synapse innervations and the full activation of all memory cells [15].

False memory is used to describe unrealistic or distorted memories [109, 110]. Memories are believed to be the reconstructed phenomenon and can often be influenced by expectation, emotions, beliefs to others, inappropriate interpretation, or desired outcome [110, 112]. Human memory may be unreliable to some degree by failing to remember or by remembering incorrectly under normal conditions. Memories can also be disturbed by some disorders, such as amnesia, Alzheimer's disease, and post-traumatic stress disorder with a dramatic loss of memory. Emotion's influences on false memory depend upon whether the mood is an embedded memory content, especially the negative natural mood which initiates false memory [17, 113]. In terms of the origin of false memory, hypnosis can be used to form false memories because some false memories are formed by the rehearsal of fantastic events. After repeated thinking and imagining these events, subjects may begin to remember these events as actual occurrences and recall them confidently. The rehearsal is the strongest mechanism for the conversion of short-term memory into long-term memory, or memory consolidation, which is applied to implanted and real memories [109, 113]. Therefore, false memory is influenced by repeated thinking and imagining, hypnosis, fantastic events in neighbors, expectations, emotions, trust to others, and desired outcome, i.e., improper associations in events stored in the brain and repeated rehearsal for memory strengthening.

In terms of cellular architectures for eidetic memory and false memory, associative memory cells and their circuits are presumably involved. As eidetic memory is influenced by the times and duration of exposure to learned objects, the relevance to persons, the consciousness and attentions, and its formation can be explained by the recruitment of associative memory cells and their upregulation by different synapse inputs. As we know, associative memory cells are innervated by different sources of synapse inputs, such as innate inputs, new inputs carrying learned signals, as well as inputs from the arousal system. The activity level of the arousal system influences the consciousness state and attention by monoamine transmitters that upregulate associative memory cells [22]. Synapse innervations that carry the signals previously acquired and memorized in relevance to current learning can work as pathways to bring attention cues. Activity-dependent positive cycle in the recruitment and refinement of associative memory cells during the prolonged learning may increase the number and the functional state of associative memory cells. These mechanisms work together to facilitate the formation of eidetic memory by upregulating the associative memory cells under the conditions of consciousness and attention cues [15]. On the other hand, false memory is affected by repeated thinking and imagining, hypnosis, fantastic events in neighbors, expectations, desired outcome, emotions, and trust to others, i.e., improper associations in events stored in the brain and repeated rehearsal for memory strengthening in a slow formation manner. As indicated in the recent studies, there are primary associative memory cells in sensory cortices and secondary associative memory cells in the brain areas relevant to cognitions and emotions [8, 22, 43, 47, 52, 53]. Their mutual innervations formed by their coactivity during learning, cognitions and emotions constitute various patterns of memory circuits [15]. As long as associative memory cells have stored the signals experienced directly by individuals themselves or learned indirectly from other persons' experiences in stories, movies, and social communications, the coactivations of these associative memory cells that encode realistic signals and indirect signals during hypnosis or dreams may induce their mutual innervations and in turn recruit new pairs of associative memory cells advanced in the including of realistic signals and indirect signals. These new pairs of the advanced association among associative memory cells that have encoded the storage of learned realistic signals and indirect signals may lead to the formation of false memory.

8.2.4 Associative Memory Cells for Working Memory and Perceptual Memory

Memory processes can also be classified as short-term working memory from sensory guidance to behavior manipulation [114–116] and long-term perceptual memory that includes the storage of the visual, auditory, olfactory, gustatory, and somatosensory signals [117, 118]. Perceptual memory refers to long-term memory of subject's perceptual experiences, i.e., memories for people faces, voices, appearance, and so on, in which these signals inputted from sight, sound, smell, taste, and tactile sensations. Working memory as a core executive function is cognitive memory buffer with a limited capacity that is responsible for the transient holding, processing, and manipulation of information, which is important for the reasoning and the guidance of decision making and behavior [119–121].

Based on the routes of information inputted from distinct modalities, perceptual memory is sorted as visual memory, verbal memory, spatial memory, and olfactory memory [118–123]. Visual memory maintains some characteristics of subject's sensations in relevance to visual experiences. After visual memory is formed, these subjects are able to consign the memorized things that are featured as similar objects, places, animals, or people in sort of a mental image for subsequent retrievals, such that visual memory seems more likely to be semantic memory [124–127].

Spatial memory refers to signal storage about organism environment and spatial orientation [128, 129]. Olfactory memory stands to the storage and retrieval of odor signals [117, 130, 131]. Neuroimaging studies show that the medial prefrontal cortex is activated during the test of these perceptual memories [132, 133], indicating the involvement of perceptual memory in high cognition.

Working memory refers to structures and processes used for temporarily storing and manipulating signals. The frontal cortex, parietal cortex, anterior cingulate, and parts of the basal ganglia are crucial for working memory [114, 116, 134–139]. Working memory is considered to possess limited capacity [140–142]. The capacity of working memory appears to be regulated by norepinephrine in the locus coeruleus and dopamine in substantia nigra of the brain stem [115, 143] as well as other psychological state [144, 145]. In terms of cellular mechanisms underlying working memory [146], the persistent firing of certain neurons in the delay period of working memory tasks shows that the brain has a mechanism of keeping representations active without external input. It has been speculated that synchronous firing of neurons involved in working memory oscillates in frequencies of theta band [146–149].

It seems to us that perceptual memory and working memory are granted by semantic memory and cognitive processes. The interactions between primary associative memory cells in the distinct sensory cortices and secondary associative memory cells in cognition-relevant brain areas play an important role in these two types of memories. In addition, the transferring of the associated signals from primary and secondary associative memory cells to memory output cells in motor cortices influences temporal maintenance from sensory cue guidance to behavior manipulation in working memory [15]. To the maintenance of perceptual and working memory as well as the skilled operations in working memory based on the rehearsal and practices, activity-dependent positive cycles in the recruitment and upregulation of associative memory cells are critically important [22].

8.2.5 Associative Memory Cells in Age-Relevant Memory

Regardless explicit memory or implicit memory, the learning efficiency and memory accuracy are largely influenced by the age. Both of them appear to a bell shape or the increment up to young adults and the decay with aging [53, 150, 151]. The unitary signals and their organized contents (topographic organization) to be learned and memorized in manners of intramodal and cross-modal are also different in postnatal development [53, 132]. Learning leads to the acquisition and storage of unitary signals as well as memories to complicated signals that are reorganized from unitary signals.

In early life, newborns or infants start to possess sensory learning based on their experience through visual, auditory, olfactory, gustatory, and tactile sensations [152]. The acquisition and memory of signals from various sensory systems occur in sensory cortices [153]. As sensory organs have dissected the complicated signals taken from environments into the unitary signals (please see Chap. 7 in detail), these unitary signals have to be integrated into the original complete styles in sensory cortices, such as visual images and verbal language, in the form of neuronal encoding [15]. The improvement of perceptual abilities through training offers important insight into the physiological basis of learning in cerebral cortices. The rapid time course and facilitation, with which perceptual capabilities can improve, suggest that the learning is an integral part of normal perception. Perceptual learning occurs in a variety of areas and involves multiple mechanisms, especially the task-specific suppression of signals that interfere with performance [154]. The contents learned from parents and their neighbors include the unitary signals, such as a sound with single frequency or tone, a simple image (e.g., blue sky), and a simple taste (e.g., milk), as well as complicated signals, such as image complexes with multiple objects and colors, which are organized from unitary signals in different ways. Each of unitary signals is detected by sensory receptors for single feature in one sensory modality and is transmitted to their innate sensory neurons in the sensory cortex by specific afferent nerves, which can strengthen the encoding of sensory signals and synapse interconnections in a given pathway. A group of sensory cortical neurons fulfills the encoding in relevance to a unitary signal. However, the complicated signals in the learning need be integrated by sensory cortical neurons to form complete neural signals correspondent to the complicated exogenous signals since these complicated signals have been dissected into the unitary signals by sensory receptors for transmitting to their sensory cortical neurons. The different frequencies of complicated sound are detected by hair cells on the different fragment of cochlea-based membrane. The complicated images are dissected to photon beams with their light strengths and colors and detected by cone cells in the retina. One way for their integrations is done by sensory cortical neurons in intramodal manner. The coactivations of these sensory cortical neurons by unitary signals from complicated signals lead to their mutual synapse innervations for the integration of unitary signals into complicated signals in the neuronal form or neuronal integrated codes. Another way of their integrations is done by cross-modal sensory cortical neurons. The coactivity of sensory cortical neurons among respective modalities by cross-modal learned signals leads to their mutual synapse innervations among these sensory cortical neurons in these modalities for the integration of signals from different modalities. The association of visual and auditory signals in learning leads to mutual synapse innervation between visual and auditory cortical neurons. With more associations, mutual synapse innervations and signal integrations will be more complicated. These mutual synapse innervations will recruit associative memory cells [15].

After growing to the age when the systemic education is given by parents and schools, children experience more cross-modal associative learning, i.e., cross-modal associative learning gradually moves forward from unitary to complicated signals, especially the association of auditory signals with visual signals, smell signals, taste signals, and tactile signals for their definitions, compared with intramodal associative learning. In the associative learning of these complicated signals reorganized from unitary signals, mutual synapse innervations that lead to the recruitment of associative memory cells become more and more enriched. In the subsequent learning, the strengthening of these mutual innervated synapses through the conversion of inactive synapses into active synapses or silent synapses into functional synapses leads to groups of associative memory cells to be activated for memory formation and retrieval [15].

Associative memory cells have memorized various unitary signals. The efficiency of associative memory and the number of associative memory cells are variable among subjects based on their experiences and repetitions during associative learning. Memory strength and maintenance at different levels influence their subsequent learning and cognition, i.e., interaction among learning, memory, and cognition in feedforward manner for these experiences influencing events or in feedback manner for new experiences influencing the strength and maintenance of previous experiences [15].

8.3 Associative Memory Cells for Emotional Memory

Emotional experiences, either good or bad, can be maintained as stronger traces in the brain, which refers to emotional memory. Memories about emotional situations and reactions are stored in explicit and implicit manners. A typical example of implicit emotional memory is Pavlovian fear conditioning, in which subjects receive the neutral conditioned stimulus (a tone) and an aversive unconditioned stimulus (electrical footshock). After a few pairings, the tone signal elicits emotional responses that are naturally present in threatening stimuli. Emotional responses include changes in behavior, autonomic nervous system, and hormones. Neural substrates for this emotional memory may include neuronal plasticity and protein kinases in the amygdala [155–157] as well as larger circuitry consisting sensory cortices, motor cortices, the hippocampus, and the prefrontal cortex [158, 159]. In terms of explicit emotional memory, the medial temporal lobe may be involved since the damage to this area, especially the hippocampus and its connected cortical areas, in human beings leads to profound deficit in the ability to store new memories and emotion-relevant memory behaviors [160]. Emotional reactions often lead to stronger explicit memories. While emotional experiences produce very powerful and vivid memories that are easily recollected, memories are not more accurate in detail than nonemotional memories, such as memories about emotional events are not always to be trusted.

Most emotional memories are cue-induced. Anything connected to individual's sensations may be a cue to ignite emotional recall, such as the association of delightful smell with someone in the perfume counter of a shopping center and the association of certain places with pleasant or unpleasant emotion. Holding onto certain properties may activate the recall of emotional memory. To precious objects as memorabilia, most of these objects are cherished for reasons other than their value as inducers of reminiscence [161]. Emotional memory is powerful and serves to guide and inform individuals who navigate the present and prepare for the future. If someone has ever had a drink or taste to something spoiled, he or she knows that emotional memory protects from doing that again.

The effect of emotional arousal on explicit memory is secondary to the activation of emotional processing system in the brain. Processing threat environment activates the amygdala, transmits information to neural networks in the hypothalamus and brain stem, and activates the arousal system by activating the medulla to release serotonin, norepinephrine, and acetylcholine throughout the brain, leading to expressions of behavior, autonomic, and endocrine responses and facilitating the formation of memories about emotions [155]. The connections from the amygdala to the paraventricular hypothalamus either directly or indirectly through other areas induce the release of ACTH from the pituitary gland. ACTH circulates to the adrenal gland to stimulate the release of glucocorticoid hormone (CORT) from the adrenal cortex. CORT has complex effects on memory. In mild stress, the low or mediate levels of CORT enhance explicit memory formation. In prolonged and intense stress, the higher level of CORT over an extended period can lead to the impairment of explicit memory. Memory failure after trauma has been attributed to memory repression due to glucocorticoid-induced amnesia. The connections from the amygdala to the brain stem and the hypothalamus that control the autonomic nervous system lead to the activation of sympathetic nerves. Epinephrine and norepinephrine released from nerve endings and adrenal medulla can act on peripheral nerves that project into the brain. For example, epinephrine directly acts on the peripheral nerve endings of sensory component in vagus nerves. The nerves innervate the brain stem and connect the locus coeruleus that releases norepinephrine in the amygdala, hippocampus, and other forebrain areas. Through these channels, peripheral catecholamines such as epinephrine and norepinephrine can alter the strength of explicit memory.

In addition, brain regions relevant to emotional memory, such as the amygdala and prefrontal cortex, have widespread mutual connections with other areas [155]. Emotional experiences and reactions often can be memorized [22, 162]. Based on activity together, connection together, and strengthening together, the brain regions involved in emotional experiences and reactions as well as sensory cortices taking in fear/pleasure information from can be mutually innervated by their coactivations during these events. So, associative memory cells that imprint emotional events are recruited in these regions. With repeated experiences and reactions, activity-dependent positive cycle in the recruitment and upregulation of associative memory cells leads to their further recruitments and upregulation, such that emotional memory is strengthened [15]. It is noteworthy that emotional memories are usually stronger in their depth and maintenance due to the upregulation of associative memory cells by the monoamine released from the arousal system.

8.4 Associative Memory Cells for Cognitive Processes

Cognitive processes include many events, such as associative thinking, logical reasoning, imagination, and computation, which are associative or integrative in nature. In other words, these cognitive processes can be thought of the association of associative memory [15]. Although studies by neural imaging have detected certain correlations between brain areas and these cognitive events, cellular architectures in relevance to associative thinking, logical reasoning, and imagination are far away from being known due to the limitations of accessing the human brain at cellular levels. Associative memory cells and their networks that have been identified in rodents may encode the association of the multiple endogenous signals based on mutual synapses innervations among secondary associative memory cells during these cognitive events. In the meantime, the secondary associative memory cells are able to memorize the association and integration of these cognitive events for their representation and recollection.

8.4.1 Associative Thinking

Associative thinking is thought of as a mental process to associate ideas and events that are seemingly unrelated and the tool to create other results in innovation, in which the brain encodes various signals by integrating pattern, finding contextual relationship, linking unrelated elements, and making three-dimensional mental modeling. Associative thinking is characterized by the brain's rapid, fluid, and cross-disciplinary capability to use the proper thinking combination. Associative thinking approach is proactive, integrative, and naturally strong in solving complex problems, finding opportunities in uncertain times, creating the future, and designing new valuable products and processes. In associative thinking, people bring the high degree of creativity to problem solving and thinking systemically. Individuals during associative thinking mentally hold the big picture to build the structure for design, development, and delivery. Associative thinkers intuitively combine art and science to develop innovative solutions and have intense focus with strong personal interest. In fact, the seemingly unrelated things do have certain internal connections in timing, features, and spatial locations for their potential to be associated in the brain [163, 164].

As illustrated in Figs. 5.2 and 5.3 based on experiment data, primary associative memory cells in sensory cortices encode the integrative storage of exogenous signals based on their mutual synapse innervations. Secondary associative memory cells with their mutual synapse innervations in cognition-relevant brain regions can encode the integration and storage of endogenous signals. Each group of the primary associative memory cells linearly connects a population of secondary associative memory cells. In addition, their connections include convergence, divergence, interconnection, and recurrent structures as well as feedforward and feedback in functional links. This widespread interconnected network confers these associative memory cells to interact with each other. In terms of the functions of these connections, learning process is usually initiated by learning objects and attention cues in relevance to these objects. These attention cues strengthen learning efficiency (please see Chap. 7). Attention cues may activate those associative memory cells that have stored events and knowledge potentially relevant to learning objects. The coactivations of ready-recruited associative memory cells for learning objects and those associative memory cells by attention cues will induce their mutual synapse innervations to form the advanced association among these associative memory cells. In other words, these two groups of associative memory cells are interconnected for their associative interactions. With the accumulation of the learning plus attention cues, numerous groups of associative memory cells are formed through their advanced associations. The connections consent these groups of associative memory cells that store different events and knowledge to be sequentially or simultaneously activated if they are active enough. That is, once the driving force of these synapse innervations efficiently shifts membrane potentials to spiking threshold potential or the excitability of these associative memory cells is upregulated, the active associative memory cells can perform their interactions for associative thinking [15].

8.4.2 Logical Reasoning

Logical reasoning is a cognitive process to look for the reasons for beliefs, conclusions, actions, or feelings, which is a branch of logical and creative thinking. In philosophy, the study of reasoning focuses on what can make reasoning efficient or inefficient, appropriate or inappropriate, and good or bad. Psychologists think which brain processes are involved and how logical reasoning is influenced by the structure of the brain. Specific forms of reasoning are also studied by mathematicians and lawyers. In scientific research, logical reasoning falls into two general areas. The biological function of the brain is studied by neuropsychologists, including the structure and function of normal brains and of damaged or unusual brains. The behaviors of logical reasoning are studied by psychologists, such as how people perform on reasoning tests (intelligence test) or how well people's reasoning matches ideals set by logic. Main divisions between forms of reasoning in philosophy are deductive reasoning and inductive reasoning. Formal logical has been described as the science of deduction. Deductive reasoning is intended to have reasoning that is valid and to determine whether the truth of conclusion can determine a rule, while inductive reasoning allows some degree of probability and attempts to support the presence of a rule by hypothesizing a rule after numerous examples are taken for a conclusion [165–167]. It seems to us that deductive reasoning is the application of the theory to each of processes, i.e., divergence from a central theory to peripheral cases, and inductive reasoning is to have the conclusion from numerous examples, i.e., convergence from peripheral examples to a central rule. Moreover, in the formal logical, a variety of forms of deductive reasoning is developed, such as abstract reasoning using symbols, logical operators, and a set of rules that specify what processes may be followed to arrive at a conclusion. These forms of reasoning include Aristotelian logic (syllogistic logic), propositional logic, predicate logic, and modal logic.

Cellular architectures for logical reasoning may be similar to associative thinking with some specials, since logical reasoning is thought of as a branch of associative thinking with logical and creative features. In addition, the well performance of logical reasoning, similarly to associative thinking, needs the accumulation of memory contents and information storage by more and advanced experiences in individual growth. For instance, deductive reasoning requires the accumulation of theories and principles, and inductive reasoning needs the accumulation of experience cases. Associative thinking and logical reasoning occur under the conditions of consciousness and attention. To meet these requirements, more associative learning, memory, and cognitions should be taken in life. Associative learning with attention cues will allow the association of relevant signals and the abstraction of their common features. The cognitions with attention cues confer the association and abstraction of relevant memory contents. In these processes, associative memory cells, especially secondary associative memory cells, are mutually interconnected. If associative memory cells encode the integrative storage of common featured signals among various relevant signals, this group of associative memory cells is commonly connected by other groups of associative memory cells. This convergence from these interconnected associative memory cells may work for inductive reasoning (Figs. 8.1 and 8.2). For instance, a few groups of associative memory cells may encode the associated signals, wet and car, wet and cloth, wet and grass, wet and road, as well as wet and something outside after raining. The convergence of these groups of associative memory cells onto a group of associative memory cells for encoding wet and something outside after raining leads to inductive reasoning about the association of the raining and wet road, car, cloth, and grass, especially their coactivations for many times. On the other hand, if the convergence onto one group of associative memory cells has been already formed, any new association from other associative memory cells to this group will be thought of as that their encoded signals should be followed to the deductive reasoning encoded by these groups of associative memory cells [15].

8.4.3 Imagination

Imagination refers to the capability to produce and simulate novel objects, peoples, and ideas in the mind without any immediate input of individual's sensations. The imagination is classified into two forms, i.e., reproductive imagination that reproduces images stored in the memory based on the suggestion of associated images and creative imagination that recombines former experiences in the creation form of new images directed by specific goal. It is also stated as the forming of experiences in the mind, which can be recreated from previous experiences in that vivid memories have been embedded by certain imagined changes, or forming of images with complete invention. The contents of imagination can be enriched by more memories to previous experiences in accumulative manners. Imagination assists to make knowledge applicable in solving problems and is fundamental to integrating experiences and other learning processes. As imagination allows to create new images and thoughts, memories to these images and thoughts with frequent rehearsals may lead to their more dominance in the mind, compared with memories to previous experiences. If the strong confidence and belief to the imagination occur, false memory may be formed. A basic training for imagination is the associations of ideas, objects, and experiences based on previous memories with freedom [168–170].

As the imagination is featured by the association and reorganization of memories to previous experiences and knowledge based on the order of associated images and specific goal, the imagination contains the regulation of associative thinking and reasoning. Imagination contents are proportional to memory enrichment accumulated from experiences and knowledge. Cellular architectures for imagination may be similar to associative thinking and logical reasoning with some specials, such that the following interconnections among associative memory cells are critically important for imagination, such as the linear connection from primary to secondary associative memory cells with convergence and divergence, the mutual synapse innervations among different groups of associative memory cells in intramodal and cross-modal manners, as well as their feedforward and feedback connections. With the driving force from their synapse inputs and the capability of their spike encoding, the excitation of these associative memory cells will lead to the retrieval of their memorized information for reorganizing and integrating reproductive and creative images. In the imagination embedded with associative thinking and reasoning, associative memory cells, especially secondary associative memory cells, are mutually interconnected. If associative memory cells encode the integrative storage of common featured signals among various relevant signals, this group of associative memory cells connects other groups of associative memory cells. The imagination will be based on these associative memory cells. A few groups of associative memory cells that encode image signals including lanes on the road, cars, traffic signs, trees, pavement, and people may work together for imaginary pictures integrated from people, vehicles, road, and traffic signs in three-dimensional manner. If these groups of associative memory cells connect one another in a sequential manner, the associative memory cells that encode image signals including these objects work together for imaginary pictures integrated from people, vehicles, road, and traffic signs sequentially. In addition, cross-modal innervations among associative memory cells linked with visual cortices and auditory cortices are able to translate the imaginary pictures into verbal languages to have the representation of these pictures quickly [8, 22].

References

1. Eichenbaum H. Still searching for the engram. Learn Behav. 2016;44(3):209–22.
2. Kandel ER, Dudai Y, Mayford MR. The molecular and systems biology of memory. Cell. 2014;157(1):163–86.
3. Lisman J, et al. Memory formation depends on both synapse-specific modifications of synaptic strength and cell-specific increases in excitability. Nat Neurosci. 2018;21(3):309–14.
4. McGaugh JL. Memory – a century of consolidation. Science. 2000;287(5451):248–51.
5. Poo MM, et al. What is memory? The present state of the engram. BMC Biol. 2016;14:40.

6. Squire LR, Wixted JT, Clark RE. Recognition memory and the medial temporal lobe: a new perspective. Nat Rev Neurosci. 2007;8(11):872–83.
7. Tonegawa S, et al. Memory engram storage and retrieval. Curr Opin Neurobiol. 2015;35:101–9.
8. Wang JH, Cui S. Associative memory cells: formation, function and perspective. F1000Res. 2017;6:283.
9. Squire LR, Knowlton B, Musen G. The structure and organization of memory. Annu Rev Psychol. 1993;44:453–95.
10. Squire LR. Declarative and nondeclarative memory: multiple brain systems supporting learning and memory. J Cogn Neurosci. 1992;4(3):232–43.
11. Squire LR, Zola SM. Episodic memory, semantic memory, and amnesia. Hippocampus. 1998;8(3):205–11.
12. Tulving E. Multiple memory systems and consciousness. Hum Neurobiol. 1987;6(2):67–80.
13. Wood F. Focal and diffuse memory activation assessed by localized indicators of CNS metabolism: the semantic-episodic memory distinction. Hum Neurobiol. 1987;6(2):141–51.
14. Kim JI, Humphreys GW. Working memory, perceptual priming, and the perception of hierarchical forms: opposite effects of priming and working memory without memory refreshing. Atten Percept Psychophys. 2010;72(6):1533–55.
15. Wang J-H. Searching basic units of memory traces: associative memory cells. F1000Res. 2019;8(457):1–28.
16. Wasinger K, Zelhart PF, Markley RP. Memory for random shapes and eidetic ability. Percept Mot Skills. 1982;55(3 Pt 2):1076–8.
17. Bookbinder SH, Brainerd CJ. Emotion and false memory: the context-content paradox. Psychol Bull. 2016;142(12):1315–51.
18. Byrne JH. Learning and memory: basic mechanisms. In: Square LR, et al., editors. Fundament neuroscience. 2nd ed. Amsterdam: Academic; 2003. p. 1275–98.
19. Kandel ER. Nerve cells and behavior. In: Kandel ER, Schwartz JH, Jessell TM, editors. Principles of neural science. 4th ed. New York: McGraw-Hill; 2000. p. 19–35.
20. Wasserman EA, Miller RR. What's elementary about associative learning? Annu Rev Psychol. 1997;48:573–607.
21. Fanselow MS, Poulos AM. The neuroscience of mammalian associative learning. Annu Rev Psychol. 2005;56:207–34.
22. Wang JH, Cui S. Associative memory cells and their working principle in the brain. F1000Res. 2018;7:108.
23. Agra G, et al. Analysis of the concept of meaningful learning in light of the Ausubel's theory. Rev Bras Enferm. 2019;72(1):248–55.
24. Baars BJ, Gage NM. In: Baars BJ, Gage NM, editors. Cognition, brain, and consciousness: introduction to cognitive neuroscience. London: Elsevier Ltd; 2007.
25. Bolhuis JJ, Honey RC. Imprinting, learning and development: from behaviour to brain and back. Trends Neurosci. 1998;21(7):306–11.
26. Cadorin L, et al. An integrative review of the characteristics of meaningful learning in healthcare professionals to enlighten educational practices in health care. Nurs Open. 2014;1(1):3–14.
27. Cake MA. Deep dissection: motivating students beyond rote learning in veterinary anatomy. J Vet Med Educ. 2006;33(2):266–71.
28. Carcea I, Froemke RC. Biological mechanisms for observational learning. Curr Opin Neurobiol. 2019;54:178–85.
29. Cardoso-Leite P, Bavelier D. Video game play, attention, and learning: how to shape the development of attention and influence learning? Curr Opin Neurol. 2014;27(2):185–91.
30. Eiriksdottir E, Catrambone R. Procedural instructions, principles, and examples: how to structure instructions for procedural tasks to enhance performance, learning, and transfer. Hum Factors. 2011;53(6):749–70.
31. Frenois F, Le Moine C, Cador M. The motivational component of withdrawal in opiate addiction: role of associative learning and aversive memory in opiate addiction from a behavioral, anatomical and functional perspective. Rev Neurosci. 2005;16(3):255–76.

32. Gewirtz JL. Conditional responding as a paradigm for observational, imitative learning and vicarious-reinforcement. Adv Child Dev Behav. 1971;6:273–304.
33. Giannini AJ, Giannini JN, Condon M. Use of tangential visual symbols to increase the long-term learning process: applications of linkage in teaching pharmacological principles of addiction. J Clin Pharmacol. 2000;40(7):708–12.
34. Hedges JH, et al. Play, attention, and learning: how do play and timing shape the development of attention and influence classroom learning? Ann N Y Acad Sci. 2013;1292:1–20.
35. Hill HM, Dietrich S, Cappiello B. Learning to play: a review and theoretical investigation of the developmental mechanisms and functions of cetacean play. Learn Behav. 2017;45(4):335–54.
36. Horn G. Imprinting, learning, and memory. Behav Neurosci. 1986;100(6):825–32.
37. Letzkus JJ, Wolff SB, Luthi A. Disinhibition, a circuit mechanism for associative learning and memory. Neuron. 2015;88(2):264–76.
38. McLaughlin B. "Intentional" and "incidental" learning in human subjects: the role of instructions to learn and motivation. Psychol Bull. 1965;63:359–76.
39. Pearce MT. Statistical learning and probabilistic prediction in music cognition: mechanisms of stylistic enculturation. Ann N Y Acad Sci. 2018.
40. Suzuki WA. Associative learning signals in the brain. Prog Brain Res. 2008;169:305–20.
41. Zentall TR. Perspectives on observational learning in animals. J Comp Psychol. 2012;126(2):114–28.
42. Squire LR, Zola SM. Structure and function of declarative and nondeclarative memory systems. Proc Natl Acad Sci U S A. 1996;93(24):13515–22.
43. Feng J, et al. Barrel cortical neuron integrates triple associated signals for their memory through receiving epigenetic-mediated new synapse innervations. Cereb Cortex. 2017;27(12):5858–71.
44. Gao Z, et al. Associations of unilateral whisker and olfactory signals induce synapse formation and memory cell recruitment in bilateral barrel cortices: cellular mechanism for unilateral training toward bilateral memory. Front Cell Neurosci. 2016;10(285):1–16.
45. Guo R, et al. Associative memory extinction is accompanied by decayed plasticity at motor cortical neurons and persistent plasticity at sensory cortical neurons. Front Cell Neurosci. 2017;11(168):1–12.
46. Liu Y, et al. Piriform cortical glutamatergic and GABAergic neurons express coordinated plasticity for whisker-induced odor recall. Oncotarget. 2017;8(56):95719–40.
47. Wang D, et al. Neurons in the barrel cortex turn into processing whisker and odor signals: a cellular mechanism for the storage and retrieval of associative signals. Front Cell Neurosci. 2015;9:320.
48. Wang JH, et al. Upregulation of glutamatergic receptor-channels is associated with cross-modal reflexes encoded in barrel cortex and piriform cortex. Biophys J. 2014;106(2):supplement 191a.
49. Wang J-H, et al. Both glutamatergic and Gabaergic neurons are recruited to be associative memory cells. Biophys J. 2016;110(3):supplement 191a.
50. Yan F, et al. Coordinated plasticity between barrel cortical glutamatergic and GABAergic neurons during associative memory. Neural Plast. 2016;2016(ID5648390):1–20.
51. Wang J-H, Guo R, Wei Z. Associative memory extinction is accompanied by decays of associative memory cells and their plasticity at motor cortex but not sensory cortex. Soc Neurosci. 2017;81(09):10385.
52. Wang J-H, et al. Prefrontal cortical neurons are recruited as secondary associative memory cells for associative memory and cognition. Biophys J. 2018;114(3):155a.
53. Wang JH, et al. Secondary associative memory cells and their plasticity in the prefrontal cortex. Biophys J. 2019;116(3):427a.
54. Squire LR. Mechanisms of memory. Science. 1986;232(4758):1612–9.
55. Gais S, Born J. Declarative memory consolidation: mechanisms acting during human sleep. Learn Mem. 2004;11(6):679–85.

56. Begg IM, Nicholson S. Semantic and episodic relations are experimentally dissociable and stochastically independent. Can J Exp Psychol. 1994;48(3):399–417.
57. Arwas S, Rolnick A, Lubow RE. Conditioned taste aversion in humans using motion-induced sickness as the US. Behav Res Ther. 1989;27(3):295–301.
58. Cohen NJ, et al. Different memory systems underlying acquisition of procedural and declarative knowledge. Ann N Y Acad Sci. 1985;444:54–71.
59. Garcia-Lazaro HG, et al. Neuroanatomy of episodic and semantic memory in humans: a brief review of neuroimaging studies. Neurol India. 2012;60(6):613–7.
60. Horner MD. Psychobiological evidence for the distinction between episodic and semantic memory. Neuropsychol Rev. 1990;1(4):281–321.
61. Moscovitch M, et al. The cognitive neuroscience of remote episodic, semantic and spatial memory. Curr Opin Neurobiol. 2006;16(2):179–90.
62. Platel H. Functional neuroimaging of semantic and episodic musical memory. Ann N Y Acad Sci. 2005;1060:136–47.
63. Renoult L, et al. Personal semantics: at the crossroads of semantic and episodic memory. Trends Cogn Sci. 2012;16(11):550–8.
64. Greenberg DL, Verfaellie M. Interdependence of episodic and semantic memory: evidence from neuropsychology. J Int Neuropsychol Soc. 2010;16(5):748–53.
65. Diamond DM, et al. The temporal dynamics model of emotional memory processing: a synthesis on the neurobiological basis of stress-induced amnesia, flashbulb and traumatic memories, and the Yerkes-Dodson law. Neural Plast. 2007;2007:60803.
66. Buzsaki G. Theta rhythm of navigation: link between path integration and landmark navigation, episodic and semantic memory. Hippocampus. 2005;15(7):827–40.
67. Janowsky JS, et al. Cognitive impairment following frontal lobe damage and its relevance to human amnesia. Behav Neurosci. 1989;103(3):548–60.
68. Janowsky JS, Shimamura AP, Squire LR. Source memory impairment in patients with frontal lobe lesions. Neuropsychologia. 1989;27(8):1043–56.
69. Nyberg L, et al. Activation of medial temporal structures during episodic memory retrieval. Nature. 1996;380(6576):715–7.
70. Nyberg L, Cabeza R, Tulving E. Asymmetric frontal activation during episodic memory: what kind of specificity? Trends Cogn Sci. 1998;2(11):419–20.
71. Tulving E, Markowitsch HJ. Episodic and declarative memory: role of the hippocampus. Hippocampus. 1998;8(3):198–204.
72. Wheeler MA, Stuss DT, Tulving E. Frontal lobe damage produces episodic memory impairment. J Int Neuropsychol Soc. 1995;1(6):525–36.
73. Gabrieli JD, Poldrack RA, Desmond JE. The role of left prefrontal cortex in language and memory. Proc Natl Acad Sci U S A. 1998;95(3):906–13.
74. McIntosh AR, et al. Differential functional connectivity of prefrontal and medial temporal cortices during episodic memory retrieval. Hum Brain Mapp. 1997;5(4):323–7.
75. Wagner AD, et al. Prefrontal cortex and recognition memory. Functional-MRI evidence for context-dependent retrieval processes. Brain. 1998;121(Pt 10):1985–2002.
76. Croxson PL, et al. Acetylcholine facilitates recovery of episodic memory after brain damage. J Neurosci. 2012;32(40):13787–95.
77. Danion JM, et al. Pharmacology of human memory and cognition: illustrations from the effects of benzodiazepines and cholinergic drugs. J Psychopharmacol. 1993;7(4):371–7.
78. Chowdhury R, et al. Dopamine modulates episodic memory persistence in old age. J Neurosci. 2012;32(41):14193–204.
79. Grogan J, et al. Dopamine and consolidation of episodic memory: timing is everything. J Cogn Neurosci. 2015;27(10):2035–50.
80. Huang M, et al. 5-HT1A parital agonism and 5-HT7 antagonism restore episodic memory in subchronic phencyclidine-treated mice: role of brain glutamate, dopamine, acetylcholine and GABA. Psychopharmacology. 2018;235(10):2795–808.

81. O'Halloran KD. Blast from the past! Phrenic motor memory of antecedent episodic hypercapnia is serotonin dependent: relevance to respiratory rehabilitation and sleep-disordered breathing? Exp Physiol. 2016;101(2):258–9.
82. Papenberg G, et al. Dopamine and glutamate receptor genes interactively influence episodic memory in old age. Neurobiol Aging. 2014;35(5):1213 e3–8.
83. Kihlstrom JF. Posthypnotic amnesia for recently learned material: interactions with "episodic" and "semantic" memory. Cogn Psychol. 1980;12(2):227–51.
84. Fletcher PD, Warren JD. Semantic dementia: a specific network-opathy. J Mol Neurosci. 2011;45(3):629–36.
85. Irish M. Elucidating a Core semantic network in the brain-implications for disorders of semantic cognition. J Neurosci. 2016;36(23):6144–6.
86. Tune S, Asaridou SS. Stimulating the semantic network: what can TMS tell us about the roles of the posterior middle temporal gyrus and angular gyrus? J Neurosci. 2016;36(16):4405–7.
87. Fisk AD, et al. Age-related retention of skilled memory search: examination of associative learning, interference, and task-specific skills. J Gerontol B Psychol Sci Soc Sci. 1995;50(3):P150–61.
88. Fransen E, Lansner A. A model of cortical associative memory based on a horizontal network of connected columns. Network. 1998;9(2):235–64.
89. Haberly LB, Bower JM. Olfactory cortex: model circuit for study of associative memory? Trends Neurosci. 1989;12(7):258–64.
90. Page RA, von Merten S, Siemers BM. Associative memory or algorithmic search: a comparative study on learning strategies of bats and shrews. Anim Cogn. 2012;15(4):495–504.
91. Mayes AR, Montaldi D. Exploring the neural bases of episodic and semantic memory: the role of structural and functional neuroimaging. Neurosci Biobehav Rev. 2001;25(6):555–73.
92. Starr JM, et al. Episodic and semantic memory tasks activate different brain regions in Alzheimer disease. Neurology. 2005;65(2):266–9.
93. McCarthy RA, Warrington EK. Disorders of semantic memory. Philos Trans R Soc Lond Ser B Biol Sci. 1994;346(1315):89–96.
94. McCarthy RA, Warrington EK. Past, present, and prospects: reflections 40 years on from the selective impairment of semantic memory (Warrington, 1975). Q J Exp Psychol (Hove). 2016;69(10):1941–68.
95. Farah MJ, et al. Category-specificity and modality-specificity in semantic memory. Neuropsychologia. 1989;27(2):193–200.
96. Farah MJ, McClelland JL. A computational model of semantic memory impairment: modality specificity and emergent category specificity. J Exp Psychol Gen. 1991;120(4):339–57.
97. Lambon Ralph MA, Patterson K. Generalization and differentiation in semantic memory: insights from semantic dementia. Ann N Y Acad Sci. 2008;1124:61–76.
98. Lambon Ralph MA, et al. Semantic memory is impaired in both dementia with Lewy bodies and dementia of Alzheimer's type: a comparative neuropsychological study and literature review. J Neurol Neurosurg Psychiatry. 2001;70(2):149–56.
99. Nyberg L, McIntosh AR, Tulving E. Functional brain imaging of episodic and semantic memory with positron emission tomography. J Mol Med (Berl). 1998;76(1):48–53.
100. Tippett LJ, Grossman M, Farah MJ. The semantic memory impairment of Alzheimer's disease: category-specific? Cortex. 1996;32(1):143–53.
101. Kable JW, et al. Conceptual representations of action in the lateral temporal cortex. J Cogn Neurosci. 2005;17(12):1855–70.
102. Thompson-Schill SL, et al. Role of left inferior prefrontal cortex in retrieval of semantic knowledge: a reevaluation. Proc Natl Acad Sci U S A. 1997;94(26):14792–7.
103. Lei Z, et al. Synapse innervation and associative memory cell are recruited for integrative storage of whisker and odor signals in the barrel cortex through miRNA-mediated processes. Front Cell Neurosci. 2017;11(316):1–11.
104. Pignatelli M, et al. Engram cell excitability state determines the efficacy of memory retrieval. Neuron. 2019;101(2):274–284.e5.

105. Chen N, Chen X, Wang J-H. Homeostasis established by coordination of subcellular compartment plasticity improves spike encoding. J Cell Sci. 2008;121(17):2961–71.
106. Wang JH, et al. The gain and fidelity of transmission patterns at cortical excitatory unitary synapses improve spike encoding. J Cell Sci. 2008;121(17):2951–60.
107. Feng J, et al. Cell-specific plasticity associated with integrative memory of triple sensory signals in the barrel cortex. Oncotarget. 2018;9(57):30962–78.
108. Kaylor CW, Davidon RS. Accuracy of recall as a function of eidetic imagery. Percept Mot Skills. 1979;48(3 Pt 2):1143–8.
109. Brainerd CJ, Reyna VF, Ceci SJ. Developmental reversals in false memory: a review of data and theory. Psychol Bull. 2008;134(3):343–82.
110. Loftus EF. Memory distortion and false memory creation. Bull Am Acad Psychiatry Law. 1996;24(3):281–95.
111. Giray EF, et al. The incidence of eidetic imagery as a function of age. Child Dev. 1976;47(4):1207–10.
112. Kurkela KA, Dennis NA. Event-related fMRI studies of false memory: an activation likelihood estimation meta-analysis. Neuropsychologia. 2016;81:149–67.
113. Reyna VF, Brainerd CJ. Fuzzy-trace theory and false memory: new frontiers. J Exp Child Psychol. 1998;71(2):194–209.
114. Baddeley A. Working memory: looking back and looking forward. Nat Rev Neurosci. 2003;4(10):829–39.
115. Constantinidis C, Klingberg T. The neuroscience of working memory capacity and training. Nat Rev Neurosci. 2016;17(7):438–49.
116. Zhou YD, Ardestani A, Fuster JM. Distributed and associative working memory. Cereb Cortex. 2007;17(Suppl 1):i77–87.
117. Granger R, Lynch G. Higher olfactory processes: perceptual learning and memory. Curr Opin Neurobiol. 1991;1(2):209–14.
118. Magnussen S, Greenlee MW. The psychophysics of perceptual memory. Psychol Res. 1999;62(2–3):81–92.
119. Hedge C, Oberauer K, Leonards U. Selection in spatial working memory is independent of perceptual selective attention, but they interact in a shared spatial priority map. Atten Percept Psychophys. 2015;77(8):2653–68.
120. Koshino H, Olid P. Interactions between modality of working memory load and perceptual load in distractor processing. J Gen Psychol. 2015;142(3):135–49.
121. Weil RS, et al. Opposite effects of perceptual and working memory load on perceptual filling-in of an artificial scotoma. Cogn Neurosci. 2012;3(1):36–44.
122. Brewin CR. Episodic memory, perceptual memory, and their interaction: foundations for a theory of posttraumatic stress disorder. Psychol Bull. 2014;140(1):69–97.
123. Schacter DL. Perceptual representation systems and implicit memory. Toward a resolution of the multiple memory systems debate. Ann N Y Acad Sci. 1990;608:543–67; discussion 567–71
124. Hampton RR. Monkey perirhinal cortex is critical for visual memory, but not for visual perception: reexamination of the behavioural evidence from monkeys. Q J Exp Psychol B. 2005;58(3–4):283–99.
125. Khan ZU, Martin-Montanez E, Baxter MG. Visual perception and memory systems: from cortex to medial temporal lobe. Cell Mol Life Sci. 2011;68(10):1737–54.
126. Miyashita Y. Inferior temporal cortex: where visual perception meets memory. Annu Rev Neurosci. 1993;16:245–63.
127. Super H. Cognitive processing in the primary visual cortex: from perception to memory. Rev Neurosci. 2002;13(4):287–98.
128. Lanke J, et al. Spatial memory and stereotypic behaviour of animals in radial arm mazes. Brain Res. 1993;605(2):221–8.
129. O'Keefe J. Hippocampus, theta, and spatial memory. Curr Opin Neurobiol. 1993;3(6):917–24.
130. Wilson DA, Stevenson RJ. Olfactory perceptual learning: the critical role of memory in odor discrimination. Neurosci Biobehav Rev. 2003;27(4):307–28.

131. Wilson DA, Sullivan RM. Cortical processing of odor objects. Neuron. 2012;72(4):506–19.
132. Mizuno K, Ueda A. Antenatal olfactory learning influences infant feeding. Early Hum Dev. 2004;76(2):83–90.
133. Pautassi RM, et al. Acute ethanol counteracts the acquisition of aversive olfactory learning in infant rats. Alcohol. 2005;36(2):99–105.
134. Bodner M, et al. Patterned firing of parietal cells in a haptic working memory task. Eur J Neurosci. 2005;21(9):2538–46.
135. Cowan N. What are the differences between long-term, short-term, and working memory? Prog Brain Res. 2008;169:323–38.
136. Cowan N. Multiple concurrent thoughts: the meaning and developmental neuropsychology of working memory. Dev Neuropsychol. 2010;35(5):447–74.
137. Ericsson KA, Kintsch W. Long-term working memory. Psychol Rev. 1995;102(2):211–45.
138. Fuster JM, Bressler SL. Cognit activation: a mechanism enabling temporal integration in working memory. Trends Cogn Sci. 2012;16(4):207–18.
139. Rhodes S, Cowan N. Attention in working memory: attention is needed but it yearns to be free. Ann N Y Acad Sci. 2018;1424(1):52–63.
140. Buszard T, Masters RS, Farrow D. The generalizability of working-memory capacity in the sport domain. Curr Opin Psychol. 2017;16:54–7.
141. Glassman RB. A working memory “theory of relativity”: elasticity in temporal, spatial, and modality dimensions conserves item capacity in radial maze, verbal tasks, and other cognition. Brain Res Bull. 1999;48(5):475–89.
142. Oberauer K, et al. What limits working memory capacity? Psychol Bull. 2016;142(7):758–99.
143. Unsworth N, Robison MK. A locus coeruleus-norepinephrine account of individual differences in working memory capacity and attention control. Psychon Bull Rev. 2017;24(4):1282–311.
144. Fukuda K, Awh E, Vogel EK. Discrete capacity limits in visual working memory. Curr Opin Neurobiol. 2010;20(2):177–82.
145. Moran TP. Anxiety and working memory capacity: a meta-analysis and narrative review. Psychol Bull. 2016;142(8):831–64.
146. Hartley AA, Speer NK. Locating and fractionating working memory using functional neuroimaging: storage, maintenance, and executive functions. Microsc Res Tech. 2000;51(1):45–53.
147. Hsieh LT, Ranganath C. Frontal midline theta oscillations during working memory maintenance and episodic encoding and retrieval. NeuroImage. 2014;85(Pt 2):721–9.
148. Jensen O. Maintenance of multiple working memory items by temporal segmentation. Neuroscience. 2006;139(1):237–49.
149. Nyberg L, Eriksson J. Working memory: maintenance, updating, and the realization of intentions. Cold Spring Harb Perspect Biol. 2015;8(2):a021816.
150. Foster TC. Dissecting the age-related decline on spatial learning and memory tasks in rodent models: N-methyl-D-aspartate receptors and voltage-dependent Ca2+ channels in senescent synaptic plasticity. Prog Neurobiol. 2012;96(3):283–303.
151. Wu WW, Oh MM, Disterhoft JF. Age-related biophysical alterations of hippocampal pyramidal neurons: implications for learning and memory. Ageing Res Rev. 2002;1(2):181–207.
152. Lipsitt LP. The study of sensory and learning processes of the newborn. Clin Perinatol. 1977;4(1):163–86.
153. Ghose GM. Learning in mammalian sensory cortex. Curr Opin Neurobiol. 2004;14(4):513–8.
154. Boldin AM, Geiger R, Emberson LL. The emergence of top-down, sensory prediction during learning in infancy: a comparison of full-term and preterm infants. Dev Psychobiol. 2018;60(5):544–56.
155. Dolan RJ, Vuilleumier P. Amygdala automaticity in emotional processing. Ann N Y Acad Sci. 2003;985:348–55.
156. Han JH, et al. Selective erasure of a fear memory. Science. 2009;323(5920):1492–6.
157. Maren S. The amygdala, synaptic plasticity, and fear memory. Ann N Y Acad Sci. 2003;985:106–13.

158. Cardinal RN, et al. Emotion and motivation: the role of the amygdala, ventral striatum, and prefrontal cortex. Neurosci Biobehav Rev. 2002;26(3):321–52.
159. Maren S, Quirk GJ. Neuronal signalling of fear memory. Nat Rev Neurosci. 2004;5(11):844–52.
160. Phelps EA. Emotion and cognition: insights from studies of the human amygdala. Annu Rev Psychol. 2006;57:27–53.
161. Sherman E. Reminiscentia: cherished objects as memorabilia in late-life reminiscence. Int J Aging Hum Dev. 1991;33(2):89–100.
162. Wang J-H, Lu W. Molecular profiles in the brain are involved in fear memory induced by physical and psychological stress. Soc Neurosci. 2018;425.19(425):III61.
163. Adler LL, Berkowitz PH. Influencing associative thinking and imagery in emotionally disturbed children. Psychol Rep. 1976;39(1):183–8.
164. Gordon R, Silverstein ML, Harrow M. Associative thinking in schizophrenia: a contextualist approach. J Clin Psychol. 1982;38(4):684–96.
165. Martins J, Mendes RV. Neural networks and logical reasoning systems: a translation table. Int J Neural Syst. 2001;11(2):179–86.
166. Procyk E, Joseph JP. Problem solving and logical reasoning in the macaque monkey. Behav Brain Res. 1996;82(1):67–78.
167. Ramsey NF, et al. Excessive recruitment of neural systems subserving logical reasoning in schizophrenia. Brain. 2002;125(Pt 8):1793–807.
168. Cevallos C, et al. Oscillations in the human brain during walking execution, imagination and observation. Neuropsychologia. 2015;79(Pt B):223–32.
169. Henkin RI, Levy LM. Functional MRI of congenital hyposmia: brain activation to odors and imagination of odors and tastes. J Comput Assist Tomogr. 2002;26(1):39–61.
170. Mullally SL, Maguire EA. Memory, imagination, and predicting the future: a common brain mechanism? Neuroscientist. 2014;20(3):220–34.

Chapter 9
The Impacts of Associative Memory Cells on Pathology

Abstract Neurological diseases and psychological disorders with cognition and mood impairment are more or less accompanied by memory deficits, since the capability and efficiency of normal cognitions, emotion, and behaviors are influenced by memory capacity. In psychiatric diseases, fear memory induced by acute severe stress is coupled with anxiety, the accumulated memories to negative outcomes induced by chronic mild stress may lead to anhedonia and low self-esteem in major depression, and weird memory is associated with schizophrenia. Memory deficits are also associated with neurodegenerative diseases, such as Alzheimer's disease and Parkinson's disease. The etiology and pathogenesis of memory deficits in neurological and psychiatric diseases remain unknown. As memory-relevant cognition and behaviors are based on the number and functional state of associative memory cells, it is hypothesized that these disease-associated memory deficits may be caused by pathological alternation in associative memory cells. Many genes and proteins in neurons are believed to result in these neurological and psychiatric diseases, and certain molecules accumulated in extracellular spaces are thought to deteriorate neuron encoding and synapse transmission. Associative memory cells are neuronal in nature prior to their recruitment for basic memory units; these intracellular and extracellular molecules that impair neurons may influence synapse innervations, synapse transmission efficiency, and spike-encoding capability at these associative memory cells and in turn take them to be abnormal. Pathological alternation in the synapse innervation, structural identity, and functional state of associative memory cells eventually results in memory deficits in these neurological and psychiatric diseases. Although this hypothesis needs to be tested experimentally, pathological alternations in neurons can be cited to associative memory cells. Here, the dysfunction of associative memory cells for memory deficits is discussed.

Keywords Memory deficit · Memory cell impairment · Anxiety · Depression · Schizophrenia and neurodegeneration

J.-H. Wang, *Associative Memory Cells: Basic Units of Memory Trace*,
https://doi.org/10.1007/978-981-13-9501-7_9

9.1 Introduction

Associative memory is a major brain function, essentially for cognitive activities, emotional reactions, and other integrative behaviors in life [1–4]. In addition to the impairment of cognition, emotion, and behaviors, the neurological and psychiatric diseases are more or less featured by memory deficit [5–12]. The suffering of neurological and psychiatric diseases follows certain obvious alternations in learning and memory in their initial stage. For instance, fear memory induced by acute severe stress is associated with anxiety, memories to negative outcomes induced by chronic mild stress lead to anhedonia in major depression, and unusual memories are associated with schizophrenia. Memory deficit is also shown in the early phase of neurodegenerative diseases, such as Alzheimer's disease and Parkinson's disease. It seems likely that these neurological and psychiatric diseases occur due to the impairment of associative memory, in which associative memory cells dysfunction and/or loss.

Psychiatric diseases, such as anxiety, major depression, bipolar disorder, and schizophrenia, appear to have the elevated memories to certain events. Anxiety based on strong fear memory is featured to be phobia to a specific event or panic to generalized events. Major depression based on strong memories accumulated from negative events is featured by hopelessness and low self-esteem. Schizophrenia based on weird memory or false memory possesses strong beliefs in hallucination and delusion. These strengthened memories may be caused by the recruitment of more memory cells unusually or the upregulation of their functional state due to an imbalance of excitation and inhibition. It has been found that GABAergic neurons are impaired or dysfunctional in these psychiatric diseases [13–22]. Associative memory formation is based on the recruitment of associative memory cells including glutamatergic and GABAergic neurons [23–28]. Though GABAergic associative memory cells are recruited, their function states are downregulated after associative memory formation [29–32]. In this regard, the more recruitment of glutamatergic associative memory cells and/or the further weakening of GABAergic associative memory cells that encode certain associated signals may be involved in these psychiatric diseases. This consequence may be reasons why smart individuals with super memory and cognition capability show more or less psychological symptoms, i.e., no clear borderline between memory and psychological problems.

Neurodegenerative diseases, such as Alzheimer's disease and Parkinson's disease, are characterized as the loss of memories gradually [5, 10, 33–39]. Pathological alternations of these diseases presumably result from loss of neurons and synapses by certain proteins, such as α-synuclein, tau-protein, and β-amyloid [34, 40–45]. These misfolding proteins may influence intracellular signaling pathways and/or building block molecules for neurons and synapses, which may cause the deterioration and/or loss of neurons and synapses. The associative memory cells possess more synapse innervations and relatively high encoding capability [3, 46], such that they are more active neurons in memory-relevant cognition and behaviors. Their elevated activities make them to be vulnerable to the metabolism and internal environment, likely GABAergic neurons [47]. Therefore, these associative memory ells are easily attacked by these misfolding proteins to have the impaired memory [48]. Based on the influence of these insoluble molecules on primary associative memory cells, secondary associative memory cells, or memory output cells, the dominant loss of memories may be characterized as episodic perceptual memory, semantic memory, memory-relevant cognitions, or entire memory output.

9.2 General Information About Associative Memory Cells

The integrative storage and the reciprocal retrieval of the associated signals are critical for the bidirectional alertness, prediction, cognition, and emotion. The multigrade integration and storage of intramodal and cross-modal exogenous and endogenous signals are based on mutual synapse innervations as well as convergence and divergence connections among primary and secondary associative memory cells [46]. With their integrative storage, one signal induces the recall of its associated signals or the other way around, as well as the signals learned from one modality can be automatically converted into another modality for their retrievals. Individuals are able to perform logical reasoning and associative thinking and to predict future events in forward and backward manners. Furthermore, associative memory cells in each of the coactivated brain regions encode the associated innate signal and newly learned signal as well as each of the associated signals is stored in multiple brain regions. These encoding properties allow associated signals for multiple uses and largely decrease the chance of memory loss [3, 26]. The storage of multiple signals in one of the associative memory cells can strengthen the efficiency of memory retrieval [23]. The storage of multiple signals in a cortical area and the recall of one signal triggered by multiple signals enable the individuals to strengthen their abilities in memory retrieval and well-organized cognitions.

Associative memory cells are able to encode the integrative storage of the associated signals in objects and environments, receive synapse innervations from coactivated brain areas, and innervate their downstream associative memory cells. Mutual synapse innervation among a group of primary associative memory cells and their convergent innervations onto secondary associative memory cells constitutes the basic neural circuits for the reciprocal retrieval of associated signals, the automatic conversion retrieval of associated signals, as well as the processes of cognition and emotional reactions. The coactivity-dependent positive cycle in the recruitment and upregulation of associative memory cells determines the memory specificity as well as the efficiency of learning and memorizing associated signals. The cue-induced or spontaneous activations and persistent activities of associative memory cells lead to the retrieval of associated signals, as well as the representation of associated signals by behaviors if they successfully activate memory output cells. Therefore, associative memory cells in neural circuits are basic units in engrams. The morphological basis for associative memory cells to encode multi-associated signals is their receptions of the innate inputs and new synapse innervations from coactivated brain areas. Based on the localizations in the cerebral brain, associative memory cells are classified into the primary group that integrates exogenous signals in sensory cortices and innervate on the secondary group in cognition- and emotion-relevant brain areas. Secondary associative memory cells integrate endogenous signals from the primary group during cognitive processes and emotional reaction. Based on the complication of integrating associated signals, associative memory cells are classified into grade one, grade two, grade three, and so on, whose activity-dependent upregulation works for the integrative storage and reciprocal retrieval of complicated associative signals. Associative memory cells plus their upregulation make them be more active, recruit more ready-recruited neurons to be associative memory cells, and cause coactivity-dependent positive cycle in the recruitment and refinement of associative memory cells for impressive memory in repetitive learnings. Primary associative memory cells are basic units for the storage of exogenous associated signals that can influence the specific contents of cognitions and emotional reactions. The consequences and processes of cognition and emotion recruit more associative memory cells for them to be stored. The recycle in memory and cognition allows individual capabilities, skills, and experiences to be strengthened. In addition, the functional state of associative memory cells is modulated by the arousal system from the midbrain raphe nuclei, cholinergic nuclei, locus coeruleus, and substantia nigra, which regulate the efficiency of learning and memory. Therefore, the impairment of associative memory cells in their morphology, function, and local environment declines memory retrievals and cognitions, which are usually associated with neurological diseases and psychiatric disorders [48].

It is widely accepted that the normal consciousness and well attention are important for memory formation [49–51], which is based on associative memory cells and their characters [46]. The arousal system maintains wakefulness. Associative memory cells under the wakefulness condition support the individuals to identify themselves and environment objects, which constitute consciousness. With the activations of ready-recruited neurons or associative memory cells that store topic-related information by attention cues, their coactivation and coactivity during associative learning will induce their mutual synapse innervations and make them to encode these associated signals. Furthermore, the consciousness supports the activation and activity of associative memory cells to execute activity-dependent positive cycle in their refinement and recruitment, so that more associative memory cells are recruited and impressive memory is formed in the mind. Thus, the lowered function of associative memory cells makes consciousness to be obscure [48].

In terms of memory maintenance versus extinction, the recruitment and refinement of associative memory cells are not significantly declined, but the activities of memory output cells in the motor cortex are decayed [30, 52]. The sustained presence of associative memory cells and the recruitment of more associative memory cells in repeated brain activities confer memorized signals to be retrieved in life span, in which the information can be retrieved as long as their innervations onto memory output cells successfully drive the latter to be functionally active. Memory retrievals show different patterns in spontaneous, cue-induced, and realistic object-triggered manner with age. Spontaneous retrievals often occur in child stage or brain excitation. Cue-induced retrievals usually occur in young and adult. Real object-induced retrievals occur in senior individuals. In addition, when the brain is highly excited in many regions, such as euphoria perception, extreme fear, and strong stimulations, more associative memory cells are recruited through their mutual innervations, and in turn impressive memories and spontaneous recalls to these experiences are generated in the life span [30, 52]. For instance, certain memories to extreme euphoria may lead to physiological and psychological dependence and even addiction, such as gambling and substance abuses. Some fear memory with extremely emotional reactions can lead to anxiety and even panic. It is hard to remove the newly formed synapse innervations and the recruited associative memory cells to relieve euphoria or fear memory. Alternative approaches are the avoidance of fear stimulations and the induction of pleasures to rebalance such two states and weaken fear memory, since lack of uses in neural circuits relevant to fear memory, especially from associative memory cells to memory output cells, may propel them to be functional silence [30, 52]. In the brains of individuals with a history of substance abuse or its addiction, primary and secondary associative memory cells relevant to these events are recruited under euphoria condition. Their large amount in the widespread areas, where the neurons contain receptor channels acted by abused drugs, will lead to much impressive memories to associated signals including environmental clues and drug-induced psychological cognitions, as well as facilitate potential relapses in their lifetime [3]. Similar to the electrical shock to activate the entire brain, addiction drugs can widely activate numerous neurons in the areas that are distributed with these chemical receptor channels, such as opioid receptors, monoamine receptors, and glutamate receptors. According to a rule of activity together, connection together, and strengthening together, mutual synapse innervations among associative memory cells and their strengthening are widely distributed in the brain in response to drug-relevant cues. There are no confident ways to erase synapse innervations and associative memory cells newly formed in drug abuse as well as to relieve memories to drug-relevant cues [53]. Two strategies can be used to weaken addiction-relevant memory in individuals. The avoidance of cues relevant to those abused drugs can reduce the activation of their associative memory cells. The establishment of alternative happiness recruits correspondent associative memory cells that innervate memory output cells. By competing the innervations from addiction memory cells, the rebalance of these two states strengthens memory output cells to serve these new associative memory cells and to be away from addiction-related associative memory cells [46].

9.3 Associative Memory Cells in Psychiatric Disorders

The deteriorated memory is often associated with psychological disorders, such as anxiety, depression, and schizophrenia [54, 55]. For instance, fear memory induced by acute severe stress is usually linked with anxiety [56, 57]. Stimulations to engram cells through optogenetic approach in the hippocampus activate fear memory recall and anxiety [58]. Memories to negative outcomes induced by chronic mild stresses are coupled with depression-like behavior [19, 59]. Stimulations to positive memory traces by optogenetic method in the amygdala suppress depression-like behaviors [60]. These data indicate that the formation of associative memory cells induced by different patterns of abnormal stimulations can lead to psychological disorders, i.e., acute severe stresses recruit associative memory cells relevant to fear memory and anxiety, and chronic mild stresses recruit associative memory cells related to negative memory and depression [48, 56, 59]. In this regard, the artificially manipulated reorganization of associative memory cells are expected to be used for memory reediting and memory rebalance to correct memory deficits in these psychological disorders.

9.3.1 Fear Memory and Anxiety

Fear memory is one type of emotion memory [61, 62]. Fear memories to specific experiences may lead to phobia to these events as well as to general events and may induce generalized anxiety. In other words, anxiety is an emotional reaction induced by sustained fear [56, 57, 63]. On the other hand, anxiety patients express odd fear memory that induces unrealistic and dangerous prediction to future events [64–66]. Emotional reactions in anxiety include unstable mood, elevated attention, negative interpretation, and social phobia under the conditions of the potential threat signs [67–69]. The involvement of the amygdala in anxiety is based on the evidences that neural imaging shows the amygdala hyperactivity in anxiety patients [70, 71], the stimulation to the amygdala induces anxiogenic somatic and autonomic responses [72], and the graft of GABAergic-rich neural tissue into the amygdala improves anxiety-like signs in animal models [73]. The abnormality of the amygdala may be major origin for anxiety pathogenesis [70, 74–82]. In addition, the hippocampus and the prefrontal cortex may be involved in the incidence of anxiety [83, 84]. Together these data with the fact that the onset of anxiety each time is induced by potential threat signs from the sensory system, anxiety results from abnormal neural circuits across many brain areas, such as the sensory cortices, prefrontal cortex, hippocampus, and amygdala, in which the amygdala initiates emotional reactions and other areas for fear memory.

In terms of cell pathology for fear memory and anxiety, the functional weakening of GABAergic neurons and synapses in the amygdala and prefrontal cortices is likely related to anxiety [15, 21, 85–87]. In addition, glutamate synapses are associated with anxiety [15, 88], since the antagonists of ionotropic glutamate receptor are potential reagents for anxiety therapy [89–91]. Moreover, the transmission of glutamatergic synapses is strengthened, and dendritic spine heads are enlarged in the prefrontal cortex under the regulation of protein kinase C [15], similar to the alternations in memory formation [92]. These data indicate that fear memory and anxiety are caused by imbalances between excitation and inhibition by glutamatergic upregulation and GABAergic downregulation toward the more excitation of neural networks. It has been found that the associative memory is featured by the upregulation of glutamatergic associative memory cells and the downregulation of GABAergic associative memory cells [24, 28, 30–32].

Based on these data, a hypothetical diagram about fear memory and anxiety is given below. Fear signals by intramodal or cross-modal sensory systems are transmitted to sensory cortices. The pairings of the associated signals coactivate sensory cortical neurons to induce their mutual synapse innervations and then recruit these interconnected neurons as primary associative memory cells. In the meantime, the coactivated neurons in the amygdala, prefrontal cortex, and hippocampus are mutually innervated with sensory cortices and themselves, so that secondary associative memory cells are recruited. These associative memory cells include glutamatergic and GABAergic neurons [23, 26–28] with the upregulation of glutamatergic associative memory cells and the downregulation of GABAergic associative memory cells [24, 28, 30–32]. With these associative memory cells, fear signals are memorized in these brain areas. When potential threatening cues appear, fear memory is retrieved from the prefrontal cortex, amygdala, and hippocampus. Emotional reactions in relevance to this fear memory are initiated by associative memory cells in the amygdala. Through their axonal innervations to memory output cells in the motor cortex and hypothalamic autonomic system and hormone-releasing neurons, symptoms in relevance to fear memory and anxiety are emerged in response to the fear signals [48].

In terms of psychological therapeutics, the reinforcement of positive memory is expected to overcome fear memory and to relieve anxiety [93]. The principle for psychological therapy is based on the views that fear memory is associated with anxiety and positive memory competes with fear memory to rebalance memory-guided cognitions and specific anxiety [94–98]. For instance, anxiety patients show odd fear memory that induces unrealistic and dangerous prediction to their future activity [64–66]. The primary goal in psychological therapy is that the use of positive memory competes with fear memory to a specific event and desensitizes fear mood to its scenarios [99–102]. Psychological desensitization by associating threat signals with the rewards or naïve signal has been achieved to reduce fear memory and relieve anxiety state [63]. The association of reward and threat signals recruits new associative memory cells whose axons compete dendritic arbors on memory output cells occupied by fear memory cells, such that fear memory and emotional reactions can be attenuated [48].

9.3.2 Depression and Memories to Negative Outcomes

Major depression is characterized as anhedonia, loss of interest, and low self-esteem, which may be originated from accumulated memories to negative outcomes [6, 103, 104]. In terms of pathogenesis, the chronic stress to genetically susceptible people leads to the dysfunctions of monoamine, brain-derived neurotrophic factor and hypothalamus-pituitary-adrenal axis [105–109], which induce neuron atrophy in brain reward circuits, such as the prefrontal cortex, nucleus accumbens, and amygdala, in depressed subjects [54, 110–115]. Since physiological coordination between excitatory and inhibitory neurons is critical for the neuron encoding to manage well-organized cognitions [116, 117], cell-specific pathology in major depression remains to be studied [19, 118, 119]. It has been found that the decreases in density and functions of GABAergic neurons and synapses in the prefrontal cortices and the nucleus accumbens are associated to major depression [17–20, 22, 120–129]. These changes may be caused by epigenetic-regulated expression of genes and proteins, in which some molecules are relevant to memory formation [56, 59, 130, 131]. In addition, the changes at glutamatergic neurons and synapses are similar to those in learning and memory, such as the increased spine size and density in the prefrontal cortex [19]. These studies indicate that memories to negative outcomes and depression are caused by more excitation in the prefrontal cortex and nucleus accumbens by GABAergic downregulation. In addition, associative memory is accompanied by the upregulation of glutamatergic associative memory cells and the downregulation of GABAergic associative memory cells [24, 28, 30–32].

Based on these data, a hypothetical diagram about negative memory and major depression is given below. In the presence of chronic mild stress in natural and social environments, these associated stressful signals through intramodal or cross-modal sensory systems are transmitted to sensory cortices. The integrations of various stressful signals coactivate sensory cortical neurons to induce their mutual synapse innervation and recruit the interconnected neurons to be primary associative memory cells. In the meantime, negative outcomes due to inability or unsuccess to treat these stressful situations will lead to negative cognitions and emotions, in which the coactivated neurons in the prefrontal cortex and nucleus accumbens are mutually innervated with sensory cortices and themselves, such that the secondary associative memory cells are recruited. Memories to negative outcomes with long-term accumulation will induce depressed mood [6, 103, 104], so that emotional reactions are negative, such as anhedonia, interest loss, and low self-esteem [132–139]. Once depressive individuals have lack of motivation to interact with their environments, loss of uses in their brains and bodies reduces capability in life, such that they may secondarily feel fatigue and difficulty to pay attention. In terms of cell types, these associative memory cells include glutamatergic and GABAergic neurons [27, 28] with the upregulation of glutamatergic associative memory cells and the downregulation of GABAergic associative memory cells [24, 28, 30–32]. These associative memory cells memorize negative outcomes, feelings, and cognitions that lead to the depressed mood and negative thoughts through their axons to memory output cells in the motor cortex [48].

In addition to pharmacological reagents, psychological therapy can be used to reinforce positive memory in order to overcome negative memory and relieve depressed mood [140, 141]. The principle for psychological therapy is based on the views that negative memory is associated with depression and positive memory competes with memories to negative outcomes to rebalance memory-guided cognitions and relieve depressed mood. This principle is supported by experiment study that the activation of positive memory suppresses depression-like behaviors [60]. As GABAergic associative memory cells are downregulated in associative memory [28, 30–32], a strengthening of GABAergic function during the establishment of positive memory may be helpful.

9.3.3 *Weird Memory and Schizophrenia*

Schizophrenia is featured by positive and negative signs in sensations, perception, thoughts, and other cognitive behaviors with loss of touching reality [142, 143]. Positive symptoms are psychotic behaviors not generally seen in healthy population. Patients with positive symptoms appear to lose touch in some aspects of reality, such as hallucination, delusion, unusual thinking, disordered speech, and disorganized routine behaviors with agitated movement. Negative symptoms are featured by disruptions to normal emotions and behaviors, such as the flat expression of emotions through countenance or voice tone, lack of pleasure feeling, difficulty beginning and sustaining activities, and social withdraw. Other cognitive symptoms in schizophrenia patients include changes in memories or other thoughts, such as poor executive function (inability to understand information and use it to make decisions), trouble paying attention, and problem in working memory [144–146]. These cognitive symptoms, such as false beliefs, seem based on weird memories to their experiences and unrealistic thinking, as well as the excessive recruitments of neural systems may subserve disorganized logical reasoning in schizophrenia [11]. Brain areas involved in schizophrenia include the prefrontal cortex, cingulate cortex, hippocampus, and association cortices in the reduced cortical thickness [147–150]. Cellular pathological changes include the dysfunctions of dopaminergic neurons and synapse transmission [151–153], the impairment and loss of GABAergic neurons [14, 154–157], as well as the dysfunction of the cholinergic system [158, 159].

The proper coactivation of actively ready-recruited neurons makes them be recruited as associative memory cells [26, 46]. The activity-dependent upregulation of associative memory cells facilitates the integrative storage and reciprocal retrieval of associated signals [28, 30, 32, 46, 160]. These neuronal events constitute coactivity-dependent positive cycle in the recruitment and refinement of associative memory cells, so that more associative memory cells are recruited. However, the further upregulation of associative memory cells, such as the dysfunction of GABAergic neurons in schizophrenia [13, 161], allows associative memory cells to be overly and widely activated. The overly upregulation of associative memory cells in the sensory cortices will lead to hallucination. The overly upregulation of associative memory cells in cognition- and emotion-relevant brain regions leads to delusion. The overly upregulation of memory output cells in motor cortex may induce disordered behaviors with agitated movements [46]. Thus, the medication and psychological therapy for schizophrenia patients have been suggested with emphasis or focus on strengthening GABAergic neurons and synapses [162–165].

9.4 Associative Memory Cells in Neurodegenerative Diseases

The efficiency of learning and memory decays in age-relevant manner [166, 167]. There is a bell-shaped pattern in the efficiency of associative learning and memory during the postnatal development [26]. Cell mechanisms underlying this postnatal development presumably include the following. Synaptic potentiation matures during postnatal development [92, 168]. Neuronal excitability in cortical neurons is upregulated up to the plateau level at postnatal weeks 3–4 [169]. These changes match dynamical changes in associative memory [26]. Neuronal plasticity and associative memory cell recruitment in postnatal development constitute the coactivity-dependent positive cycle in the recruitment and refinement of associative memory cells, such that more associative memory cells are recruited to strengthen the efficiency of learning and memory [32, 160], which may constitute neuronal substrates for eidetic memory in a young age [170–172]. In the aged mammalians, the accumulations of insoluble β-amyloid and phosphorylated tau-proteins in the brain may influence axon prolongations and synapse formations [23, 25] to suppress the recruitment and upregulation of the associative memory cells, to silence active associative memory cells and to impair the recruited associative memory cells for memory deficit [3, 40, 43, 44, 46, 173]. On the other hand, the activity of these associative memory cells can strengthen the coactivity-dependent positive cycle in the recruitment and refinement of associative memory cells, which in turn prevents the conversion of the soluble β-amyloid into its insoluble form and promotes the clearance of β-amyloids by associative memory astrocytes [23, 26, 46]. A current report supports this point in that light and sound stimulations coordinately attenuate the accumulation of β-amyloid [174].

In age-relevant neurodegeneration, such as Alzheimer's disease, insoluble β-amyloid may be accumulated and deposited in various brain areas [41]. For instance, the optogenetic stimulation to engram cells, which have a lack of increased synaptic strength and dendritic spines under the amnesia induced by protein synthesis inhibition, leads to memory retrievals [175]. The optogenetic stimulation to hippocampal engram cells induces memory retrievals in mice, though they show the amnesia under the condition of using natural cues in the transgenic mouse model of early Alzheimer's disease [38]. In addition to the indication about the wide distribution of memory traces for signal storage and retrieval, these results suggest that the locations involved in natural memory retrieval are dominantly impaired by the deposition of β-amyloid, rather than memory trace cells, as well as those regions for memory retrievals are not specific for a given memory. In this regard, synapse connections from associative memory cells to memory output cells are expected to be strengthened in the early stage of Alzheimer's disease [30, 46, 52].

In terms of pathology for episodic memory, patients who have suffered from an episode of transient global amnesia manifest memory deficits, especially problems in the retrievals of episodic memories and recent semantic information. These deficits may not result from the deterioration of memory but rather from difficulties in accessing memories [176, 177]. There may be a selective damage to the limbic-prefrontal episodic memory system in some people with autism. Autistic deficits include episodic or self-conscious memory in personally experienced activities [178–180]. Amnesia is often given to the patients with deficits in episodic memory, especially Alzheimer's disease, that tends to damage the hippocampus before other brain areas with the dominant impairment of episodic memory initially. This indicates that patients with Alzheimer's disease are often classified as amnesiacs

[181, 182]. However, the patient with an amnesic syndrome at 10 years old shows the impaired episodic memory for both verbal and visual materials. In addition, the deficit has been seen in a wide variety of semantic memory tasks including reading vocabulary and verbal fluency tests, semantic classification/lexical decision tasks as well as verbal intelligence tests. On the other hand, this patient shows normal learning and retention of two procedural tasks. This study is inconsistent with the view that the amnesic syndrome represents a selective defect of episodic memory that leaves semantic memory relatively unaffected [183].

Alzheimer's disease is thought of as a semantic memory disorder featured by category-specific impairments [36, 39], which may lead to errors in describing and naming objects. Semantic dementia is another disorder associated with semantic memory. Semantic dementia is a language disorder featured by the deteriorations in understanding and recognizing words. Its impairments include difficulty to generate familiar words and to name objects and visual recognition. The temporal lobe may be responsible for the category-specific impairment of semantic memory disorders. Modality-specific impairments are also included in disorders of semantic memory [184]. In terms of semantic memory in reference to modalities, the different components represent information from different sensorimotor systems. Modality-specific impairments are divided into separate subsystems on the basis of input modality. These input modalities include visual, auditory, and tactile input. Modality-specific impairments can also be divided into subsystems based on the type of information, such as visual versus verbal and perceptual versus functional signals [184]. Modality specificity accounts for category-specific impairments in semantic memory disorders. The damage to visual semantics primarily impairs knowledge of living things. The damage to functional semantics primarily impairs knowledge of nonliving things. Correlations between these types of dementia and the dysfunction of associative memory cells remain to be figured out. However, associative memory cells with advanced mutual synapse innervations as well as secondary associative memory cells are more likely impaired for these types of semantic dementia including Alzheimer's disease [48].

References

1. Kandel ER, Pittenger C. The past, the future and the biology of memory storage. Philos Trans R Soc Lond Ser B Biol Sci. 1999;354(1392):2027–52.
2. Wang D, et al. Neurons in the barrel cortex turn into processing whisker and odor signals: a cellular mechanism for the storage and retrieval of associative signals. Front Cell Neurosci. 2015;9(320):1–12.
3. Wang JH, Cui S. Associative memory cells: formation, function and perspective. F1000Res. 2017;6:283.
4. Wasserman EA, Miller RR. What's elementary about associative learning? Annu Rev Psychol. 1997;48:573–607.
5. Beard RL. Trust and memory: organizational strategies, institutional conditions and trust negotiations in specialty clinics for Alzheimer's disease. Cult Med Psychiatry. 2008;32(1):11–30.

6. Coutellier L, Usdin TB. Enhanced long-term fear memory and increased anxiety and depression-like behavior after exposure to an aversive event in mice lacking TIP39 signaling. Behav Brain Res. 2011;222(1):265–9.
7. Desmedt A, Marighetto A, Piazza PV. Abnormal fear memory as a model for posttraumatic stress disorder. Biol Psychiatry. 2015;78(5):290–7.
8. Orsini CA, Maren S. Neural and cellular mechanisms of fear and extinction memory formation. Neurosci Biobehav Rev. 2012;36(7):1773–802.
9. Parsons RG, Ressler KJ. Implications of memory modulation for post-traumatic stress and fear disorders. Nat Neurosci. 2013;16(2):146–53.
10. Ramanan S, Kumar D. Prospective memory in Parkinson's disease: a meta-analysis. J Int Neuropsychol Soc. 2013;19(10):1109–18.
11. Ramsey NF, et al. Excessive recruitment of neural systems subserving logical reasoning in schizophrenia. Brain. 2002;125(Pt 8):1793–807.
12. Whitton AE, Treadway MT, Pizzagalli DA. Reward processing dysfunction in major depression, bipolar disorder and schizophrenia. Curr Opin Psychiatry. 2015;28(1):7–12.
13. Benes FM, Berretta S. GABAergic interneurons: implications for understanding schizophrenia and bipolar disorder. Neuropsychopharmacology. 2001;25(1):1–27.
14. Cotter D, et al. The density and spatial distribution of GABAergic neurons, labelled using calcium binding proteins, in the anterior cingulate cortex in major depressive disorder, bipolar disorder, and schizophrenia. Biol Psychiatry. 2002;51(5):377–86.
15. Liu B, Feng J, Wang J-H. Protein kinase C is essential for kainate-induced anxiety-related behavior and glutamatergic synapse upregulation in prelimbic cortex. CNS Neurosci Ther. 2014;20:982–90.
16. Lloyd KG, et al. The gabaergic hypothesis of depression. Prog Neuro-Psychopharmacol Biol Psychiatry. 1989;13(3–4):341–51.
17. Ma K, et al. Impaired GABA synthesis, uptake and release are associated with depression-like behaviors induced by chronic mild stress. Transl Psychiatry. 2016;6(e910):1–10.
18. Maciag D, et al. Reduced density of calbindin immunoreactive GABAergic neurons in the occipital cortex in major depression: relevance to neuroimaging studies. Biol Psychiatry. 2010;67(5):465–70.
19. Xu A, Cui S, Wang J. Incoordination among subcellular compartments is associated to depression-like behavior induced by chronic mild stress. Int J Neuropsychopharmacol. 2015;19(5):pyv122.
20. Rajkowska G, et al. GABAergic neurons immunoreactive for calcium binding proteins are reduced in the prefrontal cortex in major depression. Neuropsychopharmacology. 2007;32(2):471–82.
21. Zhang F, et al. mGluR1,5 activation improves network asynchrony and GABAergic synapse attenuation in the amygdala: implication for anxiety-like behavior in DBA/2 mice. Mol Brain. 2012;5(1):20.
22. Zhu Z, et al. GABAergic neurons in nucleus accumbens are correlated to resilience and vulnerability to chronic stress for major depression. Oncotarget. 2017;8(22):35933–45.
23. Feng J, et al. Barrel cortical neuron integrates triple associated signals for their memory through receiving epigenetic-mediated new synapse innervations. Cereb Cortex. 2017;27(12):5858–71.
24. Gao Z, et al. Associations of unilateral whisker and olfactory signals induce synapse formation and memory cell recruitment in bilateral barrel cortices: cellular mechanism for unilateral training toward bilateral memory. Front Cell Neurosci. 2016;10(285):1–16.
25. Lei Z, et al. Synapse innervation and associative memory cell are recruited for integrative storage of whisker and odor signals in the barrel cortex through miRNA-mediated processes. Front Cell Neurosci. 2017;11(316):1–11.
26. Wang D, et al. Neurons in the barrel cortex turn into processing whisker and odor signals: a cellular mechanism for the storage and retrieval of associative signals. Front Cell Neurosci. 2015;9:320.

27. Wang J-H, et al. Both glutamatergic and gabaergic neurons are recruited to be associative memory cells. Biophys J. 2016;110(3):supplement 481a.
28. Yan F, et al. Coordinated plasticity between barrel cortical glutamatergic and GABAergic neurons during associative memory. Neural Plast. 2016;2016(ID5648390):1–20.
29. Feng J, et al. Cell-specific plasticity associated with integrative memory of triple sensory signals in the barrel cortex. Oncotarget. 2018;9(57):30962–78.
30. Guo R, et al. Associative memory extinction is accompanied by decayed plasticity at motor cortical neurons and persistent plasticity at sensory cortical neurons. Front Cell Neurosci. 2017;11(168):1–12.
31. Liu Y, et al. Piriform cortical glutamatergic and GABAergic neurons express coordinated plasticity for whisker-induced odor recall. Oncotarget. 2017;8(56):95719–40.
32. Zhao X, et al. Coordinated plasticity among glutamatergic and GABAergic neurons and synapses in the barrel cortex is correlated to learning efficiency. Front Cell Neurosci. 2017;11(221):1–12.
33. Das T, Hwang JJ, Poston KL. Episodic recognition memory and the hippocampus in Parkinson's disease: a review. Cortex. 2019;113:191–209.
34. Kahle PJ, et al. Physiology and pathophysiology of alpha-synuclein. Cell culture and transgenic animal models based on a Parkinson's disease-associated protein. Ann N Y Acad Sci. 2000;920:33–41.
35. Kulisevsky J. Role of dopamine in learning and memory: implications for the treatment of cognitive dysfunction in patients with Parkinson's disease. Drugs Aging. 2000;16(5):365–79.
36. Lambon Ralph MA, et al. Semantic memory is impaired in both dementia with Lewy bodies and dementia of Alzheimer's type: a comparative neuropsychological study and literature review. J Neurol Neurosurg Psychiatry. 2001;70(2):149–56.
37. Leentjens AF. The role of dopamine agonists in the treatment of depression in patients with Parkinson's disease: a systematic review. Drugs. 2011;71(3):273–86.
38. Roy DS, et al. Memory retrieval by activating engram cells in mouse models of early Alzheimer's disease. Nature. 2016;531(7595):508–12.
39. Tippett LJ, Grossman M, Farah MJ. The semantic memory impairment of Alzheimer's disease: category-specific? Cortex. 1996;32(1):143–53.
40. Fu H, et al. Tau pathology induces excitatory neuron loss, grid cell dysfunction, and spatial memory deficits reminiscent of early Alzheimer's disease. Neuron. 2017;93(3):533–541 e5.
41. Giacobini E, Becker RE. One hundred years after discovery of Alzheimer's disease. A turning point for therapy? J Alzheimers Dis. 2007;12:37–52.
42. Jakobsen LD, Jensen PH. Parkinson's disease: alpha-synuclein and parkin in protein aggregation and the reversal of unfolded protein stress. Methods Mol Biol. 2003;232:57–66.
43. Laferla FM, Green KN, Oddo S. Intracellular amyloid-b in Alzheimer's disease. Nat Rev Neurosci. 2007;8:499–509.
44. Mitsuyama F, et al. Amyloid beta: a putative intra-spinal microtubule-depolymerizer to induce synapse-loss or dendritic spine shortening in Alzheimer's disease. Ital J Anat Embryol. 2009;114(2–3):109–20.
45. Sisodia SS, George-Hyslop PH. r-secretase, notch, Ab and Alzheimer's disease: where do the presenilins fit in? Nat Rev Neurosci. 2002;3:281–90.
46. Wang JH, Cui S. Associative memory cells and their working principle in the brain. F1000Res. 2018;7:108.
47. Wang J-H. Short-term cerebral ischemia causes the dysfunction of interneurons and more excitation of pyramidal neurons. Brain Res Bull. 2003;60(1–2):53–8.
48. Wang J-H. Searching basic units of memory traces: associative memory cells. F1000Res. 2019;8(457):1–28.
49. Henke K. A model for memory systems based on processing modes rather than consciousness. Nat Rev Neurosci. 2010;11(7):523–32.
50. Reder LM, Park H, Kieffaber PD. Memory systems do not divide on consciousness: reinterpreting memory in terms of activation and binding. Psychol Bull. 2009;135(1):23–49.

51. Raffone A, Srinivasan N, van Leeuwen C. The interplay of attention and consciousness in visual search, attentional blink and working memory consolidation. Philos Trans R Soc Lond Ser B Biol Sci. 2014;369(1641):20130215.
52. Wang J-H, Guo R, Wei Z. Associative memory extinction is accompanied by decays of associative memory cells and their plasticity at motor cortex but not sensory cortex. Soc Neurosci. 2017;81(09):10385.
53. Nestler EJ. Cellular basis of memory for addiction. Dialogues Clin Neurosci. 2013;15(4):431–43.
54. Elizalde N, et al. Long-lasting behavioral effects and recognition memory deficit induced by chronic mild stress in mice: effect of antidepressant treatment. Psychopharmacology. 2008;199(1):1–14.
55. Penfield W, Milner B. Memory deficit produced by bilateral lesions in the hippocampal zone. AMA Arch Neurol Psychiatry. 1958;79(5):475–97.
56. Sun X, et al. microRNA and mRNA profiles in ventral tegmental area relevant to stress-induced depression and resilience. Prog Neuropsychopharmacol Biol Psychiatry. 2018;86:150–65.
57. Wang J-H, Lu W. Molecular profiles in the brain are involved in fear memory induced by physical and psychological stress. Soc Neurosci. 2018;425.19(425):III61.
58. Liu X, et al. Optogenetic stimulation of a hippocampal engram activates fear memory recall. Nature. 2012;484(7394):381–5.
59. Si Y, et al. microRNA and mRNA profiles in nucleus accumbens underlying depression versus resilience in response to chronic stress. Am J Med Genet B Neuropsychiatr Genet. 2018;177(6):563–79.
60. Ramirez S, et al. Activating positive memory engrams suppresses depression-like behaviour. Nature. 2015;522(7556):335–9.
61. Diamond DM, et al. The temporal dynamics model of emotional memory processing: a synthesis on the neurobiological basis of stress-induced amnesia, flashbulb and traumatic memories, and the Yerkes-Dodson law. Neural Plast. 2007;2007:60803.
62. LaBar KS, Cabeza R. Cognitive neuroscience of emotional memory. Nat Rev Neurosci. 2006;7(1):54–64.
63. Lei Z, Liu B, Wang J-H. Reward memory relieves anxiety-related behavior through synaptic strengthening and protein kinase C in dentate gyrus. Hippocampus. 2016;26(4):502–16.
64. Coles ME, Heimberg RG. Memory biases in the anxiety disorders: current status. Clin Psychol Rev. 2002;22(4):587–627.
65. Becker ES, et al. Explicit memory in anxiety disorders. J Abnorm Psychol. 1999;108(1):153–63.
66. McNally RJ. Memory and anxiety disorders. Philos Trans R Soc Lond Ser B Biol Sci. 1997;352(1362):1755–9.
67. Bishop SJ. Neurocognitive mechanisms of anxiety: an integrative account. Trends Cogn Sci. 2007;11(7):307–16.
68. Cameron OG. The differential diagnosis of anxiety. Psychiatric and medical disorders. Psychiatr Clin N Am. 1985;8(1):3–23.
69. Rickeles K, Rynn M. Overview and clinical presentation of generalized anxiety disorder. Psychiatr Clin N Am. 2001;24(1):1–17.
70. Rauch SL, Shin LM, Wright CI. Neuroimaging studies of amygdala function in anxiety disorders. Ann N Y Acad Sci. 2003;985:389–410.
71. Stein MB, Stein DJ. Social anxiety disorders. Lancet. 2008;371(9618):1115–25.
72. Davis M. The role of the amygdala in fear and anxiety. Annu Rev Neurosci. 1992;15:353–75.
73. Cunnigham MG, et al. Amygdala GABAergic-rich neural grafts attenuate anxiety-like behavior in rats. Behav Brain Res. 2009;205(1):146–53.
74. Anand A, Shekhar A. Brain imaging studies in mood and anxiety disorders: special emphasis on the amygdala. Ann N Y Acad Sci. 2003;985:370–88.
75. Bremner JD. Brain imaging in anxiety disorders. Expert Rev Neurother. 2004;4(2):275–84.

76. Davidson RJ. Anxiety and affective style: one of prefrontal cortex and amygdala. Biol Psychiatry. 2002;51(1):68–80.
77. Davis M, Rainnie D, Cassell M. Neurotransmission in the rat amygdala related to fear and anxiety. Trends Neurosci. 1994;17(5):208–14.
78. Garrett A, Chang K. The role of the amygdala in bipolar disorder development. Dev Psychopathol. 2008;20(4):1285–96.
79. LeDoux J. Emotion circuits in the brain. Annu Rev Neurosci. 2000;23:155–84.
80. Pape HC, Pare D. Plastic synaptic networks of the amygdala for the acquisition, expression and extinction of conditioned fear. Physiol Rev. 2010;90(2):419–63.
81. Neugebauer V, et al. The amygdala and persistent pain. Neuroscientist. 2004;10(3):221–34.
82. Roozendaal B, McEwen BS, Chattarji S. Stress, memory and amygdala. Nat Rev Neurosci. 2009;10(6):423–33.
83. Girardeau G, Inema I, Buzsaki G. Reactivations of emotional memory in the hippocampus-amygdala system during sleep. Nat Neurosci. 2017;20(11):1634–42.
84. Hubner C, et al. Ex vivo dissection of optogenetically activated mPFC and hippocampal inputs to neurons in the basolateral amygdala: implications for fear and emotional memory. Front Behav Neurosci. 2014;8:64.
85. Amano T, Unal CT, Pare D. Synaptic correlates of fear extinction in the amygdala. Nat Neurosci. 2010;13(4):489–95.
86. DuBios DW, et al. Distinct functional characteristics of the lateral/basolateral amygdala GABAergic system in C57BL/6J and DBA/2J mice. J Pharmacol Exp Ther. 2006;318(2):629–40.
87. Ehrlich I, et al. Amygdala inhibitory circuits and the control of fear memory. Neuron. 2009;62:757–71.
88. Bergink V, van Megen HJ, Westenberg HG. Glutamate and anxiety. Eur Neuropsychopharmacol. 2004;14(3):175–83.
89. Chaki S, Okubo T, Sekiguchi Y. Non-monoamine-based approach for the treatment of depression and anxiety disorders. Recent Pat CNS Drug Discov. 2006;1(1):1–27.
90. Cortese BM, Phan KL. The role of glutamate in anxiety and related disorders. CNS Spectr. 2005;10(10):820–30.
91. Simon AB, Gorman JM. Advances in the treatment of anxiety: targeting glutamate. NeuroRx. 2006;3(1):57–68.
92. Bliss TVP, Lynch MA. Long-term potentiation of synaptic transmission in the hippocampus: properties and mechanisms. In: Landfield PW, Deadwyler SA, editors. Long-term potentiation: from biophysics to behavior. New York: Alan R. Liss; 1988. p. 3–72.
93. Tryon WW, McKay D. Memory modification as an outcome variable in anxiety disorder treatment. J Anxiety Disord. 2009;23(4):546–56.
94. Amini F, et al. Affect, attachment, memory: contributions toward psychobiologic integration. Psychiatry. 1996;59(3):213–39.
95. Blaney PH. Affect and memory: a review. Psychol Bull. 1986;99(2):229–46.
96. Dere E, Pause BM, Pietrowsky R. Emotion and episodic memory in neuropsychiatric disorders. Behav Brain Res. 2010;215(2):162–71.
97. Derouesne C. Memory and affect. Rev Neurol (Paris). 2000;156(8–9):732–7.
98. Gillihan SJ, Kessler J, Farah MJ. Memories affect mood: evidence from covert experimental assignment to positive, neutral, and negative memory recall. Acta Psychol. 2007;125(2):144–54.
99. Aldao A, et al. Adaptive and maladaptive emotion regulation strategies: interactive effects during CBT for social anxiety disorder. J Anxiety Disord. 2014;28(4):382–9.
100. Brewin CR. Understanding cognitive behaviour therapy: a retrieval competition account. Behav Res Ther. 2006;44(6):765–84.
101. Cuijpers P, et al. Psychological treatment of generalized anxiety disorder: a meta-analysis. Clin Psychol Rev. 2014;34(2):130–40.

102. Rusting CL, DeHart T. Retrieving positive memories to regulate negative mood: consequences for mood-congruent memory. J Pers Soc Psychol. 2000;78(4):737–52.
103. Hamilton JP, Gotlib IH. Neural substrates of increased memory sensitivity for negative stimuli in major depression. Biol Psychiatry. 2008;63(12):1155–62.
104. Paolino RM, Levy HM. Amnesia produced by spreading depression and ECS: evidence for time-dependent memory trace localization. Science. 1971;172(3984):746–9.
105. Berton O, Hahn CG, Thase ME. Are we getting closer to valid translational models for major depression? Science. 2012;338(6103):75–9.
106. Brunoni AR, Lopes M, Fregni F. A systematic review and meta-analysis of clinical studies on major depression and BDNF levels: implications for the role of neuroplasticity in depression. Int J Neuropsychopharmacol. 2008;11(8):1169–80.
107. Elhwuegi AS. Central monoamines and their role in major depression. Prog Neuropsychopharmacol Biol Psychiatry. 2004;28(3):435–51.
108. Strekalova T, et al. Update in the methodology of the chronic stress paradigm: internal control matters. Behav Brain Funct. 2011;7:9.
109. Rohleder N, Wolf JM, Wolf OT. Glucocorticoid sensitivity of cognitive and inflammatory processes in depression and posttraumatic stress disorder. Neurosci Biobehav Rev. 2010;35(1):104–14.
110. Banasr M, Dwyer JM, Duman RS. Cell atrophy and loss in depression: reversal by antidepressant treatment. Curr Opin Cell Biol. 2011;23(6):730–7.
111. Bennett P, et al. Psychological factors associated with emotional responses to receiving genetic risk information. J Genet Couns. 2008;17(3):234–41.
112. Duman CH. Models of depression. Vitam Horm. 2010;82:1–21.
113. Lin LC, Sibille E. Reduced brain somatostatin in mood disorders: a common pathophysiological substrate and drug target? Front Pharmacol. 2013;4:110.
114. Pittenger C, Duman RS. Stress, depression, and neuroplasticity: a convergence of mechanisms. Neuropsychopharmacology. 2008;33(1):88–109.
115. Sandi C, Haller J. Stress and the social brain: behavioural effects and neurobiological mechanisms. Nat Rev Neurosci. 2015;16(5):290–304.
116. Ascoli GA, et al. Petilla terminology: nomenclature of features of GABAergic interneurons of the cerebral cortex. Nat Rev Neurosci. 2008;9(7):557–68.
117. Buzsaki G, et al. Interneuron diversity series: circuit complexity and axon wiring economy of cortical interneurons. Trends Neurosci. 2004;27(4):186–93.
118. Duman RS, Aghajanian GK. Synaptic dysfunction in depression: potential therapeutic targets. Science. 2012;338(6103):68–72.
119. Thompson SM, et al. An excitatory synapse hypothesis of depression. Trends Neurosci. 2015;38(5):279–94.
120. Bajbouj M, et al. Evidence for impaired cortical inhibition in patients with unipolar major depression. Biol Psychiatry. 2006;59(5):395–400.
121. Croarkin PE, Levinson AJ, Daskalakis ZJ. Evidence for GABAergic inhibitory deficits in major depressive disorder. Neurosci Biobehav Rev. 2011;35(3):818–25.
122. Hasler G, et al. Reduced prefrontal glutamate/glutamine and gamma-aminobutyric acid levels in major depression determined using proton magnetic resonance spectroscopy. Arch Gen Psychiatry. 2007;64(2):193–200.
123. Hettema JM, et al. Association between glutamic acid decarboxylase genes and anxiety disorders, major depression, and neuroticism. Mol Psychiatry. 2006;11(8):752–62.
124. Karolewicz B, et al. Reduced level of glutamic acid decarboxylase-67 kDa in the prefrontal cortex in major depression. Int J Neuropsychopharmacol. 2010;13(4):411–20.
125. Khundakar AA, et al. Cellular pathology within the anterior cingulate cortex of patients with late-life depression: a morphometric study. Psychiatry Res. 2011;194(2):184–9.
126. Levinson AJ, et al. Evidence of cortical inhibitory deficits in major depressive disorder. Biol Psychiatry. 2010;67(5):458–64.

127. Plante DT, et al. Reduced gamma-aminobutyric acid in occipital and anterior cingulate cortices in primary insomnia: a link to major depressive disorder? Neuropsychopharmacology. 2012;37(6):1548–57.
128. Price RB, et al. Amino acid neurotransmitters assessed by proton magnetic resonance spectroscopy: relationship to treatment resistance in major depressive disorder. Biol Psychiatry. 2009;65(9):792–800.
129. Veeraiah P, et al. Dysfunctional glutamatergic and gamma-aminobutyric acidergic activities in prefrontal cortex of mice in social defeat model of depression. Biol Psychiatry. 2014;76(3):231–8.
130. Ma K, et al. Molecular mechanism for stress-induced depression assessed by sequencing miRNA and mRNA in medial prefrontal cortex. PLoS One. 2016;11(7):e0159093.
131. Shen M, Song Z, Wang JH. microRNA and mRNA profiles in the amygdala are associated with stress-induced depression and resilience in juvenile mice. Psychopharmacology (Berl). 2019;236(7):2119–42.
132. Camp NJ, Cannon-Albright LA. Dissecting the genetic etiology of major depressive disorder using linkage analysis. Trends Mol Med. 2005;11(3):138–44.
133. Hamilton JP, Chen MC, Gotlib IH. Neural systems approaches to understanding major depressive disorder: an intrinsic functional organization perspective. Neurobiol Dis. 2013;52:4–11.
134. Jabbi M, et al. Investigating the molecular basis of major depressive disorder etiology: a functional convergent genetic approach. Ann N Y Acad Sci. 2008;1148:42–56.
135. Keers R, Uher R. Gene-environment interaction in major depression and antidepressant treatment response. Curr Psychiatry Rep. 2012;14(2):129–37.
136. Klengel T, Binder EB. Gene-environment interactions in major depressive disorder. Can J Psychiatr. 2013;58(2):76–83.
137. Lohoff FW. Overview of the genetics of major depressive disorder. Curr Psychiatry Rep. 2010;12(6):539–46.
138. Moylan S, et al. The neuroprogressive nature of major depressive disorder: pathways to disease evolution and resistance, and therapeutic implications. Mol Psychiatry. 2013;18(5):595–606.
139. Wilde A, et al. Implications of the use of genetic tests in psychiatry, with a focus on major depressive disorder: a review. Depress Anxiety. 2013;30(3):267–75.
140. Cuijpers P, et al. Efficacy of cognitive-behavioural therapy and other psychological treatments for adult depression: meta-analytic study of publication bias. Br J Psychiatry. 2010;196(3):173–8.
141. Fava GA, et al. Well-being therapy in depression: new insights into the role of psychological well-being in the clinical process. Depress Anxiety. 2017;34(9):801–8.
142. Lewis DA. Inhibitory neurons in human cortical circuits: substrate for cognitive dysfunction in schizophrenia. Curr Opin Neurobiol. 2014;26:22–6.
143. Royston MC, Roberts GW. Schizophrenia. When neurons go astray. Curr Biol. 1995;5(4):342–4.
144. Callicott JH, et al. Functional magnetic resonance imaging brain mapping in psychiatry: methodological issues illustrated in a study of working memory in schizophrenia. Neuropsychopharmacology. 1998;18(3):186–96.
145. Gordon R, Silverstein ML, Harrow M. Associative thinking in schizophrenia: a contextualist approach. J Clin Psychol. 1982;38(4):684–96.
146. Tek C, et al. Visual perceptual and working memory impairments in schizophrenia. Arch Gen Psychiatry. 2002;59(2):146–53.
147. Bachneff SA. Positron emission tomography and magnetic resonance imaging: a review and a local circuit neurons hypo(dys)function hypothesis of schizophrenia. Biol Psychiatry. 1991;30(9):857–86.
148. Lisman JE, et al. A thalamo-hippocampal-ventral tegmental area loop may produce the positive feedback that underlies the psychotic break in schizophrenia. Biol Psychiatry. 2010;68(1):17–24.

149. Suh J, et al. Impaired hippocampal ripple-associated replay in a mouse model of schizophrenia. Neuron. 2013;80(2):484–93.
150. Weinberger DR, et al. Prefrontal neurons and the genetics of schizophrenia. Biol Psychiatry. 2001;50(11):825–44.
151. Finlay JM. Mesoprefrontal dopamine neurons and schizophrenia: role of developmental abnormalities. Schizophr Bull. 2001;27(3):431–42.
152. Bertolino A. Dysregulation of dopamine and pathology of prefrontal neurons: neuroimaging studies in schizophrenia and related animal models. Epidemiol Psichiatr Soc. 1999;8(4):248–54.
153. Veselinovic T, Paulzen M, Grunder G. Cariprazine, a new, orally active dopamine D2/3 receptor partial agonist for the treatment of schizophrenia, bipolar mania and depression. Expert Rev Neurother. 2013;13(11):1141–59.
154. Jiang Z, Cowell RM, Nakazawa K. Convergence of genetic and environmental factors on parvalbumin-positive interneurons in schizophrenia. Front Behav Neurosci. 2013;7:116.
155. Keverne EB. GABA-ergic neurons and the neurobiology of schizophrenia and other psychoses. Brain Res Bull. 1999;48(5):467–73.
156. Lewis DA, Hashimoto T, Volk DW. Cortical inhibitory neurons and schizophrenia. Nat Rev Neurosci. 2005;6(4):312–24.
157. Volk DW, et al. Reciprocal alterations in pre- and postsynaptic inhibitory markers at chandelier cell inputs to pyramidal neurons in schizophrenia. Cereb Cortex. 2002;12(10):1063–70.
158. Miller R, Chouinard G. Loss of striatal cholinergic neurons as a basis for tardive and L-dopa-induced dyskinesias, neuroleptic-induced supersensitivity psychosis and refractory schizophrenia. Biol Psychiatry. 1993;34(10):713–38.
159. Yeomans JS. Role of tegmental cholinergic neurons in dopaminergic activation, antimuscarinic psychosis and schizophrenia. Neuropsychopharmacology. 1995;12(1):3–16.
160. Liu Y, et al. Activity strengths of cortical glutamatergic and GABAergic neurons are correlated with transgenerational inheritance of learning ability. Oncotarget. 2017;8(68):112401–16.
161. Galanopoulou AS. Mutations affecting GABAergic signaling in seizures and epilepsy. Pflugers Arch. 2010;460(2):505–23.
162. Brenner HD, Pfammatter M. Psychological therapy in schizophrenia: what is the evidence? Acta Psychiatr Scand Suppl. 2000;102(407):74–7.
163. Mueller DR, Roder V. Integrated psychological therapy for schizophrenia patients. Expert Rev Neurother. 2007;7(1):1–3.
164. Roder V, Mueller DR, Schmidt SJ. Effectiveness of integrated psychological therapy (IPT) for schizophrenia patients: a research update. Schizophr Bull. 2011;37(Suppl 2):S71–9.
165. Shattock L, et al. Therapeutic alliance in psychological therapy for people with schizophrenia and related psychoses: a systematic review. Clin Psychol Psychother. 2018;25(1):e60–85.
166. Blank T, Nijholt I, Spiess J. Treatment strategies of age-related memory dysfunction by modulation of neuronal plasticity. Mini Rev Med Chem. 2007;7(1):55–64.
167. Maillet D, Rajah MN. Age-related differences in brain activity in the subsequent memory paradigm: a meta-analysis. Neurosci Biobehav Rev. 2014;45:246–57.
168. Wang J-H, Kelly PT. Ca2+/CaM signalling pathway up-regulates glutamatergic synaptic function in non-pyramidal fast-spiking neurons of hippocampal CA1. J Physiol (Lond). 2001;533(2):407–22.
169. Zhang M, et al. Calcium signal-dependent plasticity of neuronal excitability developed postnatally. J Neurobiol. 2004;61:277–87.
170. Giray EF, et al. The incidence of eidetic imagery as a function of age. Child Dev. 1976;47(4):1207–10.
171. Miller E. The affective nature of illusion and hallucination. Part Ii: eidetic imagery. J Neurol Psychopathol. 1931;12(45):1–13.
172. Wasinger K, Zelhart PF, Markley RP. Memory for random shapes and eidetic ability. Percept Mot Skills. 1982;55(3 Pt 2):1076–8.

173. Roy DS, et al. Silent memory engrams as the basis for retrograde amnesia. Proc Natl Acad Sci U S A. 2017;114(46):E9972–9.
174. Martorell AJ, et al. Multi-sensory gamma stimulation ameliorates Alzheimer's-associated pathology and improves cognition. Cell. 2019;176(10):1–16.
175. Ryan TJ, et al. Memory. Engram cells retain memory under retrograde amnesia. Science. 2015;348(6238):1007–13.
176. Eustache F, et al. Episodic memory in transient global amnesia: encoding, storage, or retrieval deficit? J Neurol Neurosurg Psychiatry. 1999;66(2):148–54.
177. Guillery-Girard B, et al. Long-term memory following transient global amnesia: an investigation of episodic and semantic memory. Acta Neurol Scand. 2006;114(5):329–33.
178. Cooper RA, et al. Reduced hippocampal functional connectivity during episodic memory retrieval in autism. Cereb Cortex. 2017;27(2):888–902.
179. Gaigg SB, Bowler DM, Gardiner JM. Episodic but not semantic order memory difficulties in autism spectrum disorder: evidence from the Historical Figures Task. Memory. 2014;22(6):669–78.
180. Wojcik DZ, Moulin CJ, Souchay C. Metamemory in children with autism: exploring "feeling-of-knowing" in episodic and semantic memory. Neuropsychology. 2013;27(1):19–27.
181. Gron G, Riepe MW. Neural basis for the cognitive continuum in episodic memory from health to Alzheimer disease. Am J Geriatr Psychiatry. 2004;12(6):648–52.
182. Tromp D, et al. Episodic memory in normal aging and Alzheimer disease: insights from imaging and behavioral studies. Ageing Res Rev. 2015;24(Pt B):232–62.
183. Ostergaard AL. Episodic, semantic and procedural memory in a case of amnesia at an early age. Neuropsychologia. 1987;25(2):341–57.
184. Cohen G, Johnston R, Plunkett K. Exploring cognition: damaged brains and neural networks. In: Cohen G, editor. Exploring cognition: damaged brains and neural networks. Erlbaum: Psychology Press; 2002.

The manufacturer's authorised representative in the EU is Springer Nature Customer Service Centre GmbH, Europaplatz 3, 69115 Heidelberg, Germany. If you have any concerns regarding our products, please contact ProductSafety@springernature.com

Printed and bound by CPI Group (UK) Ltd, Croydon, CR0 4YY

28/07/2026

02179235-0001